Lecture Notes in Mechanical Engineering

Lecture Notes in Mechanical Engineering (LNME) publishes the latest developments in Mechanical Engineering—quickly, informally and with high quality. Original research reported in proceedings and post-proceedings represents the core of LNME. Volumes published in LNME embrace all aspects, subfields and new challenges of mechanical engineering. Topics in the series include:

- Engineering Design
- Machinery and Machine Elements
- Mechanical Structures and Stress Analysis
- Automotive Engineering
- Engine Technology
- Aerospace Technology and Astronautics
- Nanotechnology and Microengineering
- Control, Robotics, Mechatronics
- MEMS
- Theoretical and Applied Mechanics
- Dynamical Systems, Control
- Fluid Mechanics
- Engineering Thermodynamics, Heat and Mass Transfer
- Manufacturing
- Precision Engineering, Instrumentation, Measurement
- Materials Engineering
- Tribology and Surface Technology

To submit a proposal or request further information, please contact the Springer Editor of your location:

China: Ms. Ella Zhang at ella.zhang@springer.com
India: Priya Vyas at priya.vyas@springer.com
Rest of Asia, Australia, New Zealand: Swati Meherishi at swati.meherishi@springer.com
All other countries: Dr. Leontina Di Cecco at Leontina.dicecco@springer.com

To submit a proposal for a monograph, please check our Springer Tracts in Mechanical Engineering at http://www.springer.com/series/11693 or contact Leontina.dicecco@springer.com

Indexed by SCOPUS. All books published in the series are submitted for consideration in Web of Science.

More information about this series at http://www.springer.com/series/11236

Muhammad Abbas Ahmad Zaini · Mazura Jusoh ·
Norasikin Othman
Editors

Proceedings of the 3rd International Conference on Separation Technology

Sustainable Design in Construction, Materials and Processes

Springer

Editors
Muhammad Abbas Ahmad Zaini
University of Technology Malaysia
Bahru, Malaysia

Mazura Jusoh
University of Technology Malaysia
Bahru, Malaysia

Norasikin Othman
University of Technology Malaysia
Bahru, Malaysia

ISSN 2195-4356 ISSN 2195-4364 (electronic)
Lecture Notes in Mechanical Engineering
ISBN 978-981-16-0741-7 ISBN 978-981-16-0742-4 (eBook)
https://doi.org/10.1007/978-981-16-0742-4

This Springer imprint is published by the registered company Springer Nature Singapore Pte Ltd.
The registered company address is: 152 Beach Road, #21-01/04 Gateway East, Singapore 189721, Singapore

Preface

Design is a key for the present sustainable development for future generations, where cutting-edge improvement and development are sought for the betterment of life, economy and environment. 'Sustainable Design in Construction, Materials and Processes' brings papers from ICoST 2020, covering various aspects of sustainable construction, materials in engineering and design of industrial processes in line with the conference theme, 'Synergizing Sustainable Technology Towards Future Prospective'. ICoST 2020 offers a platform for researchers, scientists and engineers to gather and discuss specialized topics, each calling for disciplinary approaches to unlock the path to sustainable solutions. Amidst the ongoing global Coronavirus pandemic, the conference was conducted as a virtual event, giving the authors and presenters the flexibility from the comfort of home or office. The conference received more than 120 papers; all were peer-reviewed. Twenty-nine contributions have been selected and recommended to be published in this series of Lecture Notes in Mechanical Engineering. We hope that readers find this volume informative and useful. Lastly, we would like to extend our appreciation to Springer Nature for undertaking this publication and to the organizing committee for their hard work in making this momentous international meeting a reality.

Bahru, Malaysia

Muhammad Abbas Ahmad Zaini
Mazura Jusoh
Norasikin Othman

Contents

About the Editors

Muhammad Abbas Ahmad Zaini He is obtained a PhD in Applied Chemistry from Chiba University, Japan in 2010. He is an Associate Professor at the School of Chemical and Energy Engineering, UTM Johor Bahru. His research interests mainly revolve around design and characterization of adsorbent materials and dielectric properties of agro-wastes in microwave-assisted production. He was part of the organizing committee and chief editor for ICoST 2013, 2017 and 2020. Currently, he is a director and principal researcher at the Centre of Lipids Engineering & Applied Research (CLEAR) in Ibnu-Sina Institute for Industrial & Scientific Research (ISI-SIR). Muhammad Abbas is a professional engineer (PEng) registered with the Board of Engineers Malaysia, a professional technologist (PTech) with the Malaysia Board of Technologists, a chartered engineer (CEng) with the UK Engineering Council, and a member of the Institution of Chemical Engineers (IChemE, UK).

Dr. Mazura Jusoh He is currently an Associate Professor at the School of Chemical and Energy Engineering, Faculty of Engineering, Universiti Teknologi Malaysia. Mazura does research in Chemical Engineering, specifically in cold separation. She focuses on water removal process involving ice cystallisation called freeze concentration. She currently has 64 indexed and non-indexed publications and 64 conference papers presented. She has been in the Editorial Board for a few Journals including Jurnal Teknologi (UTM) and Materials Today: Proceeding (Elsevier).

Dr. Norasikin Othman He is currently an Associate Professor at the School of Chemical and Energy Engineering, Faculty of Engineering, Universiti Teknologi Malaysia. Norasikin does research in Chemical Engineering, specifically in Liquid Membrane Technology. She focuses on wastewater treatment of organic and inorganic compounds and also in recovery of precious metal from chemical industry wastewater. She currently has more than 100 papers in indexed, non-indexed and conference proceedings. She has been in the Editorial Board for a few Issues in Jurnal Teknologi (UTM) Special Issues and Malaysian Journal of Applied Science.

Sustainable Construction

The Implementation of the Industrialized Building System in the Malaysian Construction Industry—a Comprehensive Review

Aawag Mohsen Mohammed Alawag, Wesam Salah Alaloul, M. S. Liew, Mohammed Ali Mohammed Al-Bared, Noor Amila Wan Abdullah Zawawi, and Syed Ammad

Abstract Industrialized Building System (IBS) can be described as a potential prefabrication construction method that uses the best construction machineries, equipment, materials and extensive planning for the construction projects. Utilizing IBS in construction projects may enhance the overall performance at the construction site in terms of safety, quality, cost effectiveness, productivity and reduction of waste. Meanwhile, the level of IBS adoption in Malaysia is still low due to the very challenging barriers faced by the stakeholders in applying the new technology and shortages of skilled manpower. Even though the IBS construction method is not new to the Malaysian industry as it was introduced in the 90 s, its acceptance within the construction industry is still very low due to the high costs involved during its application. In this study, a comprehensive literature review was conducted in order to highlight the importance of adopting IBS in the Malaysian construction industry. The history of IBS, factors of implementation, barriers, benefits, advantages, and disadvantages of IBS were reviewed and analyzed thoroughly. The analysis of the literature review showed that various barriers contributes to the lack of IBS adoption in the Malaysian construction industry such as lack of skilled workers, reliance on the foreign workers, quality of materials, productivity of construction and limited enforcement of IBS requirements. The analysis also revealed that implementation of IBS in Malaysia reduced the reliance of the industry on the foreign labors. Therefore, it is recommended to adopt the IBS in a meaningful way during the pre-designing and planning stages of construction.

Keywords Adoption of IBS · IBS barriers · IBS benefits · IBS success factors · Malaysian construction industry

---

A. M. M. Alawag (✉) · W. S. Alaloul · M. S. Liew · M. A. M. Al-Bared · N. A. W. A. Zawawi · S. Ammad
Department of Civil and Environmental Engineering, Universiti Technologi PETRONAS, 32610 Seri Iskandar, Perak Darul Ridzuan, Malaysia
e-mail: alawaj711mohsen@gmail.com

M. A. A. Zaini et al. (eds.), *Proceedings of the 3rd International Conference on Separation Technology*, Lecture Notes in Mechanical Engineering,
https://doi.org/10.1007/978-981-16-0742-4_1

1 Introduction

IBS is considered as a modern technology that offers various advantages and benefits to the construction industry which are considered valuable in the construction processes compared to the conventional adopted methods and procedures. Those benefits can be observed in the reduction of the project cost, number of unskilled workers, and amount of wastage incurred during construction activities [1–3]. The industrialised building system (IBS) is considered to be a significant factor of the construction sustainability initiative [4, 5]. The adoption of IBS results in economical, sustainable and environmentally friendly projects that meet the current practices of having green planet and technology. Meanwhile, construction industry development board (CIDB) recognized the IBS as a construction system where parts are fabricated in an organized atmosphere, thereafter transferred, placed and installed into a structure with fewer extra onsite works [6–8]. It is also defined as an automated system that helps to customize and fasten the construction processes with regards to cost, management, and environmental prospective and project timeline [9, 10]. The adoption of IBS is confirmed to be helpful to increase and enhance the project value and production and save cost and duration of the project [11, 12]. IBS significantly reduces the need to construct new landfills to accommodate the waste generated during the various construction activities. The amount of waste accumulated in projects adopting IBS technology is very minimal and considered controlled compared to the conventional ones. This is due to the advanced techniques and methods used to manage the construction activities. In projects adopting IBS, the materials are prepared and fabricated in the factory, transported and installed at site in a well-controlled manner that is efficient and very fast. This process results in a minimal waste accumulation compared to conventional construction methods that usually end up with thousands tonnes of wastes for concrete, tiles, timbers, glass, etc. [13–18]. The Industrialized Building System or prefabrication is known as alternate approaches to substitute traditional construction to improve sustainable deliverables [19–21].

In Malaysia, Industrialised Building Systems technologies (IBS) are being encouraged as a promising to develop sustainability by the building manufacturing and government [22]. The government of Malaysia had significantly supporting the application of IBS to enhance the production in the construction industry through the collaboration with the Construction Industry Development Board (CIDB), IBS Center, and Malaysian Investment Development Authority (MIDA) [3, 23, 24]. From another point of view, the IBS is also known as a developed building method/methods that are being familiarized in the construction industry to accomplish the objectives of quicker accomplishment with mass fabrication of the building fundamentals in places far away from the last place of the building [25]. The IBS has established efficient approaches in literature and industry as the resolution to provide an increasing of housing demand [11]. CIDB through the IBS roadmap during 2003-2010, which can optimistically impulse the members of construction industry to perform and adopt the IBS in their every project included [11]. To the best of the author's knowledge, there is no available critical literature review on the IBS technologies and its adoption in construction industry. Hence, this paper focuses on the implementation of IBS in construction industry compared with the conventional methods. It reviews the history of IBS, factors of implementation, barriers, benefits, advantages, and disadvantages. Moreover, studying IBS extensively and providing a detailed review about it would serve as a platform to understand mechanisms, techniques, and provide a framework in order to guide future engineers and contractors on its performance during construction.

2 History of IBS Adoption in the Malaysian Construction Industry

This section is focused on the history or the timeline of the adoption of IBS technologies in Malaysia. Malaysian industry is divided into two main parts; public sector that involves residential, non-residential and construction of civil engineering projects. It also involves the particular trading works which includes various activities such as metal works, sewage and health works, cooling and conditioning works, electrical works, plumbing, tiling and glass works [5, 26]. IBS is not a modern process for construction in Malaysia. IBS is being observed in Malaysia as a valued solution in order to develop fundamental construction industry with proficiency, production and efficiency [27, 28]. Many efforts have been undertaken by the government of Malaysia to promote the application of IBS as a substitutional construction system and motivate the construction industry to move towards IBS which can generate high volume of houses at reasonable cost specially low-cost houses [9, 29]. The first IBS pilot project in Malaysia was initiated 50 years ago in the 1960s for a construction projects located at Jalan Pekeliling, Kuala Lumpur as shown in Fig. 1. Approximately 22.7 acres of land were developed for 17-storey flats (300 units of low cost flat and 40 shop lots) [30–32]. In addition, the second IBS project was built at Jalan Rifle Range, Penang in 1965 which consisted of six blocks. Three blocks consisted of 17-storey flats and the other three blocks consisted of 18-storey flats [31].

Meanwhile, IBS technologies in Malaysia became prominent in 1998; it started with an IBS strategical plan endorsed by the Cabinet of Ministers which plays a role as the scheme for the total industrialization of the construction section. Since then, CIDB has been actively promoting the usage of IBS in the Malaysian industry [31, 33]. Due to the variety of technologies involved within IBS, there is no description of IBS that could defines the entire building system. The most important, reasonable and applied description of IBS in Malaysia was introduced by the CIDB in Malaysia's roadmap of IBS [34]. CIDB defines IBS as a construction method in which the components are fabricated in a precise atmosphere (on or off site), transported, located and assembled into a structure with slight extra site works [31, 35]. The tendency of the Malaysian construction industry has shifted from the traditional system to IBS manufacturing and likewise, the production has moved from project based to product based [36].

Fig. 1 IBS implementation in Pekeliling Flats, Kuala Lumpur [49]

The extraordinary transmission of the construction industry to produce manufacturing has also happened in Australia, Hong Kong, Singapore, United Kingdom and United States [9, 37]. IBS was introduced in Malaysian industry as a resolution to the matters relevant to reliance of oversea labours, high request of houses with reasonable price and enhancing image, value and production of construction industry [38, 39]. The adoption of IBS technologies in Malaysia since 1960s helped to overcome the deficiency of houses [7]. It efficiently contributed in overcoming the increasing request of reasonable housing. Besides, it resulted in resolving the problems related to the imported workforce and enhancing image of IBS adoption, quality and output of construction industry [40, 41]. The Malaysian government begun to implement the IBS concept in construction industry due to several reasons such as the increase of population and the request of budgeted houses in the country [42]. In 2003, CIDB has completed a research about IBS implementation in Malaysia and found that there was 15% addition on the IBS resources use in domestic construction [43]. Throughout the years, private projects and the massive level of public have controlled the activities of the Malaysian construction industry driven by government and shareholders to implement next five-year plans targeted at turning into a developed country [44]. Although this method was presented four decades ago in Malaysia, so far its applications are still at low levels. Improvements of the infrastructure in the country permit the system to continue to extend in the future [45]. In Malaysia, CIDB is still boosting IBS in the private division, whilst, government division has effectively adopted IBS [7]. Presently, Malaysian government has powerfully supported the application of IBS in construction industry because of its quality assurance, shorter construction period, neater site condition, safer working environment and reduction in labour reliance [46, 47]. Recently, the usage of IBS process in numerous construction project in Malaysia is an enhancement, but the traditional technique cannot be handled with the rise of challenges [48]. According to CIDB, there are five features of IBS; prefabrication, offsite manufacture, mass production (repeatability), standardized components and design using modular coordination [49]. These characteristics are taken into account when the CIDB measures the level of IBS proportions used for construction projects in Malaysia [50].

The slow adoption of IBS in Malaysian construction industry is mainly due to the availability of cheap foreign labour which directly counteracting IBS cost advantages [51, 52]. The industry is reluctant to the adoption of IBS except if it is a client requirement. Therefore, the adoption of IBS in the Malaysian construction industry has been client-driven by those that recognise IBS benefits [52]. There is a high number of cheap foreign workers in Malaysia and contractors usually prefer to use the labour intensive conventional building method since it is easier to lay off workers during slack periods [52]. Nevertheless, there is solid pressure to adopt IBS due to government-led implementation plans and backing [52]. The adoption of the IBS methods in the Malaysian industry was implemented in many different ways such as the use of precast concrete load bearing wall system as shown in Fig. 2a, b and c for Senawang police quarter, teachers' quarters in Kuala Kangsar and government quarters in Putra Jaya, respectively [53]. In the case of steel structures, there are also various successful IBS projects. Figure 3 shows the KLCC convention centre, an icon building in Kuala Lumpur city centre. The building was built using a combination of prefabricated steel roof truss with composite steel deck flooring system [49].

The review through various studies showed that the adoption of IBS in the Malaysian construction industry had taken place decades ago. However, the contractors still prefer the usage of the conventional methods due to several reasons such as familiarity with those methods, difficulty of changing to automated based system as this new technology can be properly applied for small scale projects and therefore contractors undertaking small projects cannot afford to set up their own manufacturing plants. In addition, this process may involve very high budgets and small companies my lack the required financial support [53]. Moreover, some developer companies still prefer to use the conventional process of building system due to its simplicity to lay off workforces through slack period and this won't require retraining the workers in order to be skilled for the IBS [54].

(a) (b)

(c)

Fig. 2 IBS systems using precast concrete load bearing wall systems [53]

Fig. 3 During and after construction of KLCC Convention Centre and IBS adoption [53]

3 Implementation of IBS in Malaysian Construction Industry

Malaysia as a developing country is presently attempting for implementing the innovative or contemporary construction system of IBS. This will be a substitutional towards sustainable construction productivity and enhancement of construction performance [55]. The level of implementation of IBS is still low though it is under the governmental development strategic aims [55]. The construction industry plays an important role in the development of Malaysia and it is predictable to be a key contributor towards the Malaysian vision 2020. Although the construction industry is very competitive, it is still considered as labor-intensive [56]. According to CIDB, IBS is well-defined as a construction method with components that are factory-made in organized atmosphere, and at the same time, they are conveyed, placed and collected into a construction without excessive additional site works [21, 46]. Numerous studies examined pull and push factors in IBS implementation, the pull and push factors that inspired the usage of IBS are quality enhancement, decrease fault, reduce site reduction, housekeeping development, waste reduction, saving in the use of manual labour, inducement from government and cost saving [40].

In order to achieve a broader scale of IBS implementation in Malaysia, the government give complete exclusion on tax enforced by CIDB for developers who apply IBS components above 50 percent [32]. These wise decisions made by the government will absolutely generate lots of interest from the industry to use IBS as an alternative method of construction. The subsequent phase of IBS would be to discover the export of the system to overseas market. The achievement of these efforts will qualify the local construction industry to be competitive and penetrate the universal market [56]. The consciousness and acceptance for IBS implementation did not cover the entire elements of IBS implementation methods and not all practitioners could accept its implementation especially company and personnel with less experiences in handling IBS project [57]. For the time being in Malaysia, construction players are encouraged to move from the conventional process of construction to IBS. This process has an ability to reduce on-site works and rise the building quality in an organized environment [58].

Based on a survey conducted in Kedah and Perlis during the period between 2004 and 2005, Nasrun et al. [52] stated that in states of Kedah and Perlis, the practice of the IBS in the construction projects is still very low and considered minor. The survey also showed that amongst housing developers around Kedah and Perlis, statistic showed that majority of respondents have not used IBS in their projects. Out of 76 developers, only 24 of them have used IBS systems while the rest still used the conventional system [54]. Moreover, Hadi et al. [33] found that IBS was still considered new technology to most of the contractors in Sarawak and this indicated that they still lacked exposure to IBS technology and lack of IBS manufacturers in the industry. Jusoh et al. [59] stated that IBS needs high construction quality, accuracy and methods. Studies have indicated that most local professionals and contractors lack technical knowledge and experience in the IBS methods. Moreover, numerous local authorities are not fully aware with modular co-ordination and calibration concept related with IBS design and assembling processes. Rahim and Qureshi [7] indicated that IBS is still not widely applied in private sector and therefore the key problems are still unresolved regardless of the initiatives by the government using IBS method. The Government of Malaysia has stressed the complete utilization of IBS for governmental related projects by the inclusion of not less than 70% IBS component [60].

The critical review through various studies showed that the implementation of IBS is enforced by the government and most of the governmental projects used IBS components in their projects. However, for small private companies and some parts in Malaysia, the implementation of IBS is still very low and require further training and encouragement to be considered for future projects.

4 Critical Factors and Main Barriers of IBS Affecting IBS in Malaysia

The problem of housing is becoming increasingly critical because of the usage of conventional construction systems. The Malaysian construction authorities are trying to employ modern machinery and encourage the adoption of new practices of construction to meet the plans made by the government [61]. Thus, it is important to encourage the application of IBS through industrialised construction work and this can be achieved by increasing the use of technology and the recent practices concentrating on automation and robotics [62]. Anuar et al. [51] conducted a survey to determine the drivers, obstacles and the critical success factors of G7 contractors in adopting the IBS construction based on their past expertise and their overall awareness. Vision and obligation from the board management is the most important factor that defines the success of IBS, followed by early decision to use IBS and early assembly of the project team. Efficient communication, site administration, logistics, machineries, and finance are also considered as significant by the contractors [51, 63]. The contractors also concern factors like technology and competency, preparation and education and IT as less important [51].

A semi-structured interview was conducted by Hadi et al. [33] with 10 Small and medium sized enterprises SME's contractors in Sarawak. The study discovered that logistics and infrastructure issues, lack of reward, conventional payment ways, and lack of financial supports are amongst critical factors impacting the implementation of IBS. Mohamad et al. [41] successfully carried out a study in Selangor, Malaysia from 2011 until 2014. The percentage of the use of technologies in the governmental projects was determined after the analysis of the collected data from CIDB and coordination of implementation unit (ICU). The statistical analysis from the survey indicated that most of the respondent from numerous governmental agencies agreed that the important factors that influenced the adoption of IBS were decision and commitment to use IBS from board of director/top management and as well as lack of knowledge or information regarding IBS system [41, 64]. Another study was conducted in Malaysia by Yunus et al. [65] and its target was enhancing sustainability concerns and deliverables in IBS design and construction in developing countries, using Malaysia as the case study. The study identified the critical success factors to be constructability, fault and damages, manpower cost and material consumption. Moreover, a case study survey had been performed to check the potential critical success factors in actual construction setting in Malaysia by Anuar et al. [66]. The study emphasized a few success factors which are of significance such as pre-planning, organization, active communication, engagement in design, qualified staff, decision making, enhancement in procurement and agreement, supply series administration, participating, trade plan and information and communication technology to IBS in Malaysia.

Meanwhile, an extensive literature review was gathered by Ghazali et al. [67] who listed the critical factors and difficulties that affect the adoption of IBS formwork. The critical success factors outlined in this study were the high preliminary cost, familiarity on traditional process, uncompetitive manufacturing, failure in tendered bidding, alertness on training and short course, consciousness on product marketability, unreadiness of globalization era and readiness on business ability.

The review through various studies showed that the necessity for implementing IBS in Malaysian construction industry is significant due to the encouragement from the government and systematic implementation plan in place. Table 1 shows that the main barriers that affect the adoption of IBS are mainly related to the cost, difficulty of implementation and the easiness of the conventional system.

5 Conclusion

The critical review through several studies that were conducted on IBS can be summarized in the following points:

1. The IBS technology was adopted in the Malaysian construction industry long time ago. However, this adoption was only considered for few governmental construction projects. Hence, it can be concluded that IBS is still considered as new technology in the Malaysian construction industry.
2. The level of implementation of IBS in the Malaysian industry is low and the government considered many methods and various encouraging methods to increase the implementation such as waving the some of the taxes and enforcing the use of IBS in the governmental new projects.
3. Among the several barriers that affect the implementation of IBS in the Malaysian construction industry, cost, time, skilled workers are considered the main barriers.

Table 1 Summary of barriers to adoption of IBS in Malaysia [68]

Ref.	Barriers
[69]	Negative awareness, restriction of authorization, IBS application needs a detailed design, shortage of IBS components aesthetic, lack of knowledge and skilled workers, highly expensive, shortage of IBS manufacturer close to the site
[70]	Lack of knowledge and skilled workers, lack of incentives, negative perception, project delivery matters and extra cost
[71]	High capital investment, incompatibility of technologies, specialistaion capability, difficulty in adopting new system, lack of training, lack of implementation, financial obligations, difficutly in accepting technology by labour and increased maintenance cost,
[72]	Less attention from various parites, negative perception, readiness issues, finance and machinery, poor planning and procedures, awareness matters and weak konwledge
[73]	Lack of consciousness, lack of govrnment incentives, poor promotion, lack of workforce, weak quality, poor expertise and konwledge
[74]	Change in construction market, insufficient courses about IBS principles, bad reputation, incompetence of manufactured components, lack of materials, supply delay, bad weather, enoromous capital cost, cheapness of unskilled labour, lack of knowledge, difficulty of transportation process, limited storage of equipments, cost and time consuming
[75]	Miscommunication, poor coordination, weak integration, shortage of trust, negtive manners
[76]	Shortage of commercial support, poor knowledge of IBS, preference of client for conventioal method
[60]	Transformation from conventional to IBS, limited support of IBS requests, shortage of IBS experience, cash flow issues, increase of equpment cost
[77]	Design, financial problems, lack of familiarity with IBS, Developer perception, shoretage of determined rules
[57]	Lack of manufacturer, issues on jointing, lack of appropriate utilization, high cost, lack of joint and beam standardization and mononplogy
[78]	Uncertatinity of IBS chances, poor knowledge and shortage of IBS suppliers
[79]	Absence of fund investment, lack of volume, weakness of present IBS, overall costly price and inexpensive labour
[80]	Numourous capital cost, inadequate konwledge, module standardization, integratoin, transport, organization, on-site construction practice and preparation and application
[68]	Profit of investment, lack of unskilled workforce, nature implementation, low quality, shortage of motivations and mindfulness, wastage of material and poor information
[81]	Hig initial capital, higher design and crane and transport cost
[82]	Inflexible for modifications, highly preliminary budget, shortage of information, shortage of manpower and contractros, less demand, poor awareness, leaking issues, limited site space, required skilled workforce

References

1. Nasir NM, Nasrun M, Nawi M, Kamarul M, Abdul I, Bahaudin AY (2016) A review of delay factors in Malaysian industralised building system (IBS) construction project. ARPN J Eng Appl Sci 11:9868–9873

2. Abedi M, Fathi MS, Rawai NM (2012) A review of industrialised building system in Malaysian construction industry. In: Proceeding of Management in construction research association (MiCRA) postgraduate conference
3. Ariffin ST, Yunus R, Mohammad H, Khalijah S (2017) A preliminary review on economies of scale (EOS) towards industrialized building system (IBS) manufacturer. In: MATEC web of conferences, p 9
4. Meynagh et al MM, (2014) Main factors of contractor reluctancy to adopt IBS projects in Malaysia. Australian Journal of Basic and Applied Sciences 8:251–254
5. Amin MAM et al (2017) A review on the current issues and barriers of industrialised building system (IBS) adoption in Malaysia's construction industry. In: IOP conference series: materials science and engineering, vol 9, pp 1757–899
6. Musa MF, Mohammad MF, Yusof MR, Mahbub R (2015) The way forward for industrialised building system (IBS) in Malaysia. In: Proceedings of the international civil and infrastructure engineering conference, vol 7, pp 981–287
7. Rahim AA, Qureshi SL (2018) A review of IBS implementation in Malaysia and Singapore. Malaysia. J Malays Inst Plan 16:323–333
8. Baharuddin MN, Bahardin NF, Zaidi MA, Lokman I (2015) A barriers and challenging criteria of IBS formwork: a current scenario amongs stakeholder 39:14–21
9. Azman MN, Ahamad M, Hilmi N (2012) The perspective view of Malaysian industrialised buiding system (IBS) under IBS precast manufacturing. In: The 4th International engineering conference
10. Mohammad MF, Baharin AS, Musa MF (2016) The potential application of IBS modular system in the construction of housing scheme in Malaysia. Proc Soc Behav Sci 222:75–82
11. Nasrun M, Nawi M, Akmar F, Nifa A, Osman WN (2015) Malaysian industrialised building system (IBS): a review of studies Malaysian industrialised building system (IBS): a review of studies. Aust J Basic Appl Sci 9:99–101
12. Mohsen A, Yunus R, Handan R, Kasim N, Hussain K (2019) Determining factors for enhanced skilled worker requirements in IBS construction projects in Malaysia. In: IOP Conference series: earth and environmental science, 220 p
13. Azmi N, Bari A, Abdullah NA, Yusuff R, Ismail N (2012) Environmental awareness and benefits of industrialized building systems (IBS). In: ASEAN conference on environment-behaviour studies, vol 50, pp 392–404
14. Zaini N, Halipah IS (2015) A review on embodied energy through industrialised building system implementation in construction industries
15. Zainuddin N, Mohd Yunus NZ, Al-Bared MAM, Marto A, Harahap ISH, Rashid ASA (2019) Measuring the engineering properties of marine clay treated with disposed granite waste. Measurement 131:50–60. https://doi.org/10.1016/j.measurement.2018.08.053
16. Al-Bared MAM, Marto A, Latifi N, Horpibulsuk S (2018) Sustainable improvement of marine clay using recycled blended tiles. Geotech Geol Eng 36:3135–3147. https://doi.org/10.1007/s10706-018-0525-8
17. Al-Bared MAM, Harahap ISH, Marto A (2018) Sustainable strength improvement of soft clay stabilized with two sizes of recycled additive. Int J Geomate 15(51):39–46
18. Al-Bared MAM, Marto A, Latifi N (2018) Utilization of recycled tiles and tyres in stabilization of soils and production of construction materials a state-of-the-art review. KSCE J Civ Eng 22:3860–3874
19. Yunus R, Hamid ARA, Noor SRM (2019) An integrated approach for sustainability in the application of industrialised building system (IBS). Int J Geomate 17:115–121
20. Yunus R, Yang J (2012) An integrated approach to enhance sustainability in industrialised building systems. In: Proceedings of the joint 8th Asia pacific structural engineering and construction conference and 1st international conference on civil engineering research
21. Yunus R, Yang J (2011) Sustainability criteria for Industrialised Building Systems (IBS) in Malaysia. Proc Eng 14:1590–1598
22. Yunus R, Yang J (2011) Critical sustainability factors in industrialised building system. Proc Eng 14:1590–1598

23. Lou ECW (2012) industrialized building systems: strategic outlook for manufactured construction in Malaysia. J Archit Eng 18:69–74
24. Jabar IL, Ismail F, Aziz ARA (2018) Stakeholder's perception of industrialized building system (IBS) implementation. Asian J Behav Stud 3:8
25. Taherkhani R (2014) A strategy towards sustainable industrial building systems (IBS): the case of Malaysia. J Multi Eng Sci Technol 1:86–90
26. Razak A, Ibrahim B, Roy MH, Ahmed Z, and Imtiaz G, (2010) An investigation of the status of the Malaysian construction industry. Benchmarking An International Journal l:294–308
27. Fathi MS, Abedi M, Mirasa AK (2016) Construction industry experience of industralised building system in Malaysia. In: International congress on civil engineering
28. Nasir NM, Ismail Z, Ismail F, Nur S, Syed A (2012) enabling factors towards safety improvement for industrialised building system (IBS). Int J Archit Environ Eng 6:1115–1120
29. Alia A, Bohari M, Kipli K, Mahat N (2011) Contractor's knowledge on industrial building system (IBS) in Sarawak. In: ASEAN conference on scientific and social science research
30. Rashid MN, Abdullah MR, Ismail D, Mahyuddin MN (2018) Towards automation and robotics in industrialised building system (IBS): a literature review. In: International conference on rebuilding place
31. Ilaili Jabar I, Ismail F, Mardhiyah N, Janipha N (2013) Construction manager's competency in managing the construction process of IBS projects. In: Asean postgraduate conference, vol 105, pp 85–93
32. Mohamad-Kamar KA et al (2009) Industrialised Building System (Ibs): A review of experiences in UK and Malaysian construction. In: Construction industry research achievement international conference (CIRAIC), pp 1–12
33. Hadi NA, Muhammad WMNW, Othman MKF (2017) Critical factors of implementing industrialised building system in Sarawak: a research on SMEs 012006:1–7
34. Abedi M, Fathi MS, Mirasa AK (2011) Establishment and development of IBS in Malaysia. In: International building and infrastructure technology conference
35. Khan RA, Liew MS, Ghazali Z (2013) Malaysian construction sector and Malaysia Vision 2020: developed nation status, 000–000(May)
36. Yunus NM, Malik SA (2012) Ratios to predict business performance of IBS construction company in Selangor. In: International conference on innovation management and technology research, pp 441–445
37. Nasrun M, Nawi M, Hamid ZA (2012) Critical success factors for improving team integration in Industrialised Building System (IBS) construction projects:the Malaysian case. Malays Constr Res J 10:44–62
38. Anuar Mohamad Kamar K, Abd Hamid Z, Khairolden Ghani M, Egbu C, Arif M (2011) Collaboration initiative on green construction and sustainability through industrialized buildings systems (IBS) the Malaysian construction industry. Int J Sustain Constr Eng Technol 1:119–127
39. Sashitharan S, Jusoh MS, Amlus MH, Rahimi A, Ibrahim A, Ismail MS (2014) Industrialized Building System (IBS) performance in Malaysian construction industry analysing the cost, training and construction policies. J Appl Sci Agricul 9(13):6–13
40. Kamar KAM, Alshawi M, Hamid Z (2009) Industrialised building system: the critical success factors. In: 9th International postgraduate research conference (IPGRC), Salford, United Kingdom, pp 485–497
41. Mohamad D, Ramli MZ, Hn D, Sapuan WK (2016) Demand of the industrialized building system (IBS) implementation in Malaysian government projects. J Sci Res Develop 3:77–82
42. Wong SS, Lau LK (2015) Advantages and setbacks of industrialized building system (IBS) implementation: a case study in Sarawak. Int J Sustain Constr Eng Technol 6:52–61
43. Kassim U, Abdullah CS, Udin ZM (2015) Attraction factors in choosing industrialized building system (IBS) method over conventional building system. Int J Sci Basic Appl Res (IJSBAR) 4531:338–347
44. Rashid MNA, Abdullah MR, Ismail D, Saberi MH (2019) Automation and robotics in industrialized building system (IBS): the potential criteria for measurement. Int J Acad Res Bus Soc Sci 9:1020–1034

45. Mydin MAO, Sani N, Taib M (2014) Industrialised building system in Malaysia : a review 2014. In: Building surveying and technology undergraduate conference, vol 01002, pp 1–9
46. Muhammad WMNW, Azman MA, Othman MKF, Hadi NA, Sahimi SNS (2016) The insight of industrialised building system (IBS) by bumiputera construction players. In: MATEC web of conferences, 13 p
47. Zabihi H, Habib F, Mirsaeedie L (2013) Definitions, concepts and new directions in industrialized building systems (IBS). KSCE J Civ Eng 17:1199–1205
48. Pozin F, Nasrun M, Nawi M (2017) The communication in industrialised building system (IBS) construction project : Virtual environment. In AIP Conference proceedings 1891
49. Shamsuddin SM, Zakaria R, Mohamed SF, Mustaffar M (2012) Drivers and challenges of industrialised building system (IBS) in sustainable construction:2–4
50. Yunus R, Masrom AN, Abdullah AH, Mustakim F (2015) Conceptual Model of Contractor Satisfaction in the Industrialized Building System (IBS) Implementation. Appl Mech Mater 773–774:828–833
51. Kamar KAM, Azhari MNA (2014) IBS survey 2010: drivers, barriers and critical success factors in adopting industrialised building system (ibs) construction by G7 contractors in Malaysia. J Eng Sci Technol 9:490–501
52. Azam Haron N, Abdul-Rahman H, Wang C, Wood LC (2015) Quality function deployment (QFD) modelling to enhance industrialized building system adoption in housing projects. Total Qual Manage Bus Excell 26:703–718
53. Baharuddin A, Rahman A, Omar W (2006) Issues and challenges in the implementation of industrialised building system in Malaysia. In: Proceedings of the 6th Asia-pacific structural engineering and construction conference (APSEC), pp 5–6
54. Nasrun M et al (2007) A preliminary survey of the application of industrialised building system (IBS) in Kedah and Perlis Malaysia construction industry. In: Proceedings of the conference on sustainable building South East Asia, Johor, Malaysia, pp 411–415
55. Nawi MNM, Lee A, Mydin MAO, Osman WN, Rofie MK (2018) Supply chain management (SCM): Disintegration team factors in Malaysian Industrialised Building System (IBS) construction projects. Int J Supply Chain Manage 7:140–143
56. Setiani Y (2009) Implementation of industrialised building system (IBS) in Malaysia construction industry. In: Seminar Nasional Universitas Budi Luhur 14 Agustus 2009 Jakarta, pp 1–5
57. Mohamad MI, Zawawi M, Nekooie MA (2009) Implementing industrialised building system (IBS) in Malaysia: acceptance and awareness level, problems and strategies. Malaysian J Civ Eng 21:219–234
58. Yunus R, Abdullah AH, Yasin MN, Hanipah MH (2016) Examining performance of industrialised building system (IBS) implementation based on contractor satisfaction assessment. ARPN J Eng Appl Sci 11:3776–3782
59. Jusoh MS, Amlus MH, Rahimi A, Ismail MS (2014) Industrialized Building System (Ibs) performance in Malaysian construction industry: analysing the cost, training and construction policies. J Appl Sci Agri 9:6–13
60. Amin MAM, Abas NH, Shahidan S, Rahmat MH, Suhaini NA, Nagapan S, Abdul Rahim R (2017) A review on the current issues and barriers of Industrialised Building System (IBS) adoption in Malaysia's construction. In: IOP conference series: materials science and engineering, vol 27, no. 1, 012031p
61. Rashid MNA, Abdullah MR, Ismail D (2019) Critical Success Factors CSFs to Automation and Robotics in Industrialized Building System IBS. Int J Acad Res Bus Soc Sci 8:2207–2221
62. Abd Rashid MN, Abdullah MR, Ismail D, Mahyudddin MN (2018) Towards automation and robotics in industrialised building system (IBS): a literature review. In: International conference on rebuilding place (ICRP), pp 13–14
63. El-Abidi KMA, Ofori G, Zakaria SAS, Mannan MA, Abas NF (2019) Identifying and evaluating critical success factors for industrialized building systems implementation: Malaysia study. Arab J Sci Eng 44:8761–8777

64. Taherkhani R, Saleh A, Nekooie MA, Mansur SA (2012) External factors influencing on industrialised building system (IBS) in Malayisa. Int J Sustain Develop World Policy 1:227–231
65. Yunus R, Yang J (2015) Critical sustainability factors in industrialised building systems. Constr Innovat Inf Process Manage:12
66. Anuar K, Kamar M, Hamid ZA, Alshawi M (2010) The critical success factors (CSFs) to the implementation of industrialised building system (IBS) in Malaysia. In: 18th CIB World building congress, TG57-Special Track, May 2010 Salford, United Kingdom, pp 10–13
67. Ghazali MA, Bahardin NF, Zaidi MA, Baharudin MN, Yusof MR (2016) Literature mapping: critical factors in industrialized building system plastic formwork application. Int Rev Manage Market 6:204–208
68. Yahya MA, Shafie MNS (2012) Level of acceptance towards industrialised building system (IBS) in Malaysia. Int J Sustain Constr Eng Technol 3:96–103
69. Ali M, Abas NH, Affandi HM, Abas NA (2018) Factors impeding the industrialized building system (IBS) implementation of building construction in Malaysia. Int J Eng Technol 7:2209–2212
70. Nawi MNM, Lee A, Nor KM (2011) Barriers to implementation of the Industrialised Building System (IBS) in Malaysia. Built Hum Environ Rev 4:22–35
71. Syariazulfa S, Fadhil M, Mahbub R (2016) Barriers and impact of mechanisation and automation in construction to achieve better quality products. Proc Soc Behav Sci 222:111–120
72. kamar KAM, Alshawi A, Hamid H (2009) Barriers to industrialized building system (Ibs): the case of Malaysia. In: BuHu 9th International postgraduate research conference (IPGRC), pp 1–16
73. Ahmad M, Haseneshkorfu OO, Nigar K (2017) Th barriers to the implementation of industrialised building system (IBS) in high rise structure in Selangor. Int J Adv Mech Civ Eng 4:94–99
74. Qays M, Mustapha KN, Al-mattarneh H, Mohamed BS (2016) The constraints of industrialized building system from stakeholders' point of view. In: Proceedings of the ICSE2010
75. Nawi MNM, Lee A, Arif M, (2010) The IBS barriers in the Malaysian construction industry: a study in construction supply chain perspective. In: Proceedings of the conference on built environment in developing countries (ICBEDC)
76. Mohamed MR, Mohammad MF, Mahbub R, Ramli MA, Jamal KAA (2018) The issues and challenges of small and medium-sized contractors in adopting industrialised building system. Int J Eng Technol 7:432–436
77. Jalil AA, Nuruddin AR, Jaafar M, Mydin MAO (2015) Challenges for developers implementing IBS in Malaysian housing projects
78. Sazalli SAAH, Greenwood D, Morton D, Agnew B (2012) Industrialised building system (IBS) in Malaysia housing: an emerging innovation system explanation. In: ARCOM Conference
79. Alia A, Bohari M (2012) Industrialised building system (IBS) in Sarawak construction industry. In: International conference on innovation management and technology research, pp 21–22
80. Ilaili Jabar, Ismail F, Mustafa AA (2013) Issues in managing construction phases of IBS Projects. Proc Soc Behav Sci 101:81–89
81. Nor M, Azman A, Ahamad MSS, Majid TA, Hanafi MH (2011) Status of industrialized building system manufacturing plant in Malaysia. J Civil Eng Sci Technol 2:8–16
82. Begum RA, Satari SK, Pereira JJ (2010) Waste generation and recycling: comparison of conventional and industrialized building systems. Am J Environ Sci 6:383

Effect of Glass Powder and Egg Shell Powder on the Consistency and Setting Time of Cement Blends

Olumide Olu Olubajo and Osha Ade Odey

Abstract One alternative to the improper waste disposal and the high energy intensiveness of cement production is the use of cementitious materials such as eggshell (ESP) and glass powder (GP) as a partial cement replacement material. This paper investigates the effect of partially replacing ordinary Portland cement (OPC) with ESP and GP on the water consistency and setting times of cement blends (OPC-ESP-GP) up to 25 wt% replacement at interval of 2.5 wt%. Vicat apparatus was employed in the determination of the standard consistencies and setting times of cement blends at various proportions. Results indicated a variation in both setting times for cement replacement from 2.5 to 12.5 wt% except for 7.5 wt% which experienced a retardation, whereas beyond 12.5 wt% cement replacements led to an increase in the initial setting time as GP was replaced with ESP. This variation could be attributed to lime presence in ESP resulting in accelerated hydration rate despite clinker diminution and Magnesium hydroxide $Mg(OH)_2$ low solubility resulting in the formation of a protective layer which retards cement hydration respectively. The water consistency diminished as ESP was replaced with GP for various cement replacement except 2.5 wt% which was attributed to the wider particle size distribution and vice versa. The blended cement' consistencies were higher than OPC by 6.67–21.50% which was due to either clinker diminution or the high silica content and enhanced surface area of GP. The consistency and the setting times of the various cement blends satisfied the required standard.

Keywords Consistency · Eggshell powder · Glass powder · Setting time

O. O. Olubajo (✉)
Department of Chemical Engineering, Abubakar Tafawa Balewa University, Bauchi, Nigeria
e-mail: ooolubajo@atbu.edu.ng

O. A. Odey
Department of Chemical Engineering, University of Calabar, Calabar, Cross River, Nigeria

M. A. A. Zaini et al. (eds.), *Proceedings of the 3rd International Conference on Separation Technology*, Lecture Notes in Mechanical Engineering,
https://doi.org/10.1007/978-981-16-0742-4_2

1 Introduction

Most small-scale industries lack proper treatment facilities for industrial waste owing to huge financial implications, thus resulting in the disposal of harmful wastes [1] which could either be hazardous or unhazardous in nature. Accumulation of such solid waste such as eggshell powder (ESP) and Glass Powder (GP) can have a negative effect in our environment ranging from air pollution, water pollution, devastation of lands and the surroundings which affects every living thing (human, plant and animals. The low availability of non-renewable energy resources coupled with the large energy quantities requirement for building materials like cement, has provide the avenue for use of industrial waste which cannot be overemphasized. Similarly, carbon dioxide emissions act as a silent killer in the environment as various forms and as a result of this backdrop, the search for cheaper substitute to OPC is a needful one [2, 3]. In order to free the environment of these menace constituted by accumulated solid waste, one alternative is the use of SCMs to alleviate the environmental concerns impacted by the cement industry [4, 5].

Attempts have been made by various researchers to convert solid waste to beneficial materials for building, with a high level of success, as a way of reducing solid waste accumulation [6–9]. Many silica rich materials have been sourced from industrial and agricultural wastes such as waste glass [4] and eggshell [3], as well as calcined waste products include coal bottom ash CBA [10–12], rice husk ash [13–15], saw dust ash SDA [16], metakaolin MK [15, 17], fly ash FA [18] and groundnut shell ash [19] with significant successes. Deleterious chemical constituents such as sulfide, sulfates and alkalis increases the potentials of ASR attack and consumption of lime which drastically decreases efflorescence when cement is replaced with SCMs [20, 21], while if Magnesium oxide (MgO) in excess of the optimal rate of 2–3% could significantly influences the cement properties resulting in the reduction of absorption of free lime according to [22]. The individual components of the cement blends tend to play a significant role on the behavior of the blend [12]. as well as the combined effect of replacing cement clinker with two supplementary cementitious materials (SCMs) such as GP and ESP which could provide a synergetic interaction between them, thereby improving the cement blends' properties by eliminating some undesirable effects [23, 24]. The incorporation of ESP in GP cement blends can accelerate the setting time properties which could be attributed to the presence of calcium carbonate resulting in the formation of ettringite at the expense of monosulphate.

Glass is a non-crystalline (amorphous) silica similar in composition with OPC mainly composed of Calcium Oxide (CaO), Silicon (IV) Oxide (SiO_2), Aluminum Oxide (Al_2O_3) and Iron (III) Oxide (Fe_2O_3) with other minor oxides such as Magnesium Oxide (MgO), Lead (II) Oxide (PbO), Titanium IV Oxide (TiO_2), Manganese (II) Oxide (MnO), Copper (II) Oxide (CuO), Nickel Oxide (NiO) which is less than 5 wt% [25]. The properties of waste glass and other pozzolan in concrete are influenced by its chemical composition, fineness and the pore solution for reaction [4, 26, 27] and mainly dependent on the raw materials used and differs slightly for each glass type. Being amorphous and containing relatively large quantities of silicon, waste glass should be an excellent pozzolanic material for concrete industry if $Al_2O_3 + SiO_2 + Fe_2O_3 > 70$ wt%. An increase in the replacement of cement with glass powder from 10 to 30 wt% led to an increase in the workability according to [28] which was contrary to [29] which observed a decrease in the workability (requiring more water) and attributed the decrease to the high surface tension of the glass powder resulting in a strong capillary action. Kaveh and Prasada [30] observed similar trend of an increase in the concrete workability as GP content was increased. Vandhiyan et al. [31] also reported that the water demand slightly increased as cement was replaced with 10 wt% and above GP while the initial setting time accelerated slightly as the GP replacement was increased up to 10 wt% whereas a significant reduction was experienced beyond 10 wt%. This increase in the water demand (reduction in workability) was attributed to the increased surface area as well as the angular shape of the glass particle according to [32]. Similar trend was reported by [33] of a reduction in the workability of the mortar by 2.5 to 12.5 wt% as GP content was increased from 5 to 30 wt% respectively. The advantage of inclusion of glass powder into the cement matrix led to improved durability of concrete blended with waste glass with long term chloride permeability despite alkali silica reaction [27, 34].

The eggshell primarily contains calcium, magnesium carbonate (lime) and it is common practice that dried eggshell is a good source of calcium for animal feeds. The quality of lime in eggshell waste is influenced greatly by the extent of exposure to sunlight, raw water and harsh weather conditions [3, 35]. Dhanalakshmi et al. [36] observed that the workability decreased i.e. water required for consistency increased as ESP content was gradually increased from 0 to 12.5 wt%. The accelerated setting time effect of ESP can be regulated via the inclusion of GP which contain high silica content which tend to retard the setting time of cement. Hama et al. [51] also suggested that CaO presence in the egg shell powder which was the main component of eggshell powder was responsible for the decrease in the setting time of cement and observed that beyond 8% resulted in a decrease in the reactive cement component leading to significant physical modifications in the cement blends. In this study, the influence of eggshell and glass powder on the consistency and setting time of ternary cements were investigated. It also seeks to front the derived advantages and assist in solving the problem of construction during the raining seasons, reduce environmental pollution and increase the availability of cement at a cheaper rate. Cement was partially replaced with ESP and GP between 0 and 25 wt% and the consistency and setting times were evaluated for each replacement via Vicat apparatus as shown in Fig. 1 according to ASTMC187 [37] and ASTM C191 [38] respectively.

Fig. 1 Dried Eggshell

2 Materials and Method

2.1 Sample Preparation

Egg shells were collected from various bakeries and eateries at Yelwa in Bauchi State. The egg shells and waste glass carefully sorted and washed to avoid collecting sand. The egg shells collected were sundried for a few days are shown in Fig. 1. The egg shell and glass waste were then ground with a ball mill followed by size reduction with 90 μm sieve as shown in Figs. 2, 3 and 4 respectively. The Ordinary Portland cement (Dangote, Brand) used was obtained in Yelwa Bauchi. Sharp sand was used as fine aggregates were obtained from a local supplier in Bauchi Nigeria.

2.2 Experimental Procedure

Tests were conducted to determine the physical and chemical properties of the ESP and GP while their chemical composition were analyzed using X-ray fluorescence analyzer. The normal consistency was obtained before the commencement of the setting time test according to ASTM C 187 [37] and ASTM C191 [38]. The stop watch was employed at the instant when the water was added to the cement and the time was recorded. The initial and final setting time was conducted by using the Vicat apparatus as shown in Fig. 2 by gently lowering the needle to make contact with the surface of the cement paste and quickly released to penetrate its surface. The procedure is then repeated till the needle fails to penetrate the cement paste surface to point 5.0±0.5 mm measured from the bottom of the mould to determine the initial setting time while the final setting time was carried out by replacing the needle for initial setting time with annular attachment needle. The final setting time

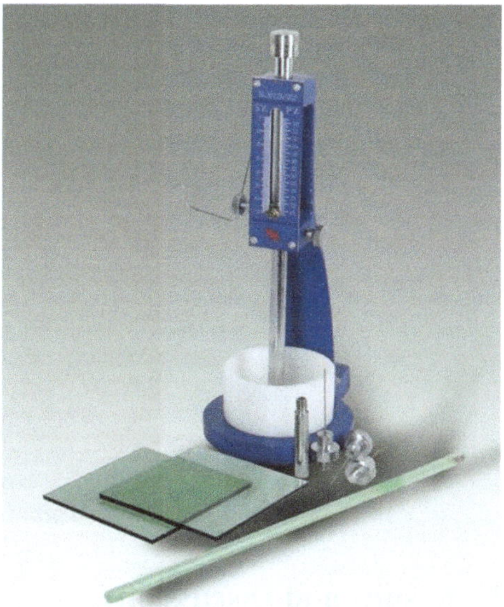

Fig. 2 Vicat Apparatus

Fig. 3 Glass Powder

is determined when the cement has finally set by the attachment needle making an impression on the top surface of the cement paste. The period elapsed between the time when the water was added to cement blend and the time when the needle makes an impression on the surface of the test block.

Fig. 4 Eggshell Powder

3 Results and Discussion

Table 1 indicates the chemical composition of OPC, ESP and GP obtained via XRF spectrometer respectively. The percentage of reactive silicates content of CEM II A-L 42.5 R contains 12.10 wt% which was lower than 69.95 wt% for GP. Thus, GP can contribute silica required for pozzolanic reaction with the available lime to produce secondary CSH. Results indicate that GP can be considered as a pozzolan as it satisfies the requirements specified by ASTM C618 [39]; $SiO_2 + Al_2O_3 + Fe_2O_3$ is greater than 70 wt% with SO_3 content of 0.29 wt% also satisfying the requirements which is less than the maximum of 5wt%. GP contains 12.39 wt% CaO which is within the range of 10-30 wt% CaO and considered a class C pozzolan. The chemical composition of the eggshell (ESP) revealed a high amount of CaO of 48.26 wt% (containing $CaCO_3$ 86.13 wt%) and other minute quantities of SiO_2, Al_2O_3, Fe_2O_3, MgO, SO_3, K_2O and Na_2O which were below 1 wt%. ESP has a high LOI of 47.38 wt% which can be attributed the decomposition of limestone to lime.

3.1 Effect of ESP and GP Content on Cement Setting Time of Cement Blends

3.1.1 Effect of ESP and GP Content on the Initial and Final Setting Time of Cement Blends

Figures 5 and 6 illustrate the effect of ESP and GP content on the initial and final setting times of ternary cement pastes respectively. Figure 5 showed a series of accelerations and retardations on the initial setting time as GP content was gradually

Table 1 Chemical composition of OPC, ESP and GP

Parameters	Ordinary Portland cement (OPC) wt%	Eggshell Powder (ESP) wt%	Glass Powder GP wt%
SiO_2 (tot)	12.10	0.99	69.95
Al_2O_3	2.97	0.15	0.52
Fe_2O_3	2.64	0.10	1.38
CaO	50.54	48.26	12.39
MgO	0.66	0.74	2.44
SO_3	1.72	0.65	0.29
K_2O	0.30	0.10	0.59
Na_2O	0.12	0.11	12.30
P_2O_5	0.08	0.41	–
Mn_2O_3	0.03	0.01	–
TiO_2	0.14	0.02	–
Cl	–	–	0.14
$CaCO_3$		86.13	
LOI	14.63	47.38	–
C_3S	90.03	187.79	
C_2S	−33.22	−138.82	
C_3A	3.40	0.23	
C_4AF	8.03	0.30	
Total	85.93	98.91	99.99
Specific gravity	3.15	1.73	2.29

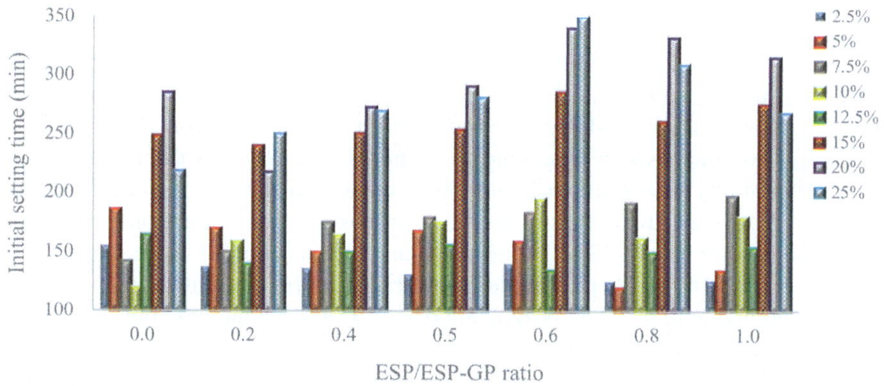

Fig. 5 Effect of ESP and GP content on the initial setting time of ternary cement blends at various cement replacement

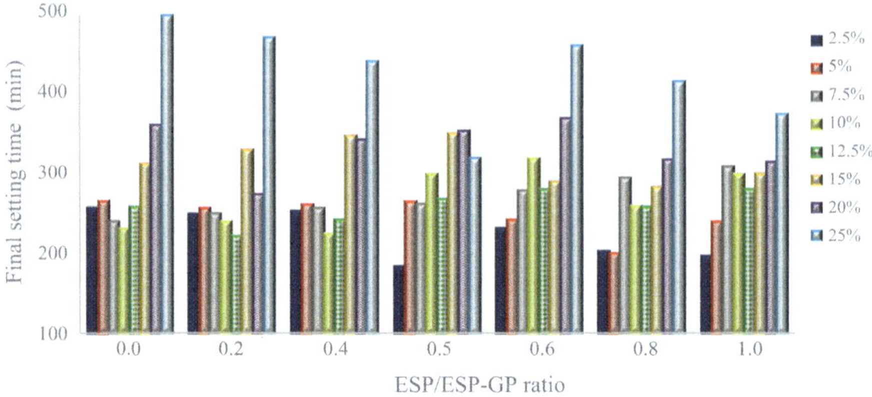

Fig. 6 Effect of ESP and GP content on the Final setting time of ternary cement blends

replaced with ESP content (i.e. an increase in the ESP/ESP-GP ratio from 0 to 1.0) for cement replacement of 2.5, 5, 10, 12.5, 20 and 25 wt% respectively except for 7.5 and 15 wt%. The initial setting time of cement blends for 2.5 and 5 wt% cement replacement experienced an acceleration followed by a retardation as GP content was replaced by the ESP content. The retardation and acceleration of the cement blends could be linked with two factors namely; clinker diminution effect and filler effect. The acceleration of the initial setting time could be attributed to the presence of calcium carbonate resulting in the formation of ettringite at the expense of monosulphate, thus, enhancing hydration rate [12, 40–43, 51] which compensates the dilution effect while the retardation of the initial setting time could be due to the diminution of the clinker content with similar trends according to [5, 44] (Table 2).

7.5 wt% cement replacement as the GP content was replaced with ESP resulted in a gradual elongation of the initial setting time from 142 to 197 min while for 10 and 12.5 wt% cement replacement, the initial setting time of cement blends experienced a series of retardations followed by accelerations as the ESP content (ESP/ESP-GP ratio) was increased respectively. A significant increase in the initial setting time (retardation) was experienced at higher cement replacement beyond 12.5 wt% cement replacement. It was also observed that most of the cement blends beyond 12.5 wt% ranging from 218 to 340 min had an elongated setting time compared to control OPC of 146 min which could be attributed to the diminution of the clinker content coupled with the slow pozzolanic reaction between the lime presence in the eggshell and the silica present in the glass powder [4, 44].

It could be observed from Fig. 6 that an increase in the ESP content by replacing GP with ESP at various cement replacements resulted in a series of retardations, followed by accelerations of the final setting time for 2.5, 5, 20 and 25 wt% cement replacement respectively. A similar steady retardation of its final setting time was observed as the GP was replaced by ESP at 7.5 wt% cement replacement whereas at 10 and 15 wt% cement replacement experienced a series of elongation and acceleration in the final setting times respectively. This retardation in the final setting time could

Table 2 Consistency, Initial and final setting for ESP and GP at 2.5, 5, 7.5, 10,12.5, 15, 20 and 25 (wt%)

S/No	OPC-ESP-GP	ESP/ESP-GP ratio	Consistency wt%	IST/FST Mins
1	100-0-0	0.0	28.00	146/189
2	97.5-0-2.5	0.0	34.00	155/257
3	97.5-0.5-2	0.2	33.33	137/249
4	97.5-1-1.5	0.4	33.33	136/253
5	97.5-1.25-1.25	0.5	33.00	131/184
6	97.5-1.5-1	0.6	32.67	140/232
7	97.5-2-0.5	0.8	31.33	125/203
8	97.5-2.5-0	1.0	31.67	126/197
9	95-0-5	0.0	31.00	186/266
10	95-1-4	0.2	31.67	170/257
11	95-2-3	0.4	32.33	150/261
12	95-2.5-2.5	0.5	33.33	168/265
13	95-3-2	0.6	32.00	159/242
14	95-4-1	0.8	31.67	120/201
15	95-5-0	1.0	33.00	135/240
16	92.5-0-7.5	0.0	30.00	142/240
17	92.5- 1.5-6	0.2	31.67	150/250
18	92.5- 3- 4.5	0.4	32.67	175/257
19	92.5- 3.75 -3.75	0.5	33.33	179/262
20	92.5- 4.5- 3	0.6	33.67	183/279
21	92.5-6 -1.5	0.8	34.00	191/295
22	92.5-7.5-0	1.0	35.00	197/309
23	90-0-10	0.0	30.33	120/232
24	90-2-8	0.2	31.33	159/240
25	90-4-6	0.4	32.00	165/225
26	90-5-5	0.5	32.33	176/300
27	90-6-4	0.6	33.00	195/320
28	90-8-2	0.8	33.33	162/260
29	90-10-0	1.0	34.00	180/300
30	87.5-0-12.5	0.0	32.67	165/258
31	87.5-2.5-10	0.2	33.00	140/222
32	87.5-5-7.5	0.4	33.67	150/242
33	87.5-6.25-6.25	0.5	34.33	156/268
34	87.5-7.5-5	0.6	34.67	135/280
35	87.5-10-2.5	0.8	35.00	150/258

(continued)

Table 2 (continued)

S/No	OPC-ESP-GP	ESP/ESP-GP ratio	Consistency wt%	IST/FST Mins
36	87.5-12.5-0	1.0	35.67	154/280
37	85-0-15	0.0	31.00	249/313
38	85-3-12	0.2	31.67	240/330
39	85-6-9	0.4	32.33	251/348
40	85-7.5-7.5	0.5	32.33	254/351
41	85-9-6	0.6	32.67	286/290
42	85-12-3	0.8	33.00	261/283
43	85-15-0	1.0	33.33	275/301
44	80-0-20	0.0	30.00	285/361
45	80-4-16	0.2	30.33	218/274
46	80-8-12	0.4	31.33	273/343
47	80-10-10	0.5	31.67	291/354
48	80-12-8	0.6	33.00	340/370
49	80-16-4	0.8	33.33	332/318
50	80-20-0	1.0	34.33	315/315
51	75-0-25	0.0	33.00	219/498
52	75-5-20	0.2	33.33	251/470
53	75-10-15	0.4	33.67	270/440
54	75-12.5-12.5	0.5	33.67	281/320
55	75-15-10	0.6	34.00	350/460
56	75-20-5	0.8	34.33	309/415
57	75-25-0	1.0	34.33	268/374

be due to the diminution of the clinker content (reduction in C_3S and C_2S content) as well as the presence of decelerators present in the glass chemical composition [12, 45], whereas an acceleration in the final setting time can be attributed to the inclusion of ESP (limestone rich content) into the matrix which improves the hydration rate thus resulting in shortened setting time.

Figure 7 and 8 indicate the effect of cement replacement on the initial and final setting time of ternary cement blend at various ESP/ESP-GP ratio respectively. In general, Fig. 7 indicated a retardation in the initial setting time results as the cement replacement was gradually increased at various ESP/ESP-GP ratios. A significant retardation in the initial setting time at higher cement replacement beyond 12.5 wt% was experienced as the ESP content was gradually increased. Whereas Fig. 8 indicated a gradual retardation of the final setting time as the cement replacement was increased from 2.5 to 25 wt%.

Figure 9 and 10 illustrate the effect of GP and ESP independently on the setting times of binary cements respectively. Figure 9 indicates a retardation in the initial setting time of GP cement blends except at 7.5 and 10 wt% which experienced

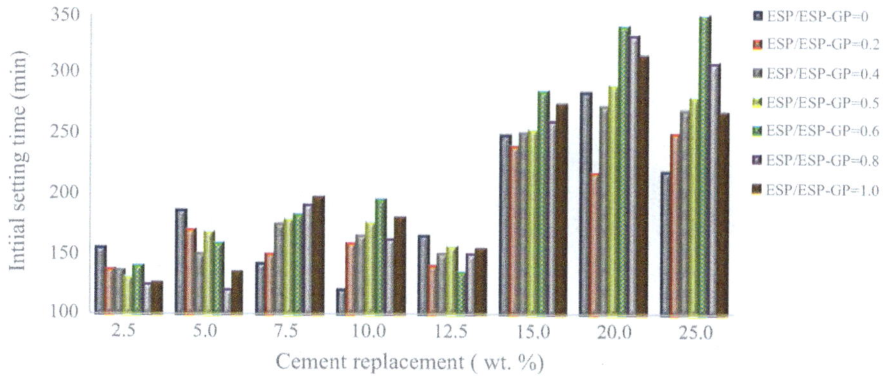

Fig. 7 Effect of Cement replacement on the Initial setting time of ternary cement blends

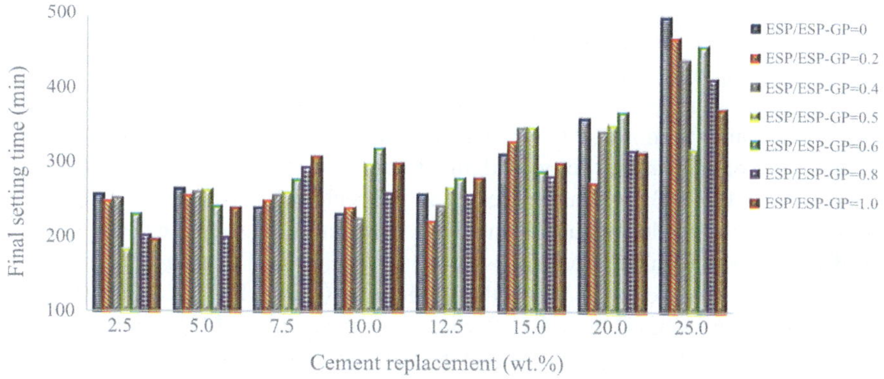

Fig. 8 Effect of cement replacement on the Final setting time of ternary cement blends

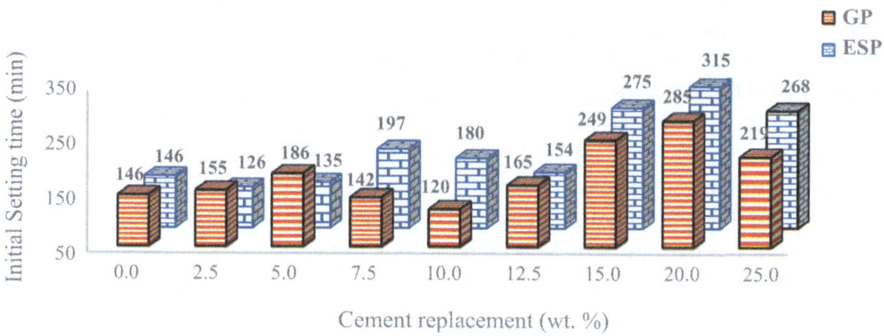

Fig. 9 Effect of GP and ESP content on the initial setting time of various binary cements

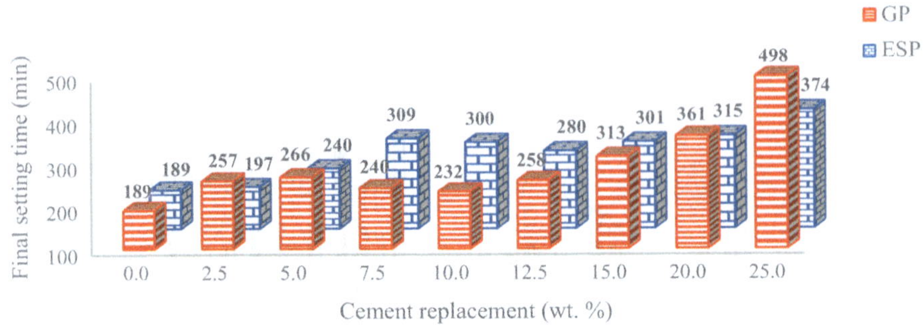

Fig. 10 Effect of GP and ESP content on the final setting time of various binary cements

an acceleration compared to OPC control. On the other hand, Fig. 10 indicated an acceleration in the initial setting of ESP cement blends between 2.5 and 5 wt% beyond which any further increment in the ESP content resulted in a retardation in its initial setting time. This acceleration could be attributed to the presence of limestone produces carboaluminate ($C_3A.CaCO_3$), which decreases the rate of ettringite transformation to monosulfate [43]. The hydration of calcium silicates is enhanced by provision of nucleation sites via calcium hydroxide crystal production at the early stage of hydration [43, 46] thereby resulting in accelerated hydration rates [12].

All cements blended with GP and ESP both had elongated final setting time results compared with OPC control as observed in Fig. 10. The elongated (retarded) setting times could be attributed to the diminution of the clinker content resulting in the reduction in the C_3S and C_2S content as reported by [44]. Péra et al. [42] reported that the presence of limestone tends to accelerate the C_3S reaction which results in the acceleration of the setting times. The inclusion of ESP tends to initate a reaction between calcium carbonate present in the ESP and tricalcium aluminate to form monocarboaluminate which stabilizes ettringite formation by increasing the volume of hydrate products and decrease porosity and promote silicate hydration by provision of nucleation sites for CSH precipitation, thus increasing the reactivity of clinker [40, 41, 45, 47, 48].

Zheng et al. [49] investigated the hydration and setting time of MgO type expansive cement and observed that the inclusion of MgO present in GP retarded the initial cement hydration and increase the setting times due to formation of $Mg(OH)_2$ which has a low solubility which affect $Ca(OH)_2$ saturation ratio negatively and $Mg(OH)_2$ formed from the hydration of MgO tend to precipitate with tiny crystallite size on the cement grains forming a protective layer which retards cement hydration. Thus, an increase in MgO content could result further retardation of both setting times of cement blends compared with OPC control as observed in Figs. 5 and 6. It could also be seen from Fig. 10 that the inclusion of either GP or ESP tend to retard the final setting time of the binary cements as the cement replacement was increased from 2.5 to 25 wt%. The GP cement blends seemed to produce a higher final setting time values compared to ESP cement blend especially between 15 and 25 wt% except 7.5 wt%, 10 wt% and 12.5 wt% cement replacement. This significant retardation could

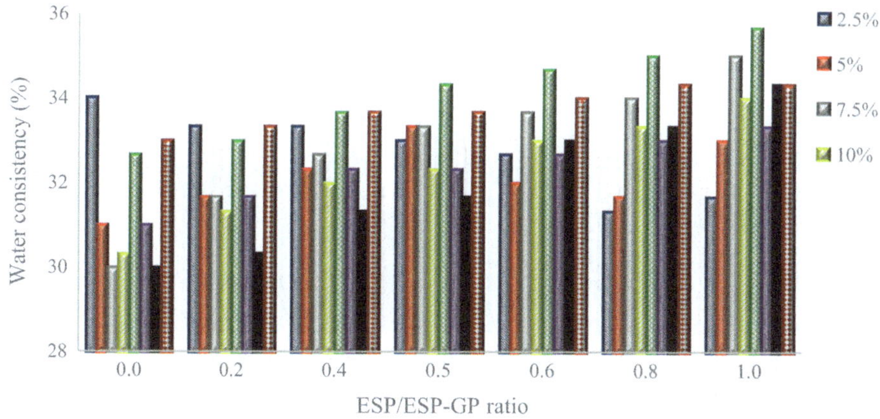

Fig. 11 Effect of ESP and GP content on the consistency of ternary cement blends

be attributed to decreased absorption of free lime [22] coupled with the formation of a protective layer by $Mg(OH)_2$ further retarding cement hydration [49]. Another reason for the elongated setting of cement blended at higher cement replacement could be attributed to excess silica from high silica present in GP.

3.2 *Effect of ESP and GP on the Water Consistency of Blended Cements*

Figure 11 illustrates the effect of ESP and GP content on the standard consistency of ternary cement blends comprising of OPC, GP and ESP at different percentages with cement replacement between 0 and 25 wt%. An increase in the water demand ranging from 90 to 108 ml was experienced as the GP content was gradually replaced with ESP especially at higher cement replacement of 7.5–25 wt% as seen in Fig. 12.

According to [50], they indicated that the particle size distribution considerably influences the water demand of cement blends. The increase in water demand observed as GP was gradually replaced with ESP could be attributed to the narrow particle size distribution while the decrease in water demand as ESP was replaced with GP could be due to the wide particle size distribution owing to inclusion of GP particles. A similar decrease in the water demand was experienced for low cement replacement at 2.5% as GP was gradually replaced with ESP owing to the wide particle size distribution of the cement blend whereas at 5 wt% cement replacement resulted in a series of increase and decrease in the water consistency which could be attributed to the variation in the particle size distribution of the cement blends.

Figure 12 illustrates the effect of cement replacement on the water consistency of ternary cement blends and it was seen that cement blends required higher water demand for normal consistency with an increase in ESP and GP content from 2.5 to 25

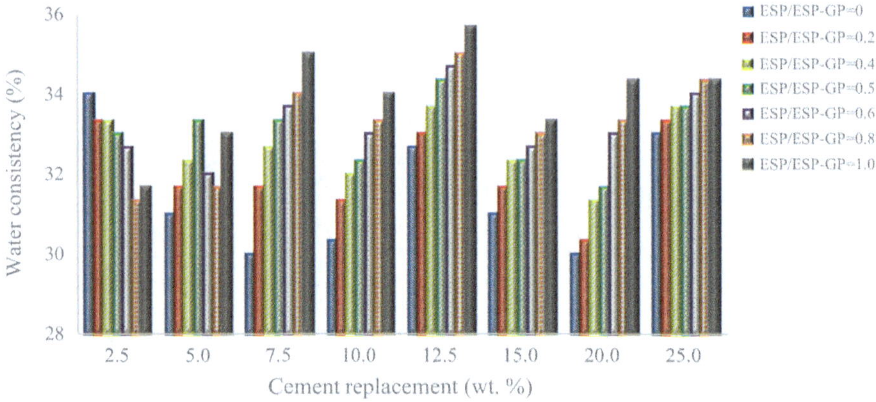

Fig. 12 Effect of cement replacement on the water consistency of ternary cement blends

wt% and their standard consistencies increments ranged from 7.14 to 27.34%. Studies revealed that replacement of OPC with GP led to an increase in the water demand according to [31] which could probably be due to porous nature of the GP particles. The glass powder chemical composition revealed high amount of silica which can imparts on the workability of the cement blends. It could also be observed that at a given ESP/ESP-GP ratio between 0 and 1.0, the water for standard consistency for the cement blends experienced a variation (i.e. a series of increases and decreases) as the cement replacement level was increased from 0 to 25 wt%.

Figure 13 illustrates the influence of cement replacement on the water consistency for various binary cement blends. All ESP and GP cement blends required more water compared to OPC control as the cement replacement was increased from 0 to 25 wt%. The increased water requirement could be due to the enhanced surface area of GP as well as the angular shape of the glass particle [32]. It was also observed that GP

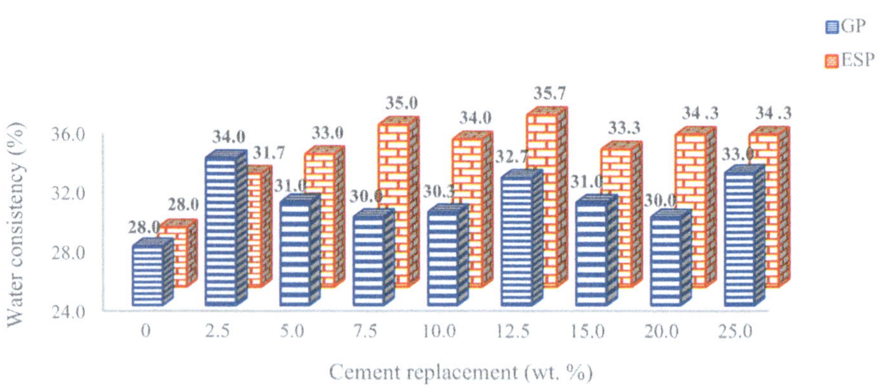

Fig. 13 Effect of cement replacement on the water consistency of various binary cements

cement blends required less water demand for all cement replacement except at 2.5 wt% compared to ESP cement blends which could be related to a wider particle size distribution of the cement blends.

4 Conclusions

In conclusion, the chemical composition of OPC, ESP and GP were obtained via XRF spectrometer and the percentage of reactive silicates content of CEM II A-L 42.5 R contains 12.10 wt% which was lower than 69.95 wt% for GP. GP could be considered as a pozzolan as it satisfies the requirements specified by ASTM C618 [39]; $SiO_2 + Al_2O_3 + Fe_2O_3$ is greater than 70 wt% with SO_3 and LOI also satisfied the requirements. GP contains 12.39 wt% CaO which is within the range of 10–30 wt% CaO and considered a class C pozzolan. On the other hand, ESP's chemical composition revealed a high amount of CaO of 48.26 wt% (containing $CaCO_3$ 86.13 wt%) and other minute quantities of SiO_2, Al_2O_3, Fe_2O_3, MgO, SO_3, K_2O and Na_2O which were below 1 wt%. ESP has a high LOI of 47.38 wt% which could be attributed to the decomposition of limestone to lime.

The initial and final setting time was influenced by two factors; dilution effect and pozzolanic/filler effect. The retardation and acceleration of the setting time of cement blends were experienced as GP content was replaced with ESP content. The retardation of the setting time could also be linked to the presence of MgO in the glass composition due to formation of $Mg(OH)_2$ which has a low solubility which affect $Ca(OH)_2$ saturation ratio negatively and $Mg(OH)_2$ formed from the hydration of MgO tend to precipitate with tiny crystallite size on the cement grains forming a protective layer which retards cement hydration. Cement replacement beyond 5 wt% for both initial and final setting times resulted in a steady retardation as the GP was replaced with ESP. Both setting times experienced significantly retardation as the cement replacement was increased. The ternary cement blends experienced a significant elongation of its initial setting time especially beyond 12.5 wt% cement replacement whereas the final setting time experienced a steady retardation as the cement replacement level was increased.

All ESP and GP cement blends required more water compared to OPC control as the cement replacement was increased from 0 to 25 wt%. An increase in the water consistency was experienced as GP gradually replaced with ESP especially at higher cement replacement between 7.5 and 25 wt% was attributed to the narrow particle size distribution while a wider particle size distribution (decrease in water demand) was observed as the ESP was replaced with GP. It could be deduced that GP particle significantly influenced the particle size distribution. Another reason for higher water demand of cement blends can be due to high silica content in glass powder owing to the enhanced surface area and angular shape of the glass particle.

Acknowledgements The authors wish to thank the Department of Chemical Engineering of Abubakar Tafawa Balewa University, Bauchi, Nigeria and Ashaka cement Plc. for providing infrastructure, facilities, and support to this research work.

Conflict of Interest The authors declared that they have no conflict of interest.

References

1. Abdel-Shafy HI, Mansour MSM (2018) Solid waste issue: sources, composition, disposal, recycling and valorization. Egypt J Pet 27(1):275–1290
2. Venkatraman S, Vinothkumar S (2018) Experimental investigation on developing low Cost concrete from paper industry waste. Int J Pure Appl Math 119(12):8999–9010
3. Gowsika D, Sarankokila S, Sargunan K (2014) Experimental investigation of egg shell powder as partial replacement with cement in concrete. Int J Eng Trends Technol (IJETT) 14(2):65–68
4. Islam SGM, Rahman MH, Kazi N (2017) Waste glass powder as partial replacement of cement for sustainable concrete practice. Int J Sustain Built Environ 6:37–44
5. Georgescu M, Saca N (2009) Properties of blended cements with limestone filler and fly ash content. Sci Bull Series B 71(3):13–14,16
6. Yukalang N, Clarke B, Ross K (2018) Solid waste management solutions for a rapidly urbanizing area in Thailand: recommendations based on stakeholder Input. Int J Environ Res Public Health 15(7):1302
7. Selin E (2013) Sustainable municipal solid waste management—A qualitative study on possibilities and solutions in Mutomo, Kenya. Department of Ecology and Environmental Science (EMG), Umeå University, Umeå, Sweden
8. Taiwo AM (2011) Composting as a sustainable waste management technique in developing countries. Sci Technol 4:93–102. https://doi.org/10.3923/jest.2011.93.102
9. Fazeera U, Kamrau SA, Zarina YHH (2016) Viability of eggshell ash affect the setting time of cement. Int Sch Sci Res Innovat 10(3):342–346
10. Kaya A (2010) A study on blended bottom ash cements MSc Thesis, Middle East Technical University, pp 1–73
11. Olubajo OO, Osha AO, El- Nafaty UA, Adamu HA (2014) Effect of water-cement ratio on the mechanical properties of blended cement containing bottom ash and limestone. Int Inst Sci Technol Educ (IISTE) 6(2):1–9
12. Olubajo OO, Osha AO, El- Nafaty UA, Adamu HA (2017) A study on coal bottom ash and limestone effects on the hydration and physico-mechanical properties of ternary cement blends. Ph.D. Thesis. Abubakar Tafawa Balewa University, Bauchi, Nigeria
13. Farah W, Ramadhansyah PJ, Badorul HA, Megat AM (2011) Effect of rice husk ash to the performance of concrete block. Int J Appl Sci Technol 1(3):53
14. Raheem A, Kareem MA (2017) Chemical composition and physical characteristics of rice husk ash blended cement akeem. Int J Eng Res Afr 32:24–35
15. Olubajo OO, Isa YM, Ayeni S, Menta S, Nwuhu W (2020) A study on ordinary Portland cement blended with rice husk ash and metakaolin. Path Sci 6(1):3001–3019
16. Marthong C (2012) Sawdust ash (SDA) as partial replacement of cement. Int J Eng Res Appl (IJERA) 2(4):1980–1985
17. Wang B, Ma H, Li M, Han Y (2013) Effect of metakaolin on the physical properties and setting time of high-performance concrete. Key Eng Mater 539:195–199
18. De Weerdt K, Kjellsen KO, Sellevold E, Justnes H (2011) Synergy between fly ash and limestone powder in ternary cements. Cement Concr Compos 33(1):30–38
19. Mujedu KA, Adebara SA (2016) The use of groundnut shell ash as a partial replacement for cement in concrete production. Int J Sci Eng Environ Technol (IJOSEET) 1(3):32–39

20. Matos AM, Sousa-Coutinho J (2012) Durability of mortar using waste glass powder as cement replacement. Constr Build Mater 36:205–215
21. Rashed AM (2014) Recycled waste glass as fine aggregate replacement in cementitious materials based on Portland cement. Constr Build Mater 72:340–357
22. Zhang J, Gong CH, Lu L, Wang S, Hou P (2015) Ceramics–Silikáty 59(2):135–144
23. Damtoft JS, Lukasik J, Herfort D, Sorrentino D, Gartner EM (2008) Sustainable development and climate change initiatives. Cem Concr Res 38(2):115–127
24. Muller C (2006) Cluster 2: Blended cements, European construction in service and society (ECO-Serve), ECO-serve seminar challenge for sustainable construction; the concrete approach, Warszawa, Poland, May 18–19
25. Al-Zubaid AB, Shabeeb KM, Ali AI (2017) Study on the effect of recycled glass on the mechanical properties of green concrete. In: International conference on technologies and materials for renewable energy, environment and sustainability, TMREES17, Beirut Lebanon, 21–24 Apr 2017
26. Imbabi MS, Carrigan C, McKenna S (2012) Trends and developments in green cement and concrete technology. Int J Sustain Built Environ 1:194–216
27. Rashad MA (2015) A brief on high-volume class F fly ash as cement replacement–a guide for civil engineer. Int J Sustain Built Environ 4:278–306
28. Manoj K, Vedpal, Ravinder (2016) Behavior of concrete with waste glass fiber powder. Int J Res Appl Sci Eng Technol (IJRASET) 4, ISSN: 2321-9653
29. Rahma A, Naber NE, Ismail SI (2017) Effect of glass powder on the compression strength and the workability of concrete. Cogent Eng Civil Environ Eng 4(1). https://doi.org/10.1080/23311916.2017.1373415
30. Kaveh A, Prasada RR (2016) Impact of combined use of glass powder and crushed glass aggregate on selected properties of Portland cement concrete. Constr Build Mater 117:263–272
31. Vandhiyan R, Ramkumar K, Ramya R (2013) Study on replacement of cement by glass powder. Int J Eng Res Technol (IJERT) 2(5):234–238
32. Shao Y, Lefort T, Moras S, Rodriguez D (2000) Studies on concrete containing ground waste glass. Cem Concr Res 30(1):91–100
33. Nwaubani SO, Poutos KI (2013) The influence of waste glass powder fineness on the properties of cement mortars. Int J Appl Innov Eng Manag (IJAIEM) 2(2):110–116
34. Taha B, Nounu G (2009) Utilizing waste recycled glass as sand/cement replacement in concrete. J Mater Civ Eng 21(12):709–721
35. Tayh SA, Razzaq AK, Yousif AR (2018) Characterization of hot mix asphalt modified by egg shell powder. Int J Eng Res Technol 11(3):481–492
36. Dhanalakshmi M, Sowmya NJ, Chandrashekar A (2015) A comparative study on egg shell concrete with partial replacement of cement by fly ash. Int J Eng Res Technol (IJERT) 4(5):1532–1538
37. ASTM C 187 (2010) Standard test method for normal consistency of hydraulic cement. Annual Book of ASTM Standards
38. ASTM C 191-19 (2010) Standard test method for Time of Setting of hydraulic cement by Vicat Needle. Annual Book of ASTM Standards
39. ASTM C 618 (2008) Standard specification for coal fly ash and raw or calcined natural pozzolan for use in concrete. Annual Book of ASTM Standards
40. Chaipanich A, Thongsanitgarn P, Wongkeo W, Sinthupinyo S (2012) Effect of limestone powders on compressive strength and setting time of Portland-limestone cement pastes. Adv Mater Res 343:322–326. Trans Tech Publications
41. Lothenbach B, Le Saout G, Gallucci E, Scrivener K (2008) Influence of limestone on the hydration of Portland cements. Cem Concr Res 38(6):848–860
42. Péra J, Husson S, Guilhot B (1999) Influence of finely ground limestone on cement hydration. Cem Concr Compos 21(2):99–105
43. Bonavetti VL, Rahhal VF, Irassar EF (2001) Studies on the carboaluminate formation in limestone filler-blended cements. Cem Concr Res 31(6):853–859

44. Olubajo OO, Osha OA (2013) Influence of bottom ash and limestone powder on the properties of ternary cement and mortar. Int J Eng Res Technol 2(7):1201–1212
45. Tsivilis S, Chaniotakis E, Kakali G, Batis G (2002) An analysis of the properties of Portland limestone cements and concrete. Cem Concr Compos 24(3):371–372
46. Soroka I, Setter N (1977) The effect of fillers on strength of cement mortars. Cem Concr Res 7(4):449–456
47. Hooton RD, Nokken M, Thomas MDAT (2007) Portland-limestone cement: state of the art report and gap analysis for CSA A 3000, SN3053. Cement Association of Canada, Toronto Ontario, Canada, pp 1–60
48. Matschei T, Lothenbach B, Glasser FP (2007) The role of calcium carbonate in cement hydration. Cem Concr Res 37:551–558
49. Zheng L, Xuehua C, Mingshu T (1992) Hydration and setting time of MgO type expansive cement. Cem Concr Res 22(1):1–5
50. Sprung S, Siebel E (1991) Assessment of the suitability of limestone for producing Portland limestone cement. ZKG Int Ed B 44(1):1–11
51. Hama SM, Hamdullah DN, Ashour HM (2019) Effects of eggshell powder as partial replacement of cement on flexural behavior of one-way concrete slabs. J Eng Sci Technol 15(5):2509–2521

A Systematic Approach to Reduce Operation and Maintenance Costs of Private Highway Projects in Malaysia

Monirul Islam, Sarajul Fikri Mohamed, Zakaria Mohd Yusof, Syamsul Hendra Mahmud, Azmi Mohamed, and Noor Azeyah Khiyon

Abstract A systematic approach in managing operation and maintenance costs of private highways is crucial to obtain sustainable long-term profit of private highways development in Malaysia. Operating and maintenance costs in road projects are a real challenge and form an significant part of the total cost of road growth. This paper is start with examining the process of private highway developments in Malaysia includes all associated funding received from Government of Malaysia since 1982. Cost data from twenty-nine private highway projects are collated and analyzed. The outcome of the research shows that majority of the highways were built with no systematic approach to managing operating and maintenance costs during the concession period agreement. A literature analysis identified five main components and twenty-two elements are contributed to the escalating of operation and maintenance costs. This paper uses a qualitative approach by analyzing secondary data from various industrial report to provide systematic breakdown of highway assets to develop a sustainable cost model. This study has clearly demonstrated the need to develop a sustainable cost model for the management of operations and maintenance of private highways in Malaysia.

Keywords Component · Cost model · Operation and maintenance · Private highways · Sustainable

---

M. Islam (✉) · S. F. Mohamed · Z. M. Yusof · S. H. Mahmud
Department of Quantity Surveying, Faculty of Built Environment and Surveying, Universiti Teknologi Malaysia, Utm Johor Bahru 81310, Johor, Malaysia
e-mail: monirl76@gmail.com

A. Mohamed
Quantity Surveying & Contract Branch, Malaysian Highway Authority (MHA), Jalan Serdang Kajang, Selangor 43000, Bandar Baru Bangi, Malaysia
e-mail: azmi.mohamed@fsmt.upsi.edu.my

N. A. Khiyon
Faculty of Society & Design, Bond University, Robina, QL 4226, Australia
e-mail: azeyah89@yahoo.com

M. A. A. Zaini et al. (eds.), *Proceedings of the 3rd International Conference on Separation Technology*, Lecture Notes in Mechanical Engineering,
https://doi.org/10.1007/978-981-16-0742-4_3

1 Introduction

Quality of a highway project is the fulfilment of client requirements and satisfaction of the end user [1–5]. Hasan et al. [6] identified labor issues and financial issues are major contributors to the increase in highway maintenance costs. According to [7] construction stakeholders handling huge quantity of projects at the similar period, material shortage, and unavailability of competent staff affects the operation and maintenance cost of highway project. Heravi and Mohammadian [8] investigated issues affecting operation and maintenance cost of highway projects in Iran found that significant factors affecting operation and maintenance cost are: poor quality of construction materials, low skill workers, lack of quality assurance, poor technical knowledge of contractors, unrealistic project cost, making slow decision, and inadequate site supervision. Other factors affecting quality of construction projects includes unclear client's requirements for design, improper material selection, lack of coordination between designer and owners, use of improper equipment [3, 9–12]. Rahman et al. [13] identified eleven factors that affect the operation and maintenance cost of highway projects in Malaysia which are: competencies of project managers, availability of equipment, availability of workforce, availability of materials, workflow for the design process, weather, site location, involvement from upper management, support from clients, efficiency of authorities and public acceptance. França et al. [14] With the emergence of more customer requirements as regards infrastructure design, maintenance costs may be even greater in future years. These rapid increases in maintenance costs, however, have serious consequences for infrastructure running costs and even researchers have raised their concerns over the drawbacks of this key question on the basis of empirical studies. Therefore, this paper is starts with a discussion on the process and development of highway assets in Malaysia since 1982. The tolls collected in earlier stage of highway privatization in Malaysia is examined with an aim to justify the budget allocation for operation and maintenance of the highway assets in Malaysia. Secondly, the paper is discussed the systematic work breakdown structure and main element of highway. It is important to established elemental breakdown structure that will facilitate the process of gathering operational and maintenance cost data throughout the concession duration. According to [15], there are some components of the highway that are difficult to operate and maintenance which includes slope and retaining structures, road surface, drainage, traffic furniture, bridges and other structures. Secondly, correlation tests were performed to determine the relationships between the operational and maintenance costs and road components. This research provides insights into key factors in the operation and maintenance of highway components' cost efficiency and maintenance complexity.

2 Literature Review

2.1 Highway Privatization Projects Funding in Malaysia

Therefore, financial aid from the government is an essential step in ensuring that the project is completed and delivered to citizens. Highway privatization projects require substantial resources. In order to execute the project on a wind-win basis, both government and private sector will cooperate. This is not necessary for the government to borrow enormous funds for repairing highways. Unless it had to rely on the traditional road building process, it would take to delay for adequate moneys to investment the schemes. This was an incredibly slow operation. The Malaysia Highway Authority (MHA) has bridged the move from the outdated technique of financing highways to open privatization. In 1980, MHA was established pursuant to Act 231. As a legal entity which was eventually meant to be independent financing, MHA allowed local and foreign lenders to obtain commercial loans. Although government guarantees all loans it intends to receive, MHA will be more like a commercial organization. In 1981, the government provided MHA with RM 500 million under the Fourth Malaysian Plan (RMK-4). It consisted of RM 150 million and RM 350 million respectively in grant and loan amounts. In the following year, the government granted a supplementary loan for a monetary organization of Japan which are RM114.6 million. Through union loan in the following year of an international financial institution, MHA secured a further loan worth RM 228,6 million. Table 1 shows the total amount of government loans for road development.

Table 2 show the government collected toll revenue over that time instead of loans worth RM2.151 billion.

In 1994, MHA had no longer gained revenue from toll road operations with the privatization of toll roads. She was then willing, with the help of the government, to continue paying her repayments on time in the method of straight allowances.

Table 1 Government's financing road projects from 1981 to 1987 [16]

Year	Government loan using amount (RM Mil)	Remarks
1987	63,820,000.00	Loan interest payment
1985	1,870,000,000.00 144,400,000.00	Loan Debt servicing
1984	1,100,000,000.00	Loan
1983	114,600,000.00	Loan (Japanese Financial Institution)
1982	500,000,000.00	RM350 million (loan) RM135 million (grant)
1981	228,600,000.00	Loan (International Financial Institution)
Total	2,151,420,000.00	

Table 2 Highway privatization toll collected for earlier stage in Malaysia [16]

Year	Toll Collected (RM Mil)	Highway
1987	65,900,000.00	Slim River–Tg. Malim Lebuhraya KL–Seremban
		Lebuhraya KL–Karak
		Labuhraya KL–Karak Johor Causeway
1985	53,000,000.00	Slim River–Tg. Malim Lebuhraya KL–Seremban &
		Lebuhraya KL–Karak
		Labuhraya KL–Karak Johor Causeway
1984	20,500,000.00	Karak Johor Causeway–Labuhraya KL
1983	2,800,000.00	Slim River–Tg. Malim
	12,300,000.00	Seremban–Lebuhraya KL
1981–1982	900,000,000/month	Lebuhraya KL–Karak

2.2 *Operating and Maintaining Highways Cost Incurred During Concession Period*

In accordance with the guiding standard of publication by MHA, the contracted company must operate and maintain the route. It has been noted that PLUS has invested 14% of its repairs and 21% of its maintenance and service budget. Table 3 shows the concessionaires' expenses during the concession periods.

2.3 *Current Toll Highways in Malaysia*

During the Fourth Malaysian Plan in 1984 the first privatization road project (North Klang Straits Bypass) began. The road was designed with two dual carriages of eight kilometers. By 2013 29 privatized road projects with a national 1,942 km total length at a cost of RM41, 34 billion have been in service. The list of existing toll roads in our country is given in Table 4.

Table 3 Highway privatization projects financing, operation and maintenance cost incurred information [16]

Highway Project	Maintenance Cost	Operation Cost	Financing Cost	Remarks
SILK				Forecasted traffic volume below 60–70%
SMART	6,000,000.00RM Mil/Yr			Maintenance
			18,000,000.00 RM Mil	Projected traffic volume below 40%
PLUS	21%	14%		4,000 culverts 5,500 slopes 50 RSA 400 bridges
LDP	1,000,000,000.00 RM Mil			Expand the toll plaza, nos. of lanes and construct interchanges
	300,000,000.00 RM Mil			New construction Year 2009–2012
	10,000,000.00 RM Mil/Year			Maintenance

2.4 *Highway Maintenance During Concession Duration*

Maintenance by highway avoids investment loss in the initial development of roads. Routine and periodic maintenance costs are estimated to be 2 to 3% of the initial investment in capital throughout the lifetime of a road [17]. Nevertheless, disused repairs could cause this amount to increase. According to [18] timely repairs expenditures of US $12 billion in Africa would save road reconstruction costs of $ 45 billion over a decade. Grimm et al. [19] an estimate of around 1–3 percent of GDP from 33 individual countries in Sub Saharan Africa is the amount of investments lost on the yearly basis from neglected maintenance. Approximately 75% of it is in shallow foreign exchange type. In 1992, equivalent figures in the Caribbean and Latin America were valued to be $1.7 trillion a year representing 1.4% of the country's GDP.

2.5 *Categorization of Highway Components for Maintenance*

According to [20] it has long been known that the roads maintained have failed to offer proper attention to the problems of maintenance associated with other designs. Aging buildings, significantly higher traffic levels, tighter budgets and workforce

Table 4 The existing toll highways lists in operations [16]

No.	Highways project	Length (KM)	Year opened	Total year of concession
1	Penang Second Bridge	24	2013	45 year
2	Eastern Dispersal Link Expressway	8.1	2012	34 year
3	South Klang Valley Expressway	52	2012	40 year
4	East Coast Expressway Phase 2	184	2011	–
5	KL–Kuala Selangor Expressway	33	2011	40 year
6	Kemunig–Shah Alam Expressway	14.7	2010	40 year
7	Kajang–Seremban Highway	44.3	2010	33 year
8	Senai–Pasir Gudang–Desaru Expressway	77	2009	33 year
9	Duta–Ulu Kelang Expressway	18	2009	34 year
10	Kuala Lumpur–Putrajaya Highway	26	2007	33 year
11	Smart Tunnel	3	2007	40 year
12	Guthrie Corridor Expressway	25	2005	33 year
13	Kuala Lumpur–Karak/East Coast Expressway Phase 1	174	2004	–
14	New Pantai Expressway	19.6	2004	33 year
15	Kajang Dispersal Link Expressway	37	2004	36 year
16	Western KL Traffic Dispersal System	26.5	2004	33 year
17	New North Klang Straits Bypass Expressway	17.5	2002	25 year
18	Ampang–Kuala Lumpur Elevated Highway	7.9	2001	33 year
19	Cheras–Kajang Expressway	11.5	1999	32 year
20	Sungai Besi Expressway	16.7	1999	35 year
21	Damansara–Puchong Expressway	40	1999	33 year
22	Second Link Expressway	47	1998	45 year
23	Shah Alam Expressway	34.5	1998	28 year 9 month
13	Kuala Lumpur–Karak/East Coast Expressway Phase 1	60	1998	32 year
24	Seremban–Port Dickson Highway	22.7	1998	50 year 9 month

(continued)

Table 4 (continued)

No.	Highways project	Length (KM)	Year opened	Total year of concession
25	Butterworth–Kulim Expressway	17	1996	32 year
26	East West Link Expressway and Kuala Lumpur–Seremban Highway	16.5	1995	23 year
27	North South Expressway	823	1994	50 year
28	Penang Bridge	13.5	1985	25 year
29	North South Expressway Central Link	48	1997	36 year

limits have served only to make the maintenance issues worse. The maintenance of pavement is degrading with age and usage irrespective of how well planned and installed, while the engineers must be conscious of the type of disability. The main cause of the decay of a maintained paved road and the consequent failure are as follows: the action of roads, with the highest detrimental effect on heavy goods vehicles; action on the weather, that is to say rain, and heat; unstable ground or poorly drained conditions; post construction trenches on bases and surfaces; etc. (Table 5).

2.6 *Component Defects*

Based on the structured breakdown of highway components and main element, it is imperative to analyze the types of defects of each highway component. This literature analysis process provides consideration of the previous studies in the area of highway maintenance management. The emphasis on empirical evidence is important in the case of practitioners, because it allows the components defects to be evaluated in a real-world environment, in addition provides an insight into the appropriate use of cost-based maintenance management as indicators. Moreover the importance of developing a structured category of maintenance cost model has demonstrated by this paper. It is now essential that a structured approach for operation and maintenance cost management is in abode to confirm the continued usability, profitability and reliability of the highway assets being achieved by Concessionaire (Table 6).

2.7 *Highway Operation and Maintenance Costs*

The structure of operations and maintenance on highway components life cycle cost should be designed for a long-term process. Meanwhile, long-term design characteristics such as maintainability, operability and human factors that ease performing

Table 5 Category of highway component and main element [21]

Category of Component	Element	Example of elements	Description
Road Surface	Shoulder/Verge	Paved or unpaved	Binder Course links the wearing course to the road base, the wearing course offers a smooth running, Saving the lower layers of abrasion and weathering of moving vehicle. The road surface' s function is to prevent the penetration of water to the ground
	Carriageway	Asphaltic concrete	
	Median	Paved or unpaved	
Slope and retaining structure	Retaining Structure	Facing wall	The functionality of the slope & retention structure portion influences both the protection and comfort of the road user. It consists of five principal elements (rough, rocky, soft), retaining system, drainage slope (surface and sub-surface), ground services and slope furniture. The components are divided into five main elements
	Slope	Rigid, soft, rock surface	
	Drainage	Surface & Subsurface Drainage	
	Above ground services furniture	Water pipes, telecommunication tower	
	Slope furniture	Handrail and facing	
Drainage	Drain Cover	R.C. Slab, M.S Grating	The efficiency of the drainage system on completed road works is assessed to make sure that there is no flooding that can impact on route operation. Each part for the purposes of this norm is divided into four parts, drain, sump and pipe/box culverts
	Drain	Roadside, median, shoulder drain	
	Sump	Concrete Sump and Brick wall sump	
	Pipe/Box Culverts	R.C., Culvert Body (Concrete & Steel), Headwall & Wing wall, C.M.P. Culvert	

(continued)

Table 5 (continued)

Category of Component	Element	Example of elements	Description
Traffic/Road Furniture	Marking	Pedestrian Crossing, pavement markings, traffic Ghost Island	The following are: traffic calming devices, road signs, pavement markings, electric devices, peat belt instalations, kerbs, barriers to traffic and delineators, traffic or road furniture elements covered by this standard
	Traffic Calming devices	Speed hump/table	
	Electrical devices	Pedestrian Push button signal, Variable Message Sign(VMS),Traffic signal, street lighting	
	Traffic Sign	Warning, information signs	
	Pedestrian facilities	Sidewalk, bus stop	
	Kerb	Median, Roadside, traffic island curb	
	Delineators	Guardrail, concrete and noise barrier, crash cushion	
	Traffic barrier	concrete and noise barrier, guardrail and crash cushion	
Bridge & Other structures	Above ground structures(bridges)	Pedestrian/Motorcycle bridge Road bridge	The components Bridge and other structures are divided into the various forms of bridge structure. pedestrian or motorcycle bridges, road bridges, and special structures like docks and tunnels for the underpass or car box
	Special structures	Box culverts, tunnels	

maintenance activities have not received much emphasis during the project design and construction phases. A major portion of the life-cycle cost of facilities is the maintenance cost and the US construction industry spends over US$200 billion each year on maintenance [36]. Despite the increasing maintenance costs, the budget allocated to carry out maintenance activities is limited, and it is a challenge for facility managers [37]. Over the years, there have been quite a few classifications of maintenance cost. These varied classifications made by different researchers and organizations prevented an agreement on a common platform. The total repair costs that a certain infrastructure causes consist different items of expenditure. According to The Finnish Property Market Technical Report, 2017 [38] maintenance and usage

Table 6 Some author categorize highway component defect groups [21]

Category of Component	Element	Defect group	Authors
Road Surface	1. Carriageway 2. Median 3. Shoulder & Verge	1. Riding Quality 2. Aesthetic 3. Safety	[22–24]
Drainage	1. Drain 2. Sump 3. Drain Cover 4. Pipe/Box Culverts	1. Joint and gap 2. Crack and damage 3. Functionality 4. Finishing	[25–27]
Slope and retaining structure	1. Slope 2. Slope drainage 3. Retaining structure 4.Aboveground services 5. Slope furniture	1. Crack and damage 2. Joints filler and gaps 3. Functionality 4. Finishing 5. Installation	[28, 29]
Traffic/Road Furniture	1. Marking 2. Traffic Calming devices 3. Pedestrian facilities Kerb 4. Electrical devices 5. Traffic Sign 6. Traffic barrier, Delineators	1. Finishing 2. Functionality 3. Installation 4. Visibility 5. Crack and damage	[30–32]
Bridge& Other Structures	1. Special Structures (Underpass/Vehicular box culverts, Tunnel) 2. above ground structures (Road bridge,Motorcycle/Pedestrian bridge)	1. Finishing 2. Crack and damage 3. Installation 4. Visibility 5. Joint and gap 6. Functionality	[33–35]

costs, outdoor maintenance costs and cleaning costs and maintenance cost can be categorized into service costs; administration costs, insurance costs, waste disposal costs and heating, water and electricity costs into energy costs and other maintenance costs into other expenditure items (Fig. 1).

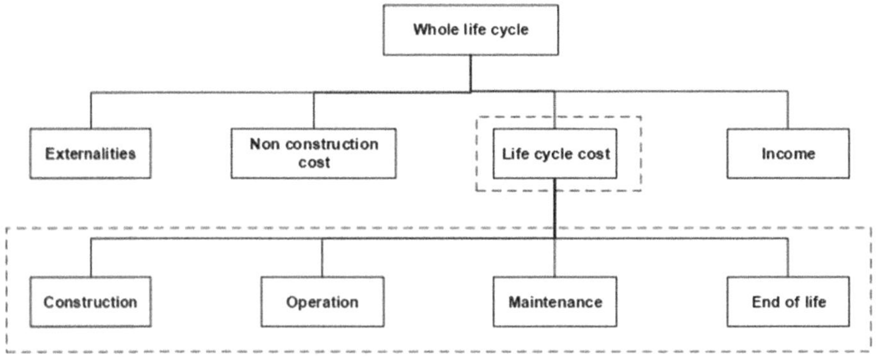

Fig. 1 Whole life cycle cost and life cycle cost structure [39]

The maintenance costs are associated with regular maintenance, replacements and outdoor (Norwegian Facility Management Association (NBEF) NS 3454). In addition, Danish Facility Management Network (DEF) focuses more on the property operation and maintenance based on DFM 19.12.2007. DEF mentioned maintenance cost that includes building exterior, building indoors, construction and installations and terrain. An appropriate maintenance management is required to control and reduce the maintenance cost systematically. Cost of maintenance can be significantly reduced if the number of maintenance activities and time are reduced. The integration of FM knowledge and experience into the design phase can facilitate cost-effective maintenance practice at post-occupancy phase [40, 41].

3 Sustainable Operation and Maintenance Cost Model

Cost of highway maintenance and operation is going up rapidly in many countries especially in developing countries (i.e. Malaysia, Vietnam and Thailand). The main reason of the increasing cost of maintenance can be regarded of the following factors: types of highway (urban, inter urban and intra urban) and characteristics, user factors, implementation of government policies and maintenance management. Wong and Chan [42] revealed that 35% from operation and establishment, 58% of infrastructure defects causes of faulty design, 12% from poor materials and 11% from unforeseen clients' necessities. Errors and omissions in changes initiated and design documentation by the end-user and client also are the key issues of increasing of maintenance cost [43]. Typically, maintenance cost for infrastructure assets not often considered during construction and design stages. As a result, crack appear on the infrastructure and highway assets due to completed construction method, types of used materials and quality. [44] find that most of infrastructure crack were the result of faulty construction and design and this led to increase repair works. The cause of design fault and its effect on maintenance also was studied by [45]. Ofori et al. [46] concluded that lack of maintenance plan, faulty designs, unavailability of skilled labor and inadequate financial support for maintenance work were the most influential factors to increase operation and maintenance cost. Based on that, within the design phase maintenance should be taken into consideration and producing appropriate design documentation should emphasize. Therefore, the study emphasize on evaluating their impact on maintenance cost of infrastructure and highway projects and classifying the types of defect. A viable and comprehensive operation and maintenance costs model shall be developed to help estimator make comparisons to the factors or risks they considered for maintenance and operation costs. It will also help to reduce uncertainty for the agreed contract terms signed by Concessionaire and the Government. By providing comprehensive cost model, it becomes possible to develop a systematic process that can guide, structure and capture their rationale during the process of generating estimate. It is important to identify the most significance infrastructure and highway assets, current estimating techniques

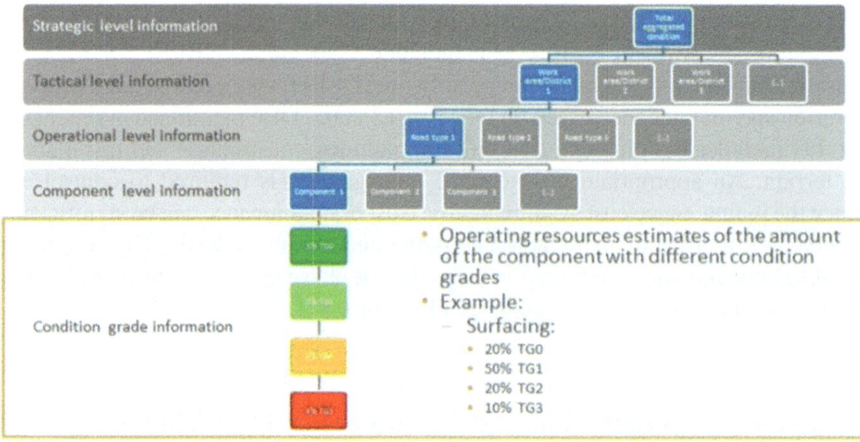

Fig. 2 Classification of highway asset infrastructure components/elements [47]

used by project team and to develop an appropriate action plan to integrate operation and maintenance costs model into privatize highway projects (Fig. 2).

4 Conclusion

Most significant category and element of highway assets has provided evidence by this study that affecting operation and maintenance cost and identified the relevant highway elements which maintenance are most difficult. Therefore, this paper has provided practical significance to facilities managers and asset it will rapid the management of private highways to attention on the systematic elemental breakdown which affecting operation and maintenance costs and thereby reduced long term profit. The problem of increasing operation and maintenance costs is not only wide but also the coverage shall not only be limited to current studies. This study recommends management system to further explore maintenance cost minimization or an innovative solution. The proposed sustainable-cost operation and maintenance model which integrate Life Cycle Costing (LCC) method are relevance to concessionaire company and relevant stakeholders the following benefits. Firstly, the proposed cost model based on LCC approach is capable to produce an acceptable toll rate for benefits to all parties involved with a systematic and strategic cost analysis and planning. Secondly, a systematic operation and maintenance cost model provides accurate cost information to Concessionaire based on elemental breakdown of maintenance components using standard format for three types of maintenance namely routine, curative and preventive maintenance.

References

1. Aoieong RT, Tang SL, Ahmed SM (2002) A process approach in measuring quality costs of construction projects: model development. Constr Manage Econ 20(2):179–192
2. Kärnä S et al (2013) Exploring project participants' satisfaction in the infrastructure projects. Eng Proj Organ J 3(4):186–197
3. Larsen JK et al (2016) Factors affecting schedule delay, cost overrun, and quality level in public construction projects. J Manage Eng 32(1):04015032
4. Walters D (2008) Demand chain management + response management = increased customer satisfaction. Int J Phys Distrib Logistics Manage 38(9):699–725
5. Yng Ling FY, Ma Y (2014) Effect of competency and communication on project outcomes in cities in China. Habitat Int 44:324–331
6. Hasan A et al (2018) Factors affecting construction productivity: a 30 year systematic review. Eng Constr Archit Manage 25(7):916–937
7. Mok KY et al (2017) Investigating key challenges in major public engineering projects by a network-theory based analysis of stakeholder concerns: a case study. Int J Project Manage 35(1):78–94
8. Heravi G, Mohammadian M (2019) Investigating cost overruns and delay in urban construction projects in Iran. Int J Constr Manage:1–11
9. Kabirifar K, Mojtahedi M (2019) The impact of engineering, procurement and construction (EPC) phases on project performance: a case of large-scale residential construction project. Buildings 9(1):15
10. Aslam M, Baffoe-Twum E, Saleem F (2019) Design changes in construction projects–causes and impact on the cost. Civil Eng J 5(7):1647–1655
11. Zidane Youcef JT, Andersen B (2018) The top 10 universal delay factors in construction projects. Int J Manag Proj Bus 11(3):650–672
12. Naoum Shamil G (2016) Factors influencing labor productivity on construction sites: A state-of-the-art literature review and a survey. Int J Prod Perform Manage 65(3):401–421
13. Rahman R et al (2020) Factors affecting the success of highway construction projects: the case of Malaysia. In: IOP Conference series: materials science and engineering. IOP Publishing
14. França CL et al (2017) An approach to business model innovation and design for strategic sustainable development. J Clean Prod 140:155–166
15. Moazami D, Behbahani H, Muniandy R (2011) Pavement rehabilitation and maintenance prioritization of urban roads using fuzzy logic. Expert Syst Appl 38(10):12869–12879
16. Bin A (2018) Project viability framework for privatized highway projects in Malaysia (Doctoral dissertation, Universiti Teknologi Malaysia)
17. Houle JJ et al (2013) Comparison of maintenance cost, labor demands, and system performance for LID and conventional stormwater management. J Environ Eng 139(7):932–938
18. Salih J, Edum-Fotwe F, Price A (2016) Investigating the road maintenance performance in developing countries. Int J Civil Environ Struct Constr Architect Eng 10(4):433–437
19. Grimm M, Krüger J, Lay J (2011) Barriers to entry and returns to capital in informal activities: evidence from sub-Saharan Africa. Rev Income Wealth 57:S27–S53
20. Oladelea AS, Adedimilab AS, Egwurubec JA (2011) Highway maintenance cost estimation modeling for developing countries: a case study of Nigeria. Botswana J Technol
21. CIDB (2011) Quality assessment system for completed road works. CIDB Malaysia, Malaysia
22. Chen F, Chen S (2011) Injury severities of truck drivers in single and multi-vehicle accidents on rural highways. Accid Anal Prev 43(5):1677–1688
23. Kropáč O, Múčka P (2011) Specification of obstacles in the longitudinal road profile by median filtering. J Transp Eng 137(3):214–226
24. Woodcock K (2014) Model of safety inspection. Saf Sci 62:145–156
25. Anderson DG (2008) Seismic analysis and design of retaining walls, buried structures, slopes, and embankments, vol 611. Transportation Research Board
26. Mo L et al (2013) Review on asphalt plug joints: performance, materials, testing and installation. Constr Build Mater 45:106–114

27. Pantha BR, Yatabe R, Bhandary NP (2010) GIS-based highway maintenance prioritization model: an integrated approach for highway maintenance in Nepal mountains. J Transp Geogr 18(3):426–433
28. Fu CC, Zhang N (2011) Investigation of bridge expansion joint failure using field strain measurement. J Perfor Constr Facil 25(4):309–316
29. Kassahun K, (2015) Causes of failure of highway cross drainage structures (A case study on Sile River Bridge-Arbaminch Ethiopia). Addis Ababa University, Addis Ababa
30. Bień J et al (2016) Quality control of road bridges in Poland. In: Proceedings of the 8th International conference on bridge maintenance, safety and management, IABMAS
31. Neves LC, Frangopol DM (2005) Condition, safety and cost profiles for deteriorating structures with emphasis on bridges. Reliab Eng Syst Saf 89(2):185–198
32. Yang S-I, Frangopol DM, Neves LC (2004) Service life prediction of structural systems using lifetime functions with emphasis on bridges. Reliab Eng Syst Saf 86(1):39–51
33. Agarwal PK, Jain V, Bhawsar U (2013) Development of a hierarchical structure to identify critical maintenance components affecting road safety. Proc Soc Behav Sci 104:292–301
34. Li D, Elberink SO (2013) Optimizing detection of road furniture (pole-like objects) in mobile laser scanner data. ISPRS Ann Photogram Remote Sens Spat Inf Sci 1:163–168
35. Terrin A, Dengo C, Meneghetti G (2017) Experimental analysis of contact fatigue damage in case hardened gears for off-highway axles. Eng Fail Anal 76:10–26
36. Fuller S (2010) Life-cycle cost analysis (LCCA). National Institute of Building Sciences, An Authoritative Source of Innovative Solutions for the Built Environment, 1090 p
37. Lavy S (2008) Facility management practices in higher education buildings: A case study. J Facil Manage 6(4):303–315
38. Pesola O (2017) Differences between real estate maintenance costs in Scandinavia, Middle East, China and Spain-The savings potential through facility management software
39. Krstić H, Marenjak S (2012) Analysis of buildings operation and maintenance costs. Građevinar 64(04):293–303
40. Islam R, Nazifa TH, Mohamed SF (2019) Factors influencing facilities management cost performance in building projects. J Perform Constr Facil 33(3):04019036
41. Yusuf GA, Mohamed SF, Binti Remeli S (2013) Factors inhibiting the adoption of a uniform pricing mechanism for building services. Int J Eng Bus Manage 5(Godište):5–3
42. Wong I, Chan A (2014) How to quantify quality architectural design from the aspect of building maintenance. In ICCREM 2014, pp 1472–1479
43. Islam R, Nazifa TH, Mohamed SF (2019) Evaluation of facilities management sustainable parameters for improving operational efficiency. Int J Constr Manage:1–17
44. Femi OT (2014) Effects of faulty construction on building maintenance. Int J Technol Enhanc Emerg Eng Res 2(3):73–79
45. Ali AS et al (2013) The effect of design on maintenance for school buildings in Penang Malaysia. Struct Surv 31(3):194–201
46. Ofori I, Duodu P, Bonney S (2015) Establishing factors influencing building maintenance practices: Ghanaian perspective. J Econ Sustain Develop 6(24):184–193
47. Rykers G (2018) Minimum levels of componentisation for road infrastructure assets: guideline

Exploring the Potentials, Impacts, and Challenges of Mobile Technology Application in Cost Management of Construction Projects

Uchenna Sampson Igwe, Sarajul Fikri Mohamed, Mohd Bin Mat Dzahir Azwarie, and Lazarus Chisom Nwankwo

Abstract Managing cost of construction projects is one of the key challenges faced during production of constructed facilities. These challenges are occasioned by diverse factors associated with the construction process which include but not limited to the dispersed nature of construction workers, complexity of sites and projects, poor communication, getting cost data from multiple sources, and challenge of tracking and keeping accurate construction information. Recent advancements of mobile technology application in the construction industry provide great innovative avenues for addressing these challenges. Mobile technology has become an essential tool for the construction industry and other engineering sectors. The potentials and current impacts of mobile technology cannot be over emphasized as the its applications spans through all stages of construction process. Most of the key tools used in the management of construction projects have been virtually translated into mobile devices such as tablets, smartphones, etc. in the form of mobile Apps. This paper explores these potentials, the impacts and the key challenges in using mobile technologies for management of construction projects, with greater emphasis on the application for cost management.

Keywords Cost management · Mobile technology · Mobile apps · Construction projects

1 Introduction

The construction industry is characterized by mobile workforce due the dynamic nature of the industry and construction process. This is attributed to the fact that construction activities are quite dispersed and frequently take place at different

U. S. Igwe (✉) · S. F. Mohamed · M. B. M. D. Azwarie
Faculti Alambina Dan Ukur, Universiti Teknologi Malaysia, 81310 Johor Bahru, Skudai, Malaysia
e-mail: usigwe@graduate.utm.my

U. S. Igwe · L. C. Nwankwo
School of Environmental Sciences, Federal University of Technology, Owerri Imo State, Nigeria

M. A. A. Zaini et al. (eds.), *Proceedings of the 3rd International Conference on Separation Technology*, Lecture Notes in Mechanical Engineering,
https://doi.org/10.1007/978-981-16-0742-4_4

changing remote site locations [1]. There is a paradigm shifts from conventional construction to a more innovative method of construction, which offers abundance of benefits, such as shortened construction duration, less waste production and lower production cost [2]. Construction activities are becoming more and more complex in recent times due to technological advancement and changing environment. The complexity of construction projects involves physical dispersion of participants such as contractors, the consultants, sub-contractors, suppliers, construction artisans and client representatives. The project-based nature of the construction industry, the dynamic system of communication on projects and the vast volume of data managed by general contractors all illustrate the need for efficiency in managing the construction process [3]. For effective management of the construction process, the dispersed project participants need to have an effective collaborative communication. Effective communication enables all the key stakeholders to be updated with the status of the ongoing project in terms of Cost, project duration, and overall performance of the project in real-time.

It is undisputable that efficiency is highly needed in managing construction project production and the current wave of technological advancement in the industry offers these opportunities to improve construction management through effective integration of fragmented systems and efficiency. One of the primary goals when carrying out a project is to ensure that the allocated budget is not surpassed at completion. Projects are time-bound and require massive capital and resource investments which translates that management of project costs is a key responsibility that cannot be managed with laxity. Yet with the advent of new procurement techniques, technologies, resources and numerous professionals involved in a project, construction cost control has become more complicated [4]. Utilizing the prevalent technologies is highly paramount and one of the greatest technologies currently being highly applied in construction management is Mobile technology. Mobile technologies are electronic devices capable of storing, accessing, producing, updating, arranging, or otherwise manipulating data in various ways without being bound to any specific location [5]. Mobile technologies include mobile and portable devices such as cell phones, personal digital assistants [PDAs], laptops, tablets, personal computers [PCs], personal media players or wireless desktop PCs. There are a range of technology-related developments in the construction industry, either driven primarily by digitization and other new technologies or impacted by technology [6]. Kapliński, [7] Highlighted recent developments transforming the construction sector from 2016 and beyond [see Fig. 1] and digital mobile technology is one of recent innovations.

Communication is the hallmark for effective project delivery, and to deliver a project within cost, communication of all cost ingredients across all project stages and project participants timely, and with ease is highly essential. In the construction industry, information and communications technology [ICT] devices are used for precise and effective information management [8]. And construction industry that has been regarded as slow adapters of technology is getting along at rates of 5G [9]. All construction processes and stages cannot function effectively without mobile technology because it is a must have in any project site by all participants including

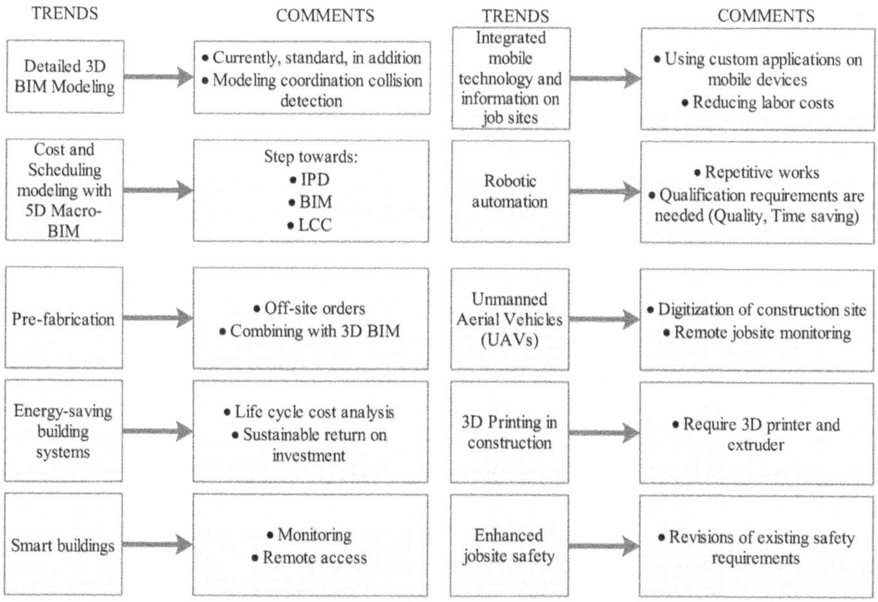

Fig. 1 Construction trends shaping the industry in 2016 and beyond [7]

unskilled labourers. Hence, since it has become inevitable to use mobile technology in all construction stages and processes, cost management process also cannot function effectively without mobile technology in this current time indicating the great potential for efficiency in construction process. However, construction companies are not taking advantage of the full potential of mobile technology [10]. There is therefore a great impact of mobile technology on construction cost management and this paper explores the key potentials, impacts and challenges of using mobile technologies in cost management of construction projects.

2 Research Methods

Systematic literature review method was used to elucidate the key aim of this paper being to articulate the potentials, the impact and the challenges of using mobile technology for managing cost of construction projects. Relevant publications in the subject matter were extensively searched for using different databases. To capture as many relevant articles as possible, a wide range of construction, engineering and scientific databases were searched to identify the potentials, and relevance of mobile technology in construction cost management, as well as the key challenges in adopting the technology. The electronic searches were supplemented by hand searching of *Scopus, Mendeley, and Google Scholar data bases*. Furthermore, various internet search engines were rifled for web pages that might provide relevant information. From the different databases, a total of 326 articles were explored and screened for inclusion for the review. The relevant articles were selected based on the criteria that the articles must have discussed the mobile technology in the context of either the potentials in the construction industry, the impacts in the construction industry, challenges of adopting the technology in the industry or the relevance in cost management. The screening process excluded a total of 282 articles which were not relevant to the subject matter. This effort resulted in inclusion and reviewing the remaining 44 articles from the different sources. Figure 2a and b indicate the number of articles referenced from the different databases and in each of referenced years respectively. The summary of the basic steps in the review process is given in Fig. 3.

The articles reviewed are articles published from 2006 to 2020. This is because Apple's iPhone was launched in 2007 and the first Android smartphone in 2008, quickly followed by third-party smartphones and tablets, has been around for just over a decade. The analysis was done chronologically using charts and tables to demonstrate and address the key findings within the three-fold area of enquiry which includes the potentials of Mobile technology for cost management, the impacts so far and the key challenges in adopting the technology.

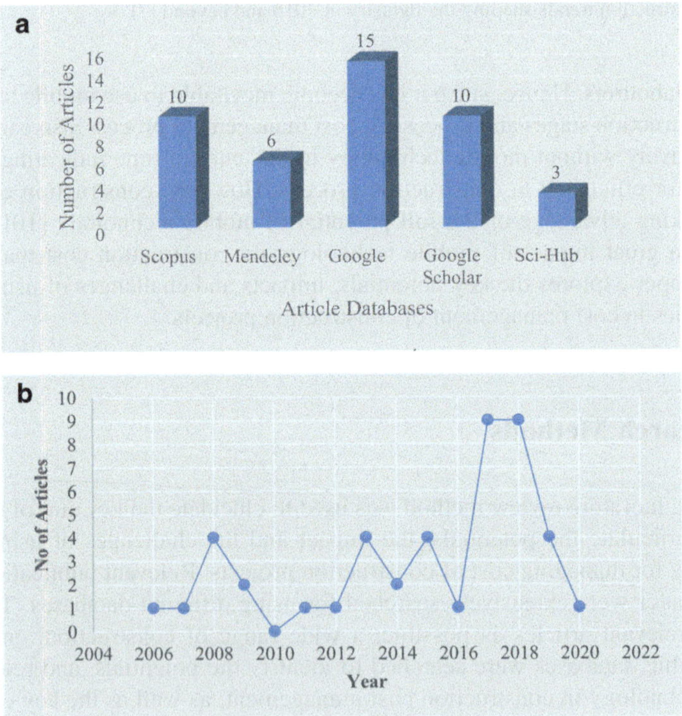

Fig. 2 **a** Articles gotten from the databases, **b** Referenced articles per year

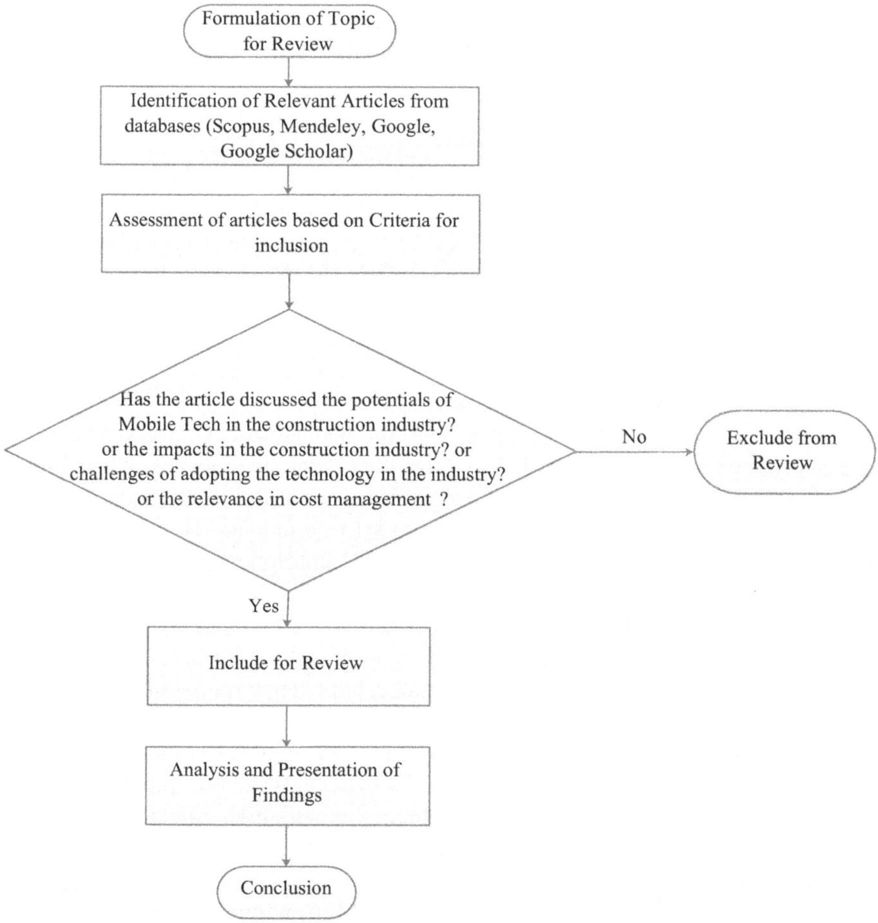

Fig. 3 Flow of the Review process

3 Potentials of Mobile Technologies in Construction

New technology involving the use of handheld devices such as smartphones and tablets has changed everyday lives and it has impacted our professional lives [11] as well. In particular, the construction industry is experiencing a new and revolutionised era with the help of technology through new technologies and resources. Such apps from software start-ups transform the way construction companies design, schedule and execute their projects [12]. The ability of construction companies to leverage and adopt current trend of technology is the great differentiator in the competitive environment where maximization of profit in order to stay in business is the hallmark of the game. Mobile technology application in the industry has experienced tremendous growth and this trend is still on the rise. In every construction site today, mobile technology must be seen being used for any construction related task. Workers clock in with mobile time cards, foremen file regular reports with their smartphones, project managers check with their tablet prototypes and requirements during site visits. So it is clear that mobile technology has a big impact on the job site.

Deployment of mobile technology is a relatively new development in the construction field. Modern smartphones have been available for just over a decade since the launch of Apple's iPhone in 2007 and the first Android handset in 2008 that was quickly followed by third-party devices and tablets [13]. A few years later 4 G mobile wireless was launched with higher Internet speeds and 5 G network is currently on the rise. Until then, the only thing you might recognise as mobile tech was a laptop equipped with an AirCard and a cell phone or PDA with limited connectivity compared to today's smartphones and tablets [13].

There are mobile applications that support the management of every aspect of a pre-construction project, from project management and field reporting to back office management. Mobile technology is providing more opportunities for faster information gathering and management to improve workflow efficiency. By that the amount of time spent on other related support functions, smartphones have great potential to improve productivity. The mobile applications available in the construction sector range from project management, calculators, tracking, integrated building and accounting costs, construction site operations, computer-aided design, 2D and 3D designs and drawings, estimating and designing knowledge modelling [BIM] [11]. These mobile apps have been developed to improve productivity and performance as well as to control construction workflow and this is closely related to controlling construction costs as all cost components (materials, labour, plant and equipment and overhead costs) are organised and handled with high efficiency. Applying mobile technologies such as personal digital assistants [PDAs], smart phones and palm-tops along with state-of-the-art networking infrastructure generation such as third-generation mobile communications system [3G], wireless local area network [WLAN] and general packet radio service [GPRS] could bring great benefits to the construction process and contribute to the re-evaluation of construction [8].

The capabilities of mobile technologies with respect to construction project delivery cannot be over emphasized. These potentials/capabilities in the industry is applicable in diverse construction activities. Some key potentials as gathered from the reviewed articles are presented in Table 1. The tools that help collaborate and remove bottlenecks in the three construction phases-planning, design and construction-are embedded in tablets, smartphones and smart mobile hotspots [14]. These tools are embedded in the mobile devices in form of mobile applications. It is estimated that nearly 13,000 construction-related development and design applications are available on the market today, and these industry-related applications have many functionalities ranging from simple calculations to detailed architectural renderings [11].

The mobile tech capabilities as enumerated in the table is undoubtedly directly or indirectly related to cost management. Every functionalities of the mobile technology utilized in any construction site is geared towards completing the project within estimated budget, scheduled period and specifications. Hence, maintaining

Table 1 Identified Potentials/Capabilities of Mobile Technology in Construction Industry

S/N	Key Potentials/Capabilities	Reference
1	Record keeping/cost data capturing	[8, 13, 15–17]
2	Real-Time Collaboration capabilities	[13, 18, 19]
3	Field Inspection and tracking of site activities	[8, 17, 20, 21]
4	Progress Monitoring	[8, 17, 18, 22]
5	Supply Chain Management	[8]
6	Construction Resource management	[14, 23]
7	Real-time information management	[11, 15, 23, 24]
8	Project Management	[11, 15, 23, 25]
9	Integrated construction cost and accounting	[11, 15, 26–28]
10	Estimating and budgeting	[5, 29]
11	Building information modelling [BIM]	[11, 14, 18, 29–31]
12	Construction safety	[18, 21, 29, 32]
13	Site operations	[11, 33]
14	Computer aided design	[1, 11, 31]
15	Augmented and virtual reality	[34]

balance between cost, time, quality and appearance becomes the hallmark of most construction undertaking. The current advances in mobile applications in the industry has greatly impacted on the way projects are delivered in terms of cost, time, and specifications.

4 The Impact of Mobile Technologies in Construction

Mobile technology has greatly transformed the construction industry as its application is presently inevitable at all stages of construction process. Mobile technology advancements allow for a whole new range of capabilities. Current 5G mobile is gradually being introduced which has a significant advantage that can boost mobile capabilities and its utilization in the construction industry. Table 2 articulated the core impacts of mobile technology in the construction industry.

Areas of application of mobile technology in construction project delivery is numerous. Table 2 gave the key areas of mobile application. Mobile technology has a significant impact on the completion of projects within the projected budget, scheduled time and specifications. Project information is obtained in real time with the current advances in mobile technology. Due to the presence of mobile technologies, less time is spent in query and approval processes. It also allows for efficient management of changes and reduces the risk of errors and rework during project delivery. It has improved efficient procurement and management of resources, effective contract management, customer satisfaction, decreased administrative costs of

Table 2 The impacts of mobile technology in the construction industry

S/N	Impact/Relevance of Mobile Technology	References
1	Mobile Technology has become tool for logging project information and reporting in real-time [Example SCADA system, RFID]	[6, 8, 16, 33, 35, 36]
2	It has Provided opportunities for improved construction communication	[6, 33, 37]
3	Improves the productivity and profit of construction projects	[6, 33]
4	Enables physical separate hardware devices to connect and share data in project delivery	[16, 38]
5	It has enhanced the real-time construction process monitoring and aids an early decision making and process control	[6, 37, 39]
6	It has become Valuable tool for Clients to execute real-time control of time and cost performance	[16, 27]
7	Allows for the real-time visualization of the locations of workers, materials, and equipment	[6, 21, 40]
8	Provides potential for greater efficiency and time and cost savings	[6, 32, 36, 41]
9	Works together and removes bottlenecks in the three construction stages-planning, design and construction	[14]
10	Improves on-site health, logistics, improves efficiency and helps deter theft and wastage	[6]
11	Reduction in operation and maintenance costs	[32]

handling and delivery of documentation to many project participants and also enables project managers to spend more time on managerial work. Despite the great functionalities and potentials of mobile technologies, full advantage of the technology for cost management and project delivery is yet to be fully harnessed due to certain barriers and challenges.

5 Barriers/Challenges of Adopting Mobile Technologies for Cost Management

Several studies have revealed that certain barriers oppose full utilization of information and communication technologies in the construction industry. Davies, [42] categorized these challenges and barriers into cultural or organizational based, and the technological base barriers. Table 3 itemised the barriers/challenges that hinders the full adoption or adequate utilization of the potentials of mobile technology for cost management.

It is not disputable that the availability of mobile technology has made most of the cost management processes easier but it is obvious that most of the mobile technologies and applications are yet to gain full entry into the construction business due the aforementioned barriers and challenges. Remote construction sites and

Table 3 Challenges/barriers to mobile technology application for construction cost management

S/N	Challenge/Barrier	Category	Reference
1	Conservatism and fear of change from old method	Culture/Organization	[14, 36, 41]
2	Limited time and money available for investment in IT	Culture/Organization	[14, 36, 41]
3	Lack of interoperability between many Mobile tools	Technological base	[36, 41]
4	Lack of or poor internet availability	Technological base	[14]
5	Cost of Mobile devices	Culture/Organization	[14]
6	Lack of ICT training and technical assistance to professionals	Culture/Organization	[14, 36]
7	Management's lack of attention to ICT	Culture/Organization	[36, 42]
8	Low benefits/Poor return on ICT investment	Culture/Organization	[36]
9	High cost of hiring experts in ICT	Culture/Organization	[36]
10	Insufficient ICT content of Construction Education	Culture/Organization	[36]
11	Insufficient information about the return on investment in ICT	Culture/Organization	[36]
12	Lack of personnel with sufficient ICT expertise and knowledge	Culture/Organization	[36, 42]
13	Fear of employment loss/being redundant staff.	Culture/Organization	[36]
14	Lack of legal support for use of ICT	Culture/Organization	[36]
15	Lack of standardization	Technological base	[43]

medium scale construction business owners seem to be the most affected. It is therefore highly necessary to bridge all necessary barriers to maximize the potentials of mobile technology as the benefit in cost management is awe-inspiring.

6 Mobile Technology Application in Construction Cost Management

Construction cost management is information orientated, meaning that any lapses in data collection, processing, reporting and documentation mars the efficiency of the system and the entire construction process will be affected. Managing construction cost is a method and its implementation requires people's innovation, technology, learning, engagement and commitment. The'mobile age' is in full swing and construction companies are slowly but steadily making the transition [23]. Although the construction sector has long been regarded as slow technology adapters, the sector is catching up at 5 G speeds [9]. Practically, Mobile devices are used at all project development phases. The smartphone, combined with mobile computing technology, has provided a versatile and interactive platform for on-site construction management and the paradigm change in traditional construction management practises is expected to lead the shift [34]. Currently available applications in the construction industry have different functionalities, from basic calculations to detailed architectural renderings [11]. In the construction industry the specific areas of mobile application include but not limited to;

i. Project planning
ii. Project designs
iii. Cost management and control
iv. Construction safety
v. Project management and control

Mobile technology helps all stakeholders to interact quickly in any project outfit. This ensures that every Project Member would be on the same page while carrying out a task. Most construction projects need constant contact between multiple parties and any delay can be detrimental to a construction process's productivity [24]. Mobile technology offers parties the ability to quickly address any problems related to information sharing using mobile email or sending documents in real time. Access to the data in real time means less costly mistakes [43]. Capturing, documenting and processing of Cost information in real-time is essential for effective management of project cost. Mobile technology allows design teams and contractors to access construction documents, take pictures and publish field reports with the help of smartphones, tablets, etc. promoting efficiency and immediacy in delivering tasks. With so many industry professionals spending significant time on job sites, transporting materials and meeting clients on the go, it is crucial that construction companies embrace mobile technology to ensure they remain competitive.

Mobile technology enables real-time data processing and transfer between the jobsite and back offices. Cloud-based systems allow construction workers to upload timetables, routine reports, requests for information [RFIs], job records, order changes and other documents reviewed on site. Keeping documents on paper is clearly time consuming, energy intensive and inefficient. Documents can be easily lost or misreported, and obtaining vital data and documents like daily reports and project managers is a slow way to rely on this information to keep a project on track and under budget. Mobile apps allow data to flow in real time and documents with date and time stamps can be captured along with GPS location data and digital signatures to ensure accuracy and authenticity.

One main aspect of any productive construction project is being able to communicate and collaborate effectively. As team leaders and stakeholders have access to the same information in real time, the collaboration process is streamlined and conversations are made more effective. Mobile technology and cloud-based software enable anyone in a project to access and update in real time documents, reports, plans and requirements, while keeping everyone on the same page as the work progresses. If there are problems on the job site, a worker can easily take few pictures of the problem, generate a report and send it off immediately to be quickly reviewed and solved. No more waiting to obtain revised plan sheets, they can be sent directly to smartphones or tablets. This boosts delivery speed, and saves costs.

With the right mobile cloud based tools, workflows and processes are easily handled. The ability to schedule labour and equipment, conduct field checks, manage order modifications, monitor progress and review plans and requirements from one single computer will greatly improve performance on a job site. The ability to easily access and review all project data collected from anywhere eliminates downtime, helps make better decisions more efficiently and increases the efficiency of the project. Using the right mobile apps one can be aware of problems or possible delays as soon as they happen. It mitigates risks and prevents costly rework in a way that was not feasible just a decade ago.

In recent time, mobile technologies such as drones, wearable mobile devices have become trending means of automated progress monitoring. Keeping accurate track of information is crucial for efficient cost management. All construction resources are better managed with the use of mobile technologies as the advent of Mobile technologies has the capabilities of tracking data on equipment usage, material management, labour management, and all cost components such as overheads, enabling cost and project managers to be updated with status of project in terms of cost performance. Currently there are software and mobile apps that can help handle different aspects of the project. There is already an integrated software framework from preconstruction to scheduling, from project management and field reporting to back office operations, which can help streamline construction processes and increase productivity. Many software systems are cloud-based, allowing for real-time adjustments and improvements to records, schedules and other management resources, enabling improved communication and collaboration [44]. The top mobile applications that are used in construction in recent times are as listed in Table 4.

Most of the mobile applications are used for cost control and construction project management. the applications are categorized into Project design management applications, Cost management applications and Project management applications. The summary of the identified applications is shown in the doughnut (Fig. 4).

It is obvious that most of the available mobile applications in the industry can be greatly utilized for cost management seeing that 40% of the identified applications has cost management functionalities. Total cost management of any construction project begins from project initiation stage up to practical completion stage and subsequently maintenance or running cost management. At every stage of construction process, there should be a mobile application relevant for cost management at that stage. Figure 5 is the Royal Institute of British Architects (RIBA) plan of work for construction projects and the cost management activities for each stage. Using play store app which is available in almost all mobile devices, mobile applications relevant for each cost management activity in every stage were articulated as shown in the Fig. 5. The various cost management related mobile applications were extracted by keying the appropriate words/phrase for each cost management activity into the play store app.

Table 4 Top Mobile Technologies Used in Construction

S/N	Technology	Type	Uses	Areas applicable
1	BIM 360 Docs	Mobile application	Manage all project files and schedules from one unified device including BIM	Project Design, Cost Management, Project Management
2	Construction Master Pro App	Mobile application	Useful for conversions of construction math problems such as length, field, and volume	Cost Management [Quantity Measurement and Estimating]
3	DroneDeploy	Mobile application	DroneDeploy is designed to draw up site plans, track job progress and compare BIM models with real site conditions to improve safety and monitor projects	Project Management
4	eSUB	Mobile application	Focused solely on the process and the needs of subcontractors	Project Management, Cost Management
5	Fall Safety Pro	Mobile application	Active control of falls and monitors location in construction sites	Project Management
6	GasBuddy	Mobile application	It helps consumers find the cheapest gas near their place	Cost Management
7	PlanGrid	Mobile application	Users enjoy the opportunity to share plans, mark-ups, photographs and feedback from their mobile device with the entire project team	Project Design, Project management, and Cost Management
8	Safety Culture iAuditor	Mobile application	Provides a flexible Checklist feature. Paper checklists are digitised so that the team can perform inspections even more easily	Project Management

(continued)

Table 4 (continued)

S/N	Technology	Type	Uses	Areas applicable
9	SmartBid	Mobile application	SmartBid provides a cloud-based platform for subcontractors to send invitations to bid, share project papers and collaborate. The mobile app allows SmartBid users to easily access their subcontractor network, so they can bid projects from anywhere quickly	Cost Management, Project Management
10	TrueLook	Mobile application	Users can view, save and share webcam images live on your work site, and more	Project Management
11	JobFLEX		JobFLEX is an app that offers forecasts on-the-spot, with or without a cellular connection	Cost Management, Project Management
12	e-Builder	Like mobile devices, available on all platforms	e-Builder is a build project management platform with real-time feedback dashboards	Project Management
13	DEWALT Mobile Pro	Mobile application	DEWALT Mobile Pro is a robust construction calculation and comparison tool	Cost Management
14	Bridgit Closeout	Available in different platforms including in mobile devices	Bridgit Closeout is a cloud-based task management system designed to assign tasks, monitor and take notes. Information about the project can be shared by teams through the Closeout app	Project Management

(continued)

Table 4 (continued)

S/N	Technology	Type	Uses	Areas applicable
15	Canvas	Mobile application	Reporting with Canvas is easy, and available. On mobile devices, you can fill in figures, forms, inspections, work orders and audits	Cost Management, Project Management
16	SAGE Estimating SQL	Available in all platforms including mobile device	Used for taking off quantities, estimating and payrolling	Cost Management and Project Management
17	SKYSITE	Mobile application	Allows users to view, monitor, distribute and collaborate with safe access on'any smartphone or desktop computer' in real time on construction documents. Apps can be downloaded on ios, Android and Windows	Project Management, and Cost Management
18	UDA	Mobile Application	Used for Estimating, Costing, and project management	Cost management and Project Management
19	Procore	Mobile Application	Allows creating and sharing project information. The coolest feature is the ability to add or make changes offline	Project management and Cost management
20	CoConstruct	Available in different platforms including in mobile devices	Useful App for project monitoring, enables easy collaborations among project participants	Project management

7 Conclusion

Mobile technologies and applications available for use in the construction industry as articulated in this paper are very vital tools for all stages of construction processes. The potentials as identified points out that current construction processes and, in the future, cannot effectively deliver without the use of mobile technologies. The potentials ranges from sketch up drawings, to detailed design, to total cost management, to project delivery on site, to maintenance of constructed facilities. It is therefore evident

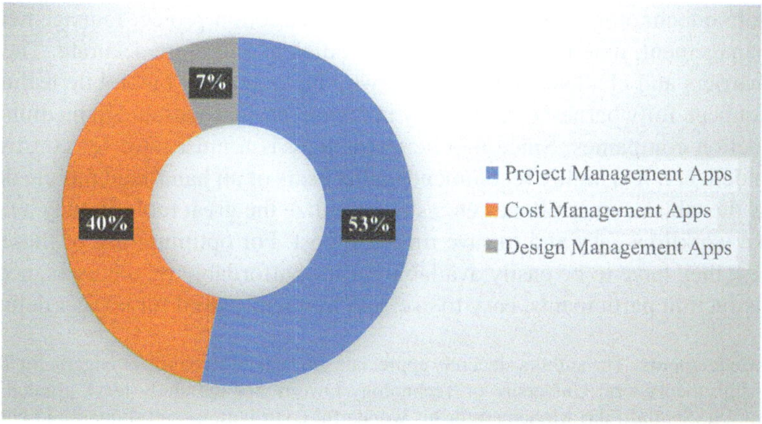

Fig. 4 Summary of top 20 Identified Mobile Applications in the Construction Industry

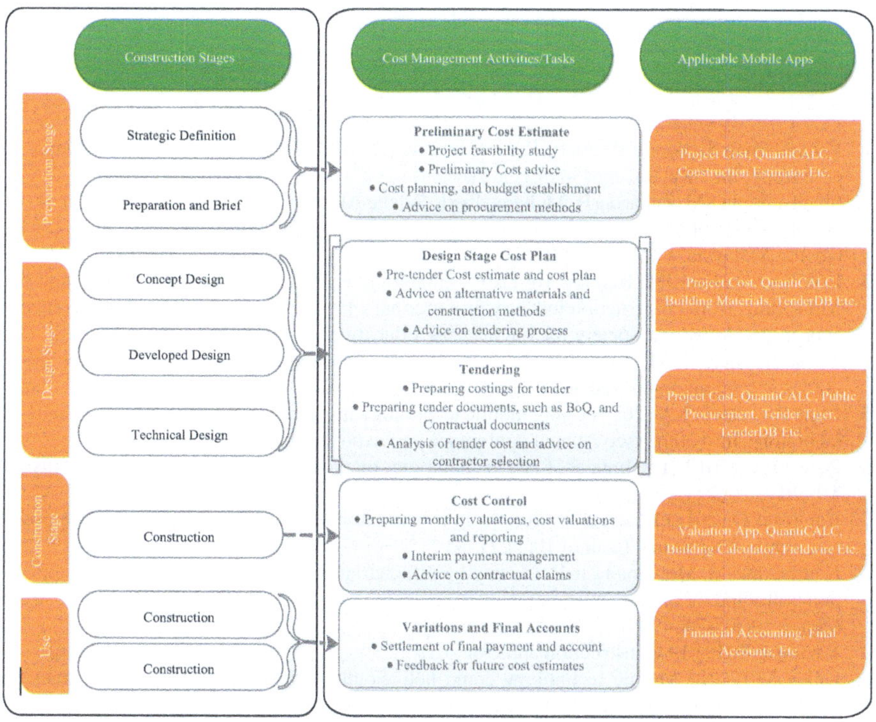

Fig. 5 Cost Management Activities/Task summarized in the context of RIBA Plan of Work and applicable Mobile Apps

that mobile technology has great impact on construction project management and cost management, making the process more efficient and more accurate. The identified barriers and challenges are reasons why the potentials of Mobile technology have not been fully harnessed especially in remote project sites and in medium scale construction companies. Since mobile technology is a must have for construction project delivery, it is highly recommended that users of all hand-held mobile devices such as the tablets, smartphones etc., should utilize the great tools already available to make workflow easy and to save time and cost. For optimum use of these technologies, they have to be easily available for use, affordable by the users, accepted by construction participants, easy to use, and frequently used for project delivery.

Acknowledgements The authors sincerely appreciate the Federal Republic of Nigeria for TetFund sponsorship, the Federal University of Technology Owerri, for the study leave granted, Associate Prof. Sr. Sarajul Fikri Mohamed for his wonderful contributions and support and Universiti Teknologi Malaysia for making available the platform that made this article possible.

References

1. Venkatraman S, Yoong P (2009) Role of mobile technology in the construction industry—A case study. Int J Bus Inf Syst 4(2):195–209
2. Shamsuddina SM, Zakariaa R, Mohamedb SF, Salehc AL, Utomod C, Majide MZA et al (2015) Developing methodology for cradle to grave cost planning for industrialised building system (Ibs) In Malaysia. J Teknol 77(16):37–42
3. Holt EA, Benham JM, Bigelow BF (2015) Emerging technology in the construction industry: perceptions from construction industry professionals. In: 122nd ASSEE annual conference & exposition. American Society for Engineering Education, Seattle, WA
4. Girma A, Alemu MG (2018) Improving project cost management practice and profitability of domestic contractors in Vadodara. J Emerg Technol Innov Res 5(5):51–58
5. Cline RC, Davis KA (2013) Using mobile technology in a construction management hands-on laboratory. In: 120th Asee Annual conference and exposition, pp 1–7
6. Pistorius C (2017) The impact of new technologies on the construction industry. Innov Insight 4:1–70
7. Kapliński O (2018) Innovative solutions in construction industry review of 2016–2018 events and trends. Eng Struct Technol 10(1):27–33
8. Nourbakhsh M, Mohamad Zin R, Irizarry J, Zolfagharian S, Gheisari M (2012) Mobile application prototype for on-site information management in construction industry. Eng Constr Archit Manag 19(5):474–494
9. Yovino J (2019) 19 Construction apps for 2019
10. Kibby M (2018) Mobile technology could help South African construction companies. IT NEWS AFRICA, Africa's Technology News Leader
11. Liu T, Mbachu J, Mathrani A, Jones B, McDonald B (2017) The perceived benefits of apps by construction professionals in New Zealand. Buildings 7(4):111
12. Santos JMD (2018) Technology and software trends in the construction industry
13. Jones K (2018) Going mobile: the benefits of mobile construction technology. construction technology. Blog Posts
14. Sattineni A, Schmidt T. (2015) Implementation of mobile devices on jobsites in the construction industry. In: Procedia Engineering. Elsevier B.V., pp 488–95
15. Riddell T (2017) Why mobile technology is the next step for the construction industry

16. El-Omari S, Moselhi O (2011) Integrating automated data acquisition technologies for progress reporting of construction projects. Autom Constr 20(6):699–705
17. Deolikar A, Neware DS (2018) An approach for site progress monitoring using mobile crowd sensing. Int J Comput Sci Eng 10(4):92–96
18. Abanda FH, Mzyece D, Oti AH, Manjia MB (2018) A study of the potential of cloud/mobile BIM for the management of construction projects. Appl Syst Innov 1(2):9
19. Ahsan S, El-Hamalawi A, Bouchlaghem D, Ahmad S (2007) Mobile technologies for improved collaboration on construction sites. Archit Eng Des Manag 3(4):257–272
20. Yamaura J, White G, Katara S, Willoughby K, Garcia R, Beer M (2015) Project inspection using mobile technology—phase II. WA-RD 840.2, Texas; 2015
21. Nasr E, Shehab T, Vlad A (2013) Tracking systems in construction: applications and comparisons. In: 49th ASC Annual international conference proceedings, pp 1–9
22. Kiganda A (2017) How mobile technologies are boosting construction site efficiency
23. Riddell T (2017) How mobile technology is transforming the construction industry
24. Deibert S, Hemmer E, Heinzl A (2009) Mobile technology in the construction industry—the impact on business processes in job production. In: 15th Americas conference on information systems. AMCIS 2009, pp 5326–35
25. Igwe US, Mohamed SF, Azwarie MBMD, Okoro CO, Adediran AO (2020) Recent advances in mobile application for real time cost management in construction projects. Int J Innov Sci Res Technol 5(5):414–418
26. MC Solutions (2019) Mobile technology trends for heavy construction. Modern Contractor Solutions
27. Andrzej F, Paweł K, Aleksander N (2014) Design and implementation of automated, mobile construction projects monitoring system (MEVMS) based on Earned Value Management as an element of BIM in the execution stage. In: 3rd Creative construction conference proceedings, pp 180–5
28. Redden L, Collins W, Kim J (2017) Integration of construction mobile technologies into construction management curriculum: a case study. Proc Eng 196:535–542
29. Hong SH, Yu JH (2018) Identification of external variables for the technology acceptance model (TAM) in the assessment of BIM application for mobile devices. In: IOP Conference series: materials science and engineering, p 012027
30. Chu M, Matthews J, Love PED (2017) Integrating mobile building information modelling and augmented reality systems: an experimental study. Autom Constr 2018(85):305–316
31. Essay U (2019) Impact of mobile communications on construction industry, vol 67, pp 291–315
32. Khelifi A, Hyari KH (2016) A mobile device software to improve construction sites communications (MoSIC). Int J Adv Comput Sci Appl 7(11):51–58
33. Kim C, Park T, Lim H, Kim H (2013) On-site construction management using mobile computing technology. Autom Constr 35:415–423
34. Abbaszadegan A (2017) Instantaneous project controls: current status, state of the art, benefits, and strategies, vol 77, Dissertation Abstracts international: section b: the sciences and engineering, pp. no-specified
35. Naik S, Bobade S (2018) The benefits of information and communication technology adoption in construction project management for small and medium enterprises Sourabh. Int J Res Appl Sci Eng Technol 6(5):107–110
36. Rebolj D, Babič NČ, Magdič A, Podbreznik P, Pšunder M (2008) Automated construction activity monitoring system. Adv Eng Inform 22(4):493–503
37. Hore A (2006) Use of IT in managing information and data on construction projects: a perspective for the Irish construction industry. Inf Tehnol Constr Proj Manag Eng Irel Proj Manag Soc:1–15
38. Srewil Y, Scherer RJ (2013) Effective construction process monitoring and control through a collaborative cyber-physical approach. IFIP Adv Inf Commun Technol. 408:172–179
39. Costin A, Pradhananga N, Teizer J (2014) Passive RFID and BIM for real-time visualization and location tracking. In: Proceedings of the 2014 Construction research congress construction research congress 2014. Construction in a Global Network, pp 169–78

40. Davies K (2008) Barriers or constraints? A review of development issues as they apply to construction IT. In: International Conference on information technology in construction santiago, pp 239–45
41. Bohn JS, Teizer J (2008) Benefits and barriers of construction project monitoring using hi-resolution automated cameras. J Constr Eng Manag. 136:1–17
42. Linderoth HCJ, Jacobsson M (2008) Understanding adoption and use of ICT in construction projects. In: CIB W78 2008 International conference on information technology in construction. Santiago, Chile, pp 1–10
43. Riddell T (2017) Why mobile technology is the next step for the construction industry [Internet]. construction technology. Accessed 29 2020.: https://www.constructconnect.com/blog/mobile-technology-next-step-construction-industry
44. Jones K (2019) How technology is disrupting the construction industry. Visual Capitalist

Develop Interlocking Concrete Block Pavement from Portland Cement, Polystyrene and Bottom Ash on Pedestrian Road

Mariah Awang, Mohamad Luqman Hakkim Idris, Azman Bin Ja'afar, Noraini Marsi, Muhammad Haikal Mohd Fodzi, Kamaruzaman Musa, Faridahanim Ahmad, M. M. Syafiq Syazwan, Fatimah Yusop, and Adibah Aiman Jumali

Abstract This study presents to utilize agriculture waste (palm oil bottom ash) and polymer material (polystyrene) as a replacement of sand and aggregate a new sustainable interlocking concrete block pavement (ICBP). The bottom ash takes from palm oil mill incinerator at palm factory Belitong, Kluang, Johor. The expanded polystyrene is collected from electric shop and garbage collection center nearest Simpang Renggam, Johor. The bottom ash through the screening process in 4.75 mm sieve complying the BS 410. Only particle passing though the 4.75 mm sieve was be used in interlocking concrete block pavement production. Expanded polystyrene size from 2.5 to 4.0 mm. Concrete mix design for this interlocking concrete block sample with size 220 mm × 110 mm × 60 mm and divided by five (5) ratios where Portland cement is a consistent ratio with (40%) and different percentage ratio for bottom ash and polystyrene. Bottom ash percentage ratio is 0%, 20%. 20%, 40%, 60%, and polystyrene percentage ratio is 60%, 40%, 30%, 20%, 0%. According to the results, the optimum ratio of interlocking concrete block pavement is 30% bottom ash and 30% expanded polystyrene with average compressive strength 4.77 N/mm^2 at days 28. The compressive strength is not archived the standard strength based on MS 1380:1995. From the result, percentages of bottom ash are significant value with compressive strength. However, water absorption and density are not significant with percentages of expanded polystyrene. The use of bottom ash in interlocking concrete block pavement in an alternative to encourage environmental protection in order to minimize waste material.

M. Awang (✉) · M. L. H. Idris · A. B. Ja'afar · N. Marsi · M. H. M. Fodzi · K. Musa · M. M. Syafiq Syazwan · F. Yusop · A. A. Jumali
Department, of Civil Engineering Technology, Universiti Tun Hussein Onn Malaysia (UTHM), Pagoh, Muar, Malaysia
e-mail: mariah@uthm.edu.my

F. Ahmad
School of Civil Engineering, Faculty of Engineering, Universiti Teknologi Malaysia, Skudai, Johor, Malaysia

M. A. A. Zaini et al. (eds.), *Proceedings of the 3rd International Conference on Separation Technology*, Lecture Notes in Mechanical Engineering,
https://doi.org/10.1007/978-981-16-0742-4_5

Keywords Bottom ash (BA) · Compressive strength · Expanded polystyrene (EPS) · Interlocking concrete block pavement (ICBP) · Water absorption

1 Introduction

Interlocking concrete block pavement is a once of construction material especially in road application. The usage of interlocking concrete block pavement has increased markedly in many countries not only for making the road, but it's also commonly used in urban roads, parking areas, industrial roads and pedestrians and the use of this material is widespread throughout the world [1]. Due to the shape and segmental nature of the block, the interlocking concrete block is easy to remove, and the underlaying utilities could be easy to maintenance because block pavers use interlocking system and easy to access. It's preferred as a great choice to implement in urban areas which are consist of playgrounds, pedestrians and parking areas. The quality and performance of block pavement are similar to the conventional flexible pavement and the capacity structure of various block dimensions and parameters was consistent the performing and the functional role as a flexible pavement [2].

Agriculture and solid waste are some of the causes of environmental pollution problem in Malaysia. Polymer products such as polystyrene are widely used in Malaysia because the manufacture is easy, and the cost is very low. Besides that, the large amount of waste packaging also is a problem to manage the disposing and increase the amount of solid waste at dumping area. Another environmental problem is agriculture waste where Malaysia is known for its palm oil-producing countries. Normally, wastes from palm oil plantation are used to produce steam and electricity for mill processes with the combustion of a boiler or an incinerator. From the combustion, fly ash and bottom ash was generated and every 200 palm oils mills in Malaysia are produce a thousand tons of fly ash and bottom ash and simply disposed without any commercial return in any industry [3].

A combination of agriculture and solid waste can give advantages to construction industry especially in concrete design. The bottom ash and expanded polystyrene can be a good combination in production of interlocking concrete block pavement because expanded polystyrene has long life span for through the decomposition process and the low specific gravity of expanded polystyrene also can reduce the weight of normal concrete blocks [4]. Bottom ash is coarser material where the particle almost same as sand and this material was suitable to replace sand in a concrete mixture in making of interlocking concrete block pavement. The combination of this waste material can improve the mechanical properties and workability of interlocking concrete block pavement.

As a whole, this chapter review about producing the new sustainable product by using waste material from expanded polystyrene and bottom ash to produce the interlocking concrete block pavement. In this study, expanded polystyrene are function as replacement of aggregate and bottom ash are as a replacement to sand in mixture of Portland cement. Waste material management is a global challenge and continues to increase if it is not properly managed and impacts on the environment and human health.

2 Literature Review

2.1 Cement

Cement is a substance that binds together other material by a combination of a chemical process known collectively as setting. Cements are dry powder and should not be confused with concretes or mortars, but they are an important constituent of both of these materials, in which they act as the 'glue' that gives strength to structures. Cement is an extremely important construction material. The constant demand for all of these structures, increasingly from the developing world, means that cement is the second most consumed commodity in the world after water [5].

Cement characteristics used in construction can be divided by two categorized either hydraulic or non-hydraulic [6]. Non-hydraulic cement will not set in conditions or underwater because the cement was set is it dries and reacts with carbon dioxide in the air. Non-hydraulic cement can be attacked by some aggressive chemicals after setting [6]. Hydraulic cement (e.g., Portland cement) was set and become adhesive due to a chemical reaction between the dry ingredients and water. This will allow the hydraulic cement to setting in wet conditions or underwater and further protect the hardened material from chemical attack. The chemical process for hydraulic cement found by ancient Romans used volcanic ash [6].

2.2 Bottom Ash

Bottom ash is the coarser material, which drops into the bottom of the furnace in latest large thermal power plants and constitute about 20% of gross ash content of the coal fed in the boilers where raw bottom ash is a granular material that consists of a mix of inert materials such as sand, stone, glass, porcelain, metals and ash from burnt materials [7]. Bottom ash particles are much coarser than the fly ash and this material is composed of silica, alumina and iron with small amounts of calcium, magnesium and sulfate. Chemical composition of bottom ash is similar to the fly ash but typically contain greater quantity of carbon [8].

In comparison to sand, an elementary raw material, the black sandy material obtained is a subsidiary raw material. Using this material is more endurable and environmentally, amicable and eludes the use of natural resources such as sand and gravel. Bottom ash can used in the following activities: [7]

- Road construction.
- Foundation material.
- Aggregate.
- Noise barrier.

In addition, American Coal Ash Association (ACAA) lead a survey on the application of bottom ash and found that usually bottom ash can be used for several construction purposes such as construction, transportation and agriculture. It also can be used as soil modification or improvement, road-based material and structural fillers [9]. In the making of brick, bottom ash also can be a replacement material to sand in concrete. Three different source of bottom ash are used in concrete mix each 30%, 40% and 50% to replacement by weight of sand to making concrete blocks. During the investigation, concrete block having 30% value of bottom ash replaced by sand are more suitable for use in the manufacture of concrete blocks [10].

2.3 *Expanded Polystyrene*

Polystyrene is a versatile polymer where it has been used for many purposes such as packaging and consumer goods. Polystyrene are widely used plastic made from styrene where a chemical compound is produced by combining the ethylene and benzene to become polystyrene and the final product are made up of 95 to 98% air [11].

The expanded polystyrene particle is uniform in size and shape, the concrete can range from that of no-fines concrete with density 300 kg/m3 or less to that of fully compact concrete with density 1000 kg/m3 or more [12]. Expanded polystyrene (EPS) is one of the most common used material for lightweight concrete in construction industry. It uses low strength materials with good energy absorbing characteristics. Expanded polystyrene (EPS) is good in thermal and acoustic insulation properties and as a result, it is mainly used in non-structural applications such as precast roof, precast wall panels and lightweight infill blocks.

The compression test was conducted a study on lightweight concrete made from waste polystyrene and fly ash proved that with the compressive strength of concrete tends to decrease when amounts of expanded polystyrene and fly ash used to replace natural sand and Portland Cement increased [13]. The study found that expanded polystyrene (EPS) in concrete is useful to absorb energy and decrease the contact loading loads during hard impact at low density [14].

3 Methodology

This study starts with studying the optimum ratio of expanded polystyrene and bottom ash as replacement of sand and aggregate in interlocking concrete pavement block (ICBP). A series of test was conducted to investigate the strength of the interlocking concrete block pavement (ICPB) by compression test, water absorption test and bulk density test. The design of interlocking concrete block pavement (ICBP) almost same with existing paver block in Malaysia. From the testing of interlocking concrete block pavement (ICPB), the sustainable and workability of the ICBP can be identified to the construction industry. Figure 1 show the detail flow chart of this study.

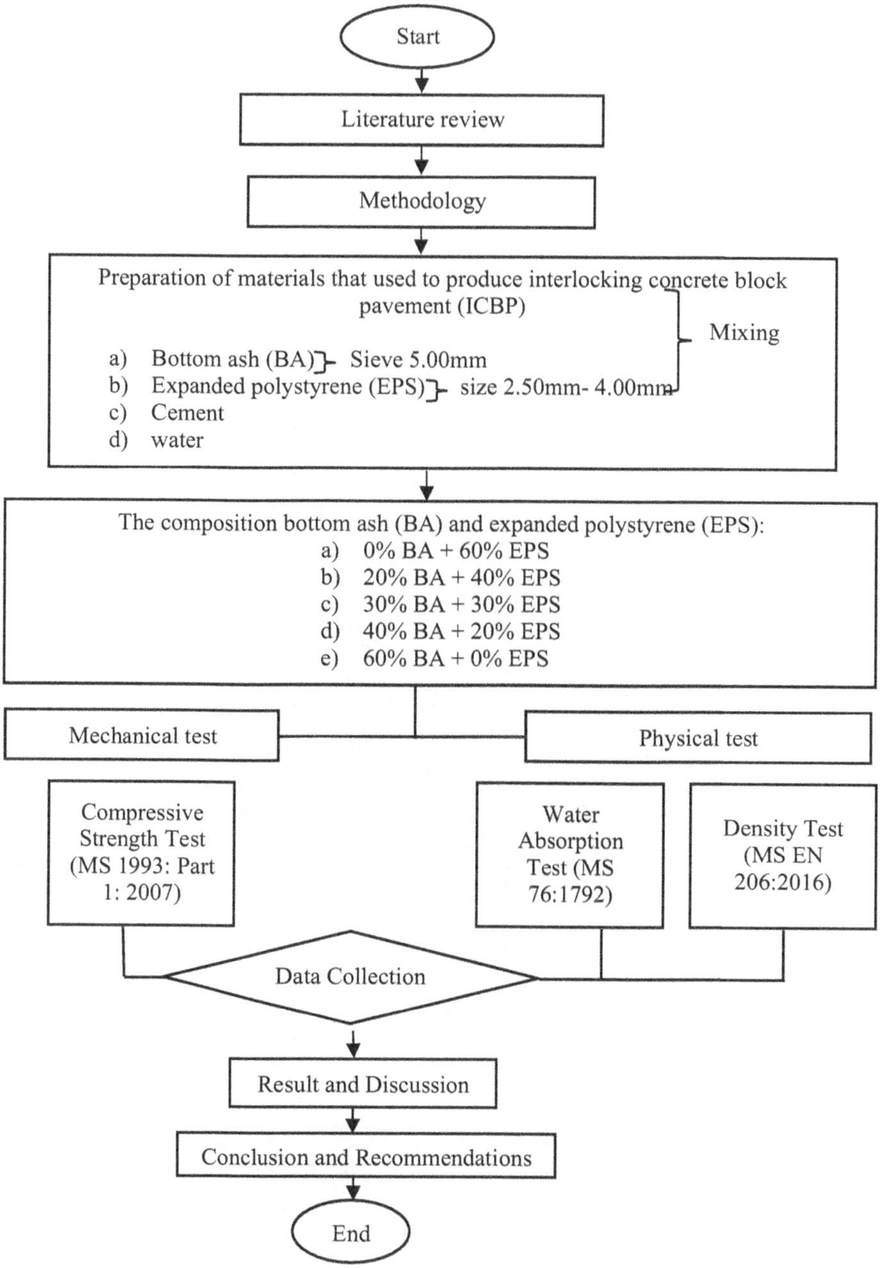

Fig. 1 Flow chart of Methodology for interlocking concrete block pavement

3.1 Bottom Ash

Palm oil bottom ash is a by-product material obtained in the form of ash when burning palm oil husks or fibers as a fuel in palm oil mill boilers or incinerator. Palm oil bottom ash will use in this study and collected from a factory processing palm oil situated at Felda Belitong at the area of Kluang, Johor Darul Takzim. The ash was found in the below of the incinerator where all heavy ash falls to the bottom during the combustion process inside the incinerator (Figs. 2 and 3).

Palm oil bottom ash will be screening to remove a coarse particle. The sample will be sieved manually using the 4.75 mm sieve complying with the BS 410 and

Fig. 2 Incinerator of the palm oil husk of fiber burning

BEFORE AFTER

Fig. 3 Bottom ash before sieve and after sieve

Fig. 4 Expanded polystyrene beads with a size in range of 2.5–4.0 mm

only the particles passing through the 4.75 mm sieve will used because the bottom ash will be sand replacement in this research.

3.2 Expanded Polystyrene

Expanded polystyrene is a naturally transparent thermoplastic where it's available as a typical solid plastic as well in the form of a rigid foam material. Besides that, polystyrene also is vinyl polymer and it is a long hydrocarbon chain with a phenyl group attached to every other carbon atom. Polystyrene in raw beads which being steam-heated and causing it to expand is namely as expanded polystyrene. There have many sizes of polystyrene beads in production and can be used in certain applications.

In this study, polystyrene in raw beads was being steam-heated and causing the polystyrene expands and is namely as expanded polystyrene. It is used to produce lightweight concrete and the size of polystyrene beads used in this study is in range of 2.5–4.0 mm as shown in Fig. 4.

3.3 Concrete Mix Design

The preparation of interlocking concrete block pavement (ICBP) was done at Concrete Laboratory in University Tun Hussein Onn Malaysia (UTHM), Pagoh. Every sample of interlocking concrete block pavement (ICBP) has different

Table 1 Mix of bottom ash (BA) and Expanded Polystyrene (EPS)

Concrete code	Portland cement (%)	Bottom ash (%)	Expanded polystyrene (%)	Total sample
B1	40	0	60	12
B2	40	20	40	12
B3	40	30	30	12
B4	40	40	20	12
B5	40	60	0	12

percentage mix proportion were the Portland cement was a control percentage is 40%. The different percentage of mix proportion of bottom ash (BA) and expanded polystyrene (EPS) is shown in Table 1.

3.4 *Laboratory Test*

Three types of testing will be conducted in this study to get in order to determine the optimum ratio, mechanical properties, and workability of interlocking concrete block pavement (ICBP). The testing is compressive strength, water absorption and density test.

(a) Compressive Strength (MS 1933: Part1: 2007)

The compressive strength test is a mechanical test measuring the maximum amount of compressive strength load material before fracturing. The crushing strength of interlocking concrete block pavement (ICBP) was brought to light by locating it in the compression testing machine and it was pressed until it cracks and read the value of the compressive strength of the sample. A size of sample is 220 mm × 110 mm × 60 mm and the test will be done at days 7 and 28.

(b) Water Absorption (MS 76:1972)

The water absorption test is conducted to determine the pore volume and porosity in the interlocking concrete block pavement (ICBP). The total water absorption sample is defined where the increment in the weight of the sample due to moisture in air. In this testing, 3 sample of interlocking concrete block pavement (ICBP) were selected for each percentage and water absorption test was performed after 7 days and 28 days.

(c) Density test (MS EN 206:2016)

Density test was carried out to determine the density of interlocking concrete block pavement (ICBP) and to know the difference on density value of each percentage of bottom ash (BA) and expanded polystyrene (EPS) as a replacement of sand and aggregate in each sample. The test was performed at days 7 and 28, whereby each of mass of sample was weighed and the data was recorded.

4 Results and Discussions

4.1 Compressive Strength

The compression test is an important test to determine the strength, quality and workability of interlocking concrete block pavement either it meets engineering strength stated in MS1993: Part 1: 2007. The compression test was performed on 6 specimens with 220 mm × 110 mm × 60 mm where the specimen is divided into 5 categories according to different ratios of BA and EPS Every specimen's ratio was tested 3 samples at days 7 and 3 sample at days 28. The compression machine test that has been used to obtain data of each specimen is a universal testing machine as shown in Fig. 4.1 and the data of each specimen were calculated in Table 2 and the graph has illustrated the average of compressive strength of each specimen at days 7 and days 28.

The table shows the percentage were 30% of BA 30% of EPS have the higher compressive strength value than other percentage. This is because the percentage ratio of material is balance and the values of compressive strength maybe increase day by day.

Based on Table 3 shows the average value compressive strength of interlocking concrete block pavement (ICBP) for 28 days. The highest value of compressive strength is sample B3 with 4.77 N/mm^2. The highest value reading of compressive strength at days 28 increased slightly from the highest value of compressive strength

Table 2 Result for compressive strength average for 7 days

Sample no.	Percentage of bottom ash (%)	Percentage of expanded polystyrene (%)	Average compressive strength (N/mm^2)
B1	0	60	0.5
B2	20	40	3.95
B3	30	30	4.62
B4	40	20	4.11
B5	60	0	4.42

Table 3 Result for compressive strength average for 28 days

Sample no.	Percentage of bottom ash (%)	Percentage of expanded polystyrene (%)	Average compressive strength (N/mm^2)
B1	0	60	2.76
B2	20	40	3.83
B3	30	30	4.77
B4	40	20	4.46
B5	60	0	4.59

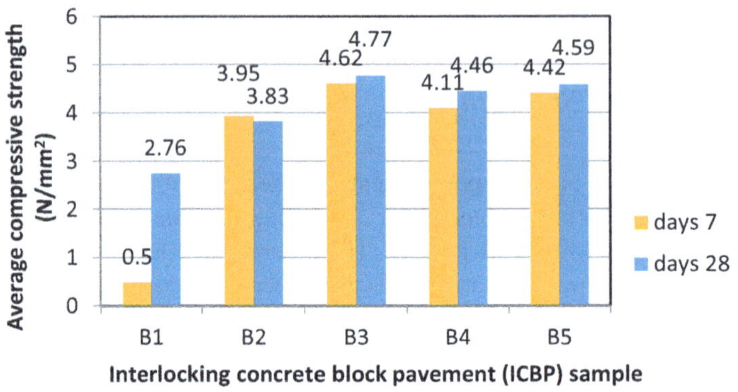

Fig. 5 Graph on the average of interlocking concrete block pavement (ICBP) compressive strength for different percentage at age 7 and 28 days

at days 7. It shows days can give effects in increasing the value compressive strength of interlocking concrete block pavement (ICBP).

Figure 5 shows the different of value compressive strength on days 7 and days 28 where observation from the graph, sample B3 have high value than another sample. The percentage of sample B3 is 30% bottom ash and 30% expanded polystyrene. The conclusion from this testing based on graph, sample B3 is optimum ratio in making interlocking concrete block pavement (ICBP) which is made from BA and EPS. This is because the percentage of BA and EPS was balance and the combination of two replaced material produced high compressive strength to interlocking concrete block pavement (ICBP). Furthermore, graph shows the lowest value of compressive strength is sample B1 because the samples have too much void. The expanded polystyrene is a lightweight and soft material; it was giving effect to the strength of interlocking concrete block pavement sample. This relationship is significantly affected by expanded polystyrene parameter and cannot be interrupted alone as example for a given proportion of expanded polystyrene, the more cement content the more concrete strength and for a given cement content, the more proportion of expanded polystyrene the less the compressive strength of interlocking concrete block pavement sample [15].

In compliance with MS 1380:1995; for 80 mm and 100 mm thick of paver with average minimum compressive strength is 45 N/mm^2 and follow another compliance with MA 20; for 60 mm thick paver with average minimum compressive strength is 30 N/mm^2. Based the result from the Fig. 7, the highest average value of compressive strength is does not pass the standard MS 1380:1995. The value average of compressive strength low because of the effect of expanded polystyrene and bottom ash. Different percentage of the expanded polystyrene particle volume had different failure mode which is shown in. The specimen fracture after the compressive strength test and the scale of crack were smaller along with the increase of the expanded polystyrene particle volume ratio. This phenomenon was caused by the characteristics of energy absorption of expanded polystyrene particles, and the appearance was still intact even if the expanded polystyrene concrete was under destruction. Based on the result, bottom ash also gives effect to average value of compressive strength because bottom ash produces the porous surface structure and have high absorptivity of the bottom ash such the hydration to all cement particle may not occurs and less paste is possible for bonding [16].

4.2 *Water Absorption Test*

The water absorption test is important to conduct to determine the pore volume and porosity of the specimen. The calculation rate of water absorption to the interlocking concrete block pavement (ICBP) is important to determine the percentage of minimum and maximum rate of water absorption was allowed in the construction. Tables 4 and 5 show the result of water absorption test on each sample.

Based on Tables 4 and 5, the result was showing the average percentage of water absorption test which is the different percentage of bottom ash and expanded polystyrene in each sample. From the table, it can be seen the water absorption rate for a different sample where the EPS percentage is affecting the average value of water absorption rate of sample. This is because the EPS has hollow properties and consequently, the EPS absorbs more water.

Table 4 Average result of water absorption test for each interlocking concrete block pavement (ICBP) with different percentage of bottom ash and expanded polystyrene on 7 days

Sample no.	Percentage of bottom ash (%)	Percentage of expanded polystyrene (%)	Average water absorption (%)
B1	0	60	42
B2	20	40	26
B3	30	30	20
B4	40	20	20
B5	60	0	17

Table 5 Average result of water absorption test for each interlocking concrete block pavement (ICBP) with different percentage of bottom ash and expanded polystyrene on 28 days

Sample no.	Percentage of bottom ash (%)	Percentage of expanded polystyrene (%)	Average water absorption (%)
B1	0	60	45
B2	20	40	22
B3	30	30	18
B4	40	20	20
B5	60	0	17

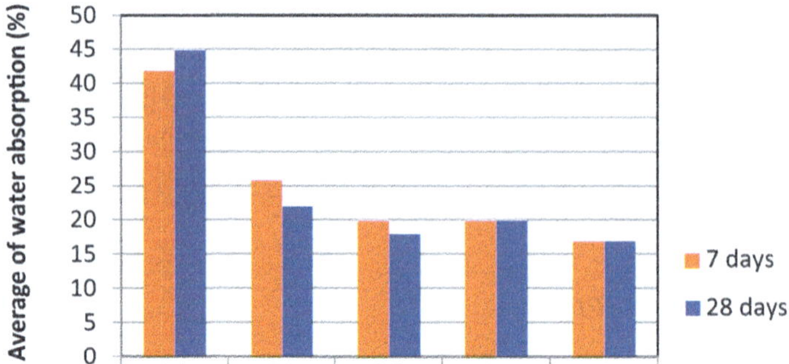

	B1	B2	B3	B4	B5
7 days	42	26	20	20	17
28 days	45	22	18	20	17

Fig. 6 Graph average value of water absorption (%) for each sample interlocking concrete block pavement (ICBP) for 7 and 28 days

Figure 6 shows the average of percentage (%) of water absorption for each sample of interlocking concrete block pavement (ICBP). From the result, the higher value of water absorption rate is sample B1 with percentage of material is 0% BA and 60% EPS. Average value of sample B1 at days 7 is 42% slightly increase to 45% on days 28. For sample B4 and B5, the average percentage (%) of water absorption is constant which 20% for sample B4 and 17% for sample B5.

From the whole observation, percentage (%) water absorption in sample B5 has the lowest value of compared to sample B1 with the percentage material for sample B5 is 60% BA and 0% EPS. This is because the EPS has pores in every particle and the pores keep the water in the sample. According to BS 3921:1985, the average value for the water absorption test for concrete is not less than 20%. The result shows, sample B5 is beyond the standard BS 3921:1985 because sample B5 doesn't have a lot of pores and a solid structure for the sample.

4.3 Density Test

The density of concrete is an important aspect in produce design of concrete. This is to determine whether the sample is of high quality in terms of strength but of optimum weight for ease of transport work. In this study, average value of density was obtained from 3 interlocking concrete block pavement (ICBP) samples for every percentage of material at the age 7 days and 28 days. Table 6 show the result of density of interlocking concrete block pavement (ICBP) for each sample.

Through the study conducted, cab be concluded that percentage of polystyrene in interlocking concrete block pavement (ICBP) is not significant with density of

Table 6 Result of density test on interlocking concrete block pavement (ICBP) at the age 7 days and 28 days

Sample no.	Percentage of bottom ash (%)	Percentage of expanded polystyrene (%)	7 days	28 days
			Average density (kg/m^3)	Average density (kg/m^3)
B1	0	60	373.3	464.2
B2	20	40	812.7	874.1
B3	30	30	957.9	985.5
B4	40	20	908.4	936.6
B5	60	0	1117.1	1165.2

interlocking concrete block pavement (ICBP). This is because EPS have low specific gravity compared to BA.

Bases on Fig. 7, the average density of interlocking concrete block pavement (ICBP) sample at the age 7 days is 373.3 kg/m^2, 812.7 kg/m^2, 957.9 kg/m, 936.6 kg/m^2 and the highest is 1165.2 kg/m^2. The density of sample increases constantly at age 28 days to 464.2, 847.1, 985.5, 936.6 and 1165.2 kg/m^2. The sample which high percentage of EPS has low density compared to the sample which high percentage of BA. From the result, it can be concluded that percentage (%) of BA was affecting to increase the average density value of sample. This is because sample B1 has high percentage of EPS and the sample become light weight, the lighter the sample, the density value become low. Besides that, the density also become increase

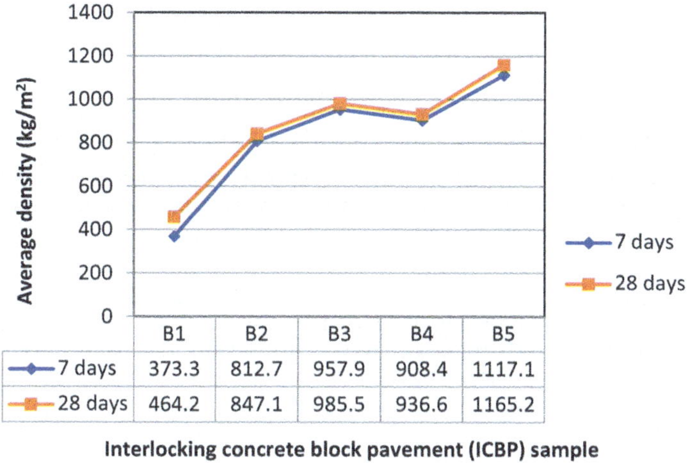

Fig. 7 Graph of average density of interlocking concrete block pavement (ICBP) for different percentages of bottom ash (BA) and expanded polystyrene (EPS)

days by days which is the value of average density sample B5 is 1117.1 kg/m^2 at days 7, the density value increase to 1165.2 kg/m^2. In addition, different weights for each sample can also cause temporary density values and this may be due to the concrete's current compaction to the mold.

5 Conclusions

From the study analysis 30% of BA and 30% EPS containing in interlocking concrete block pavement (ICBP) shows the highest compressive strength and the percentage of water absorption is 18% but still lowest percentage compare to another ratio. Interlocking concrete block pavement (ICBP) with 30% of BA and 30% EPS has higher compressive strength of 4.77 N/mm^2 compare to other percentages and at the same time reduce the use of sand and aggregate in production of interlocking concrete block pavement (ICBP) as well as reducing the environment pollution cause by BA and expanded polystyrene EPS. The compressive strength starts increase when the percentages of BA increase. From the analysis of compressive strength, when the EPS percentage increase in the sample, the average compressive strength become lower which is 2.76 N/mm^2 compare to the other percentage of EPS. However, increasing of BA in sample has increase the average compressive strength of interlocking concrete block pavement (ICBP) which is 4.59 N/mm^2 and above. So, EPS is not considerably useful to replace aggregate for construction field because the strength decrease when the used percentage of EPS high and BA can replace the sand with correct optimum percentage in concrete. The compressive strength decreases as the percentage of EPS beyond 60% and 40%. This may be attributed to the decrease in the adhesive strength between EPS and the sample become lightweight. The water absorption rate higher compares to another sample when the percentage of EPS high. Furthermore, the density of sample becomes lower when the EPS percentage high in concrete. In conclusion, the highest average value compressive strength sample is 4.77 N/mm^2 still cannot achieving the objective to develop new sustainability and workability product because the compressive strength under the standard specification according Malaysian Standard (MS 1380:1995).

Acknowledgements This work would not have been possible without the financial support of the grant project Vot. H251 and "Geran Penyelidikan Pascasiswazah" (Vot. H595) by Universiti Tun Hussein Onn Malaysia (UTHM). We are grateful to all of those with whom we have had the pleasure to work during this and other related projects.

References

1. Teiborlang LR, Mazumdar M, Pandey BB (2005) Structural behaviour of cast in situ concrete block pavement. J Transp Eng 131(9):662
2. Mokaddes DR, Ahmed A, Singhi B (2013) Overview on structural behavior of concrete block pavement. Int J Sci Res 4 (ISSN 2229-5518)
3. Awal ASMA, Hussin MW (1997) The effectiveness of palm oil fuel ash in preventing expansion due to alkali-silica reaction. Cem Concr Compos 19(4):367–372
4. Choi YW, Moon DJ, Chung JS, Cho SK (2005) Effects of waste PET bottles aggregate on the properties of concrete. Cem Concr Res 35(4):776–781
5. Ridi F (2010) Hydration of cement: still a lot to be understood. La Chemical Industrial Societal, Chimica Italiana
6. Blezard RG (2004) The history of calcareous cements in chemistry of cement and concrete. Elsevier Butterworth-Heinemann
7. Nadig VR, Sanjith J, Ranjith A, Kiran BM (2015) Bottom ash as partial sand replacement in concrete. IOSR J Mech Civil Eng (IOSR-JMCE) 12:148–151
8. Kumar D, Gupta A, Ram S (2014) Use of bottom ash in the replacement of fine aggregate for making concrete, Int J Curr Eng Technol (E-ISSN 2277-4106, P-ISSN 2347-5161)
9. Syahrul M, Sani M, Muftah F, Muda Z (2010) The properties of special concrete using washed bottom ash (WBA) as partial sand replacement. Int J Sustain Constr Eng Technol 1(2):65–76
10. Sharma S (2012) Study of usage of bottom ash as part replacement of sand for making concrete blocks
11. Laukaitis A, Zurauskas R, Kerien J (2005) The effect of foam polystyrene granules on cement composite properties. Cem Concr Compos 27 (1):41e47
12. Neville A (1981) Properties of concrete. Pitman Press Publisher, London
13. Herki B, Khatib J, Negim E (2013) Lightweight concrete made from waste polystyrene and fly ash. World Appl Sci J 21:1356–1361
14. Bischoff PH (1990) Polystyrene aggregate concrete subjected to hard impact. Proc Inst Civil Eng Part 2 Res Theory 89
15. Zahir K (2001) Polystyrene lightweight concrete (polyconcrete). Najah Univ J Res 15
16. Cadersa AS, Rana J, Ramjeawon T (2014) Assessing the durability of coal bottom ash as aggregate replacement in low strength concrete. J Emerg Trends Eng Appl Sci (JETEAS) 5(5):344–349 (ISSN: 2141-7016)

The Effectivess of Photoventi Under Malaysian Climate

Mariah Awang, Haikal Akmal Bin Jasni, M. A. A. Rahman, Nuramidah Hamidon, Alia Abdullah Saleh, Mohd Shahril Abdul Rahman, Mohd Kamaruzaman Musa, Fatimah Yusop, and M. M. Syafiq Syazwan

Abstract Malaysia relies on non-renewable sources like fossil fuel and coal by 90%. The energy industry itself contributes about 77% of total carbon dioxide gas emission in 2011. Due to that, the government of Malaysia published Green Technology Master Plan Malaysia 2016–2030 to help reducing the carbon dioxide gas emission. This shows how serious the problem caused by electricity generation. Thus, the end user is encouraged to reduce electricity usage by switching to sustainable product. Turbine ventilator and tubular skylight has been invented a long time ago, which offers natural ventilation and natural lighting respectively. The usage of these devices could reduce the dependency on electricity use especially for a windowless space or a space located far from the window. Both of these devices require an area as they are commonly placed on the roof. This research aims to overcome these problems by designing a device that can provide natural lighting and ventilation without the usage of electricity. The device is called PhotoVenti which represents the photon and ventilator. The performance of this product is analyzed through experimentation on a test house located in Panchor, Muar. The average of illuminance obtained from the experiment are 232 lx and 208 lx and the average of indoor temperature obtained are 27.6 and 28.3 °C. MS 1525 and MS 2680 and Industry Code of Practice of Indoor Air Quality 2010 to are used to describe its conformity under Malaysian climate.

Keywords Daylighting · Green product · Tubular skylight · Turbine ventilator

M. Awang (✉) · H. A. B. Jasni · M. A. A. Rahman · N. Hamidon · M. K. Musa · F. Yusop · M. M. Syafiq Syazwan
Faculty of Engineering Technology, Universiti Tun Hussein Onn Malaysia, Pagoh Campus, 84600 Muar, Johor, Malaysia
e-mail: mariah@uthm.edu.my

A. A. Saleh
Faculty of Achitecture, Planning & Surveying, UiTM Perak Campus, Seri Iskandar Branch, Perak, Malaysia

M. S. A. Rahman
Faculty of Built Environment and Surveying, Universiti Teknologi Malaysia, Skudai, Johor, Malaysia

M. A. A. Zaini et al. (eds.), *Proceedings of the 3rd International Conference on Separation Technology*, Lecture Notes in Mechanical Engineering,
https://doi.org/10.1007/978-981-16-0742-4_6

1 Introduction

Energy saving programs were conducted to urge the people to reduce energy consumption in daily life. Malaysia relies on the fossil fuels the most in order to generate electricity. Fossil fuels are non-renewable energy source, which emits a large amount of CO2 gas that can affect the environment by the inclination of temperature and greenhouse effect [1]. The most worrying part of excessive emission of CO2 gas is climate change. In order to slow down the effect, the dependency on non-renewable energy source must be reduced whereby creating a limit for electricity consumption [2]. Energy, Science, Technology, Environment and Climate Change Minister Yeo Bee Yin aimed to save up to 137,775 GwH of electricity consumption from 2016 to 2030. Buildings in Malaysia contributes more than 50% of total electricity consumption [3].

It is crucial to take into consideration of building design and the technical systems as it could help to improves energy efficiency as well as reducing the impact on the environment [4]. Green technology has been applied into our daily use products for a long time and there are variety methods used to increase the sustainability of an existing product. Researchers are working on their way on green initiatives involving energy efficiency, resources, health, environment, recycling, renewable and durability approach to create a better way of living and safer environment. Some of green technology being applied in our daily life products including the televisions, refrigerators, lighting devices and washing machines. Converting the daily products to different power source, the dependency on fossil fuels can be reduced as well as the bills the building owners need to pay [5].

The existing green devices that shares the similar characteristics for improving the lighting and ventilation are tubular skylight and roof ventilator respectively. These are separate device that provide only one function each and both of these devices needs to be placed on the roof where there will be enough sunlight and wind energy in order for them to operate efficiently [6]. By taking consideration of a building owner that wants to use these devices for natural ventilation and lighting at the same time, it would take two different spots on the roof. For a small building, these devices might have to be place next to each other due to the limited roof area. The major problem with this positioning is that the shadow from turbine ventilator would reduce the efficiency of the tubular skylight and the tubular skylight might block the wind from driving the turbine ventilator.

Thus, this research aims to design a single unit device that helps to reduce electricity consumption by implementing the concept of green technology focusing on both ventilation and lighting which is among the important parameters to be considered for a building called PhotoVenti. The next goal after designing the device is to test its performance under Malaysian climate and compare it with MS 1525 and MS 2680 and Industry Code of Practice of Indoor Air Quality 2010. MS 2680 is the Energy efficiency and use of renewable energy for residential buildings—Code of practice while MS 1525 is the Energy efficiency and use of renewable energy for non-residential buildings—Code of practice (Second Revision).

1.1 Literature Review

Tubular Skylight

Transport of light developed in 1978 describes the use of prismatic material combined with high reflection efficiency of total internal reflection in a hollow cylinder. The concept remains the same until today. Light pipe systems are linear devices that channel daylight into the core of a building [7]. Its basic component consists of dome, tube and diffuser. There are two types of light tube which are the horizontal and vertical tubular skylight. The concept of bringing natural light into a building for both types is still the same, which guides the light through reflective materials. It is the positioning of the device will highly affect its performance. Other than that, the sky condition will affect the interior illuminance [8].

Vertical tubular skylight is the most common type of tubular skylight. The dome that act as a collector is usually placed on the roof. The hemispherical shape of the dome increases its ability to capture sunlight from various angle. High reflective material is attached to the dome base to help the light travel to the diffuser. The diffuser will then illuminate the interior space. The installation of this type of tubular skylight is a lot easier than the horizontal type. There are numerous mathematical formulae derived by researchers to predict the performance of tubular skylight such as the determination of daylight penetration factor (DPF) by Zhang and Muneer [9]. The new models are continuously being derive to obtain more accurate result by adding more factors under consideration. A study by Yun et al. [10] have proposed 12 new models derived from the DPF formula to suit the Korean climate condition, resulting 5 of the new models provides more accurate data.

Turbine Ventilator

Turbine ventilator is considered as mechanical and active ventilation due to the fact that this device has mechanical parts that can move without using electricity. There are two modes of turbine ventilator which are passive and active technique. They usually rotate in its vertical axis to create updraft inside the turbine which extracts air and in the absence of wind, they can still facilitate ventilator using the stack effects [11]. As for the active technique, the ball bearing at the center of turbine ventilator improves the rotation of the ventilator. The movement of the blades create pressure difference between the outdoor air and the indoor air. The indoor air will be negatively pressured and the air will be forced to exhaust out of the space through the pipeline. Passive technique occurs without the presence of wind. The exchange of air happens with the help of stack effect. Source of heat gain such as solar, occupants and electrical appliances produces heat in a building. The hot air will be trapped in the building if there is no proper ventilation. Turbine ventilator has gaps between the blades that allows air exchange. Theoretically, cold air has higher density than warm air and it will create air buoyancy. Warm air of air in the building raise and exhaust out of the building through the gaps while the cold air sinks into the building.

Malaysian Climate

This country experience hot and humid climate throughout the year with diurnal temperature around 16 °C and varies slightly between months [12]. The highest average temperature is 27.8 °C in March and the lowest temperature is 25.9 °C in December. The average hours of sunshine in Malaysia ranges from 4.7 to 7.2 h daily with average of 6.2 h per year. The sky condition in Malaysia is always cloudy, which creates diffusion of solar radiation causing only 30–50% of the global solar radiance reaches the surface of the earth. The daily mean cloud cover in Subang is 7 octas every month and varies from 6 to 7 octas every hour. The global illuminance of Malaysian sky can reach up to 1121 m/W while the mean diffuse luminous efficacy is 1201 m/W. This makes Malaysia a suitable country for daylighting purposes. The highest illuminance can reach up to 80,000 lx during hottest month and minimum peak of 60,000 lx in December and mean of 70,000 lx in June. A study [13] prove that the illuminance could reach up to 100,000 lx at Shah Alam and 140,000 at Bangi. However, the lux value can vary as low as below 5000 lx to 80,000 lx under intermediate sky.

The mean surface winds that occur in peninsular Malaysia are actually considered as gentle. The highest speed of wind can reach up to 8.8 m/s. The minimum mean of wind speed is 0.1 m/s and the minimum mean speed is about 1.9 m/s. March and April, together with September and October receives the highest speed of wind with no large difference throughout the year.

Design Concept

Turbine ventilator and tubular skylight are devices that is commonly placed on the roof in order to make it work. The concept of both devices is combined together to create a device that serve 2 functions in one unit which helps to reduce space usage on the roof. Tubular skylight has a transparent dome to be place on the roof. Dome shape is more preferred as it can collect sunlight from various angles. Furthermore, turbine ventilator has dome shaped vanes but it is usually made out of aluminium [14]. These two devices have the same shape but different materials. Thus, Photoventi applies the concept of tubular skylight on turbine ventilator by making every part that act as collector transparent, which is the dome. This includes the top cover, vanes and the internal structure components like rod and supporter. These transparent components help to increase the transparency and improve the transmission of light through the collector into the target area [7].

Turbine ventilator is a well-known attic heat remover and for some of the cases, ducting is needed if ventilation is needed for the interior. Tubular skylight has the same characteristics where it can light up the attic or a tube might be needed to channel the light into the building. Hence, the design of the channel tube can be combined as well. In this research, the tube used to channel the light is aluminium sheet covered with high reflective aluminium foil. Similar to the existing use, the channel will be only needed when the building has an attic that separates the indoor space [14]. The channel tube is not required for attic daylighting and ventilating. The prototype will not be focusing on the strength durability but will be testing on its function as it is designed with the original scale and it works similar to the main purpose.

2 Methodology

The main point of this research is to create a prototype of a transparent that can improve the indoor environment in term of lighting and ventilation by using natural resources such as the wind and solar. The process starts from product designing, assembling, data collection, analysis and conclusion. The product will be tested under Malaysian weather and the result from the data collection will determine if the data has to be recollected. Then, the data will be analyzed and will be concluded as shown in Fig. 1.

2.1 Product Design

Materials and Design Selection

The properties of materials and the designed are selected based on the previous study to increase the efficiency. The most common material used for the dome is clear polycarbonate and acrylic. Higher transmittance of light allows more light passes through the diffuser. The collector should be placed on unobstructed roof area where it can receive the highest amount of daylight available [15]. Ensuring the shadows from nearby buildings do not falls on the dome is crucial for maximum daylighting. Robertson et al. [16] reveals that the removal of liner from the dome increases the illuminance by 15% while the flat dome reduces the illuminance by 5%.

The pipe should be made of high reflective material. Typically alloy aluminium material is used to construct the hollow tube. Higher specular-reflectance tubes prevent loss of illuminance. Bending of the pipe will affect the internal space illuminance. Robertson et al. [14] in their study analyzed that pipe with 98% reflectance results in higher illumination than the pipe with 95% reflectance.

The most common material used for Turbine Ventilator is mainly made up of aluminium. Researchers are working on their way to increase the efficiency of turbine ventilator. According to Lien and Ahmed [11], the performance of turbine ventilator depends largely on prevailing winds. Thus, the studies are mainly on the design but only few on material selection. There are wide range of design especially the rotary parts. A study by Khan et al. [17] revealed that the curved vanes ventilator had up to 25% greater performance than the straight vanes. The same study indicates that the performance of 300 mm height of vanes had 75% greater air flow than the 250 mm height of vanes. Bigger size of turbine ventilator results in increasing of ventilation rate [18]. The installation of inner vanes also helps to increase its efficiency. Table 1 summarizes the details of the materials used for each component.

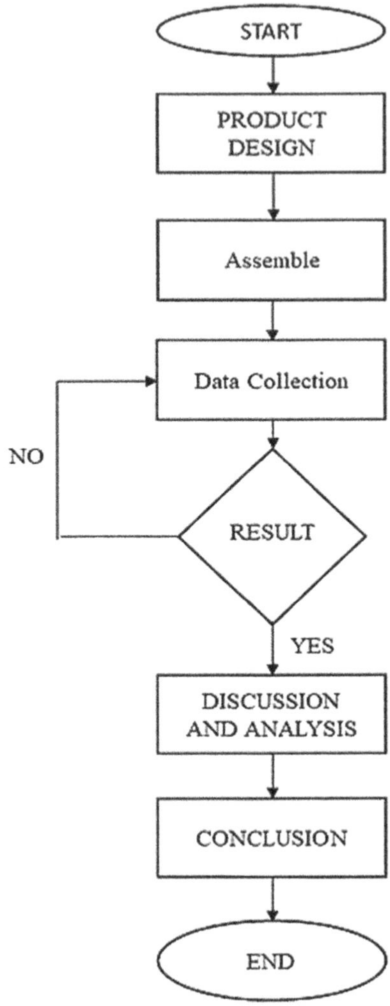

Fig. 1 Flow Chart Methodology

Assemble Process

The main components of PhotoVenti are the top cover, vanes, base and tube. These components are what makes the device to function properly. The 24 holes on the top cover are used to attach the vanes in the right place. Each vane is the same size and it is rectangular shape with two 5 mm holes at the top and bottom of the sheet. Each hole must be intercept with next and previous sheet. Thus, it will form an overlap arrangement with a little gap to exhaust the air from the inside. The same process is done for the bottom except that it is attached to a hollow cylindrical sheet with larger diameter to as a pathway for air flow. Two bearing are inserted in the 12 mm rod at a distance to hold the top cover and the bottom of the vanes. Then the bottom of the vanes is attached with four units of 150 mm acrylic stick to hold it shape as well as

Table 1 List of materials

No.	Element	Type of material	Dimension	Unit	Description
1	Top cover	Acrylic	280 mm dia. with 24 X 4.9 mm dia. holes	1	High transparency
2	Vanes	PVC plastic	80 mm (W) X 300 mm (L)	24	High transparency
3	Base	Aluminium	500 mm (W) X 500 mm (L)	1	–
4	Tube	Aluminium	300 mm dia. X 250 mm (H)	1	High transparency
5	Ball bearing	Steel	28 mm dia. with 12 mm dia. hole X 8 mm (H)	2	Reduce friction
6	Rod	Acrylic	12 mm dia. X 280 mm (L)	1	Central support
7	Stick	Acrylic	3 mm (W) X 3 mm (H) X 150 mm (L)	4	Axial support
8	Base support	Wood	25 mm (W) X 6 mm (H) X 300 mm (L)	2	Axial support
9	Fastener	Metal	10 mm dia. X 10 mm (H)	48	Joint

allowing the vanes to spin at the same time as the top cover. At another end of the rod, there is a wooden support the weight of the turbine. The wooden support is wrapped with aluminium foil to increase the reflectivity and reduce shadow appearance. The design for the upper part of PhotoVenti is as shown in Fig. 2.

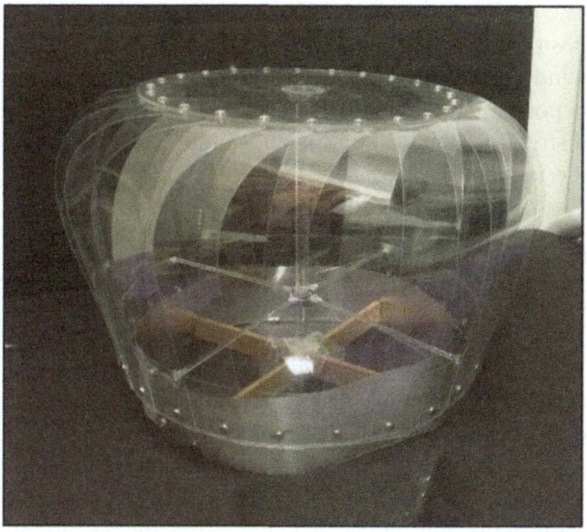

Fig. 2 PhotoVenti design

Fig. 3 Test room

Test Room

The device is tested in real condition on a test room located at Panchor, Muar. The building is used as a storage and made up of the combination of brick and wood plank with corrugated zinc roof. There is no tall trees near the building that may block the wind or sunlight especially during peak hour where the sun provides the highest outdoor illuminance. There is a window at one side the building that allows air to enter. The house is ensured to be as dark as possible to improve the accuracy of this experiment. Figure 3 below shows the test room.

Installation

The device is installed at the spot where there is less obstacle. The most preferably position is at the highest possible point so the roof itself will not block the incoming wind or cause shadow when the sun rises or sets. It is crucial to make sure the device is positioned perpendicular to the ground to make sure the PhotoVenti is stable by distributing the load uniformly so that it can spin freely, reducing the friction. Thus, the inclination of the roof is taken into consideration before designing the base. Figure 4 shows the PhotoVenti is being installed on the roof.

Data Collection

PhotoVenti helps to improve the indoor air quality and enhance vision in a building. Thus, the two parameters to be considered are the temperature and the illuminance. The experiment was conducted on 23rd, 27th and 28th November 2019 and labelled as Day 1, Day 2 and Day 3 respectively. On Day 1, the device is not installed to obtain the performance of existing building on its lighting and ventilation. The data is taken every 15 min from 7.00 am to 7.00 pm for the outdoor temperature, indoor temperature and indoor illuminance. The device is installed starting at Day 2 and the same parameters are taken to determine the performance of indoor environment after

Fig. 4 Installation of PhotoVenti on an inclined roof

an improvement is made. The device used to obtain the data is the 4 in 1 environmental meter which be able to measure the illuminance, relative humidity, temperature and wind speed. The parameters are measured at working plane which is 800 mm above the ground and vertically below the position where PhotoVenti is installed. Figure 5 illustrates the position where the data should be located.

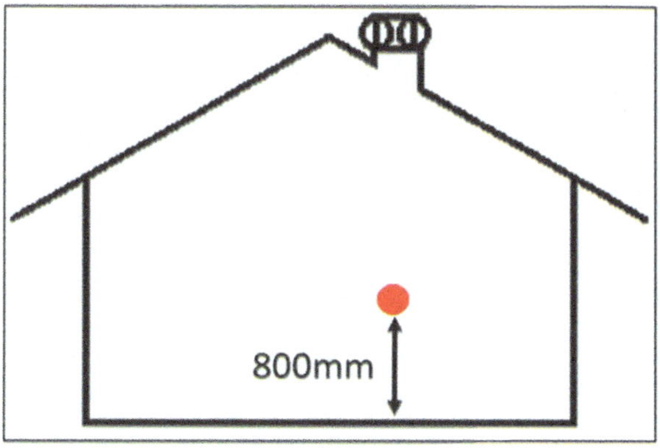

Fig. 5 The position where the data should be collected

3 Results and Analysis

3.1 Temperature Performance

The indoor and outdoor temperature is measured for three different days where the device is installed starting from Day 2 and Day 3. On Day 1, the temperature was measured to study the existing condition of the building to make a comparison with the experimental data after the PhotoVenti is installed. The data is measured from 7.00 a.m. to 7.00 p.m. with 15 min interval. The experimental result for Day 1, 2 and 3 for temperature performance are shown in Figs. 6, 7 and 8.

Day 1

The readings on Day 1 shows that the indoor temperature increases gradually with the outdoor temperature until it reaches an interception at 11:30 where the temperature is 28.4 °C. Then the indoor temperature continues to rise above the outdoor temperature until 19:00. At the peak point of outdoor temperature which is 31.5 °C at 14:15, the indoor temperature continues to rise even though the outdoor temperature started to drop. At 19:00 the outdoor temperature reduces to 26 °C while the indoor temperature is higher which is at 28 °C. The highest temperature recorded for the indoor temperature is 32.1 °Cwhile the lowest temperature for indoor and outdoor are 24.3 °C and 24 °C respectively.

On Day 1, the building is enclosed and there is no air flow in the building. Thus, heat from the sun is slowly trapped in the building and increasing the temperature. The stagnant air stores the heat and there is no escape or anything that helps the air in the building to flow except for diffusion. As the time approaches the noon, where

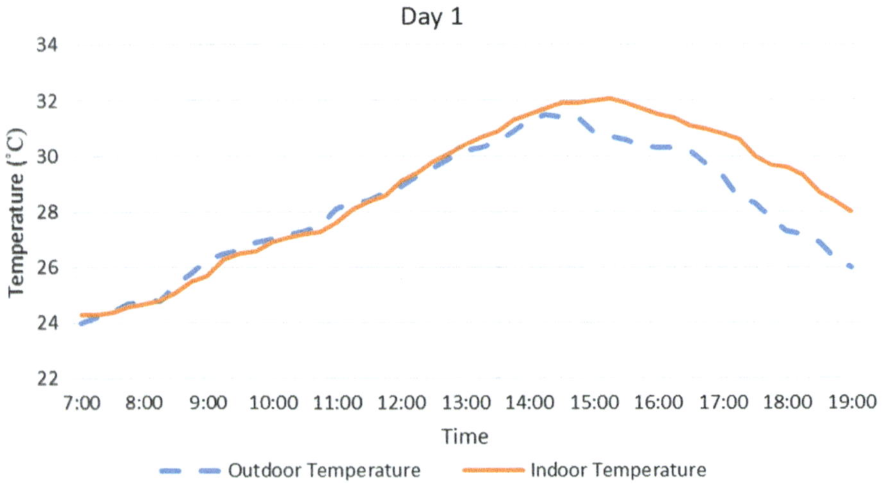

Fig. 6 Graph of Temperature vs Time on Day 1 (Without PhotoVenti)

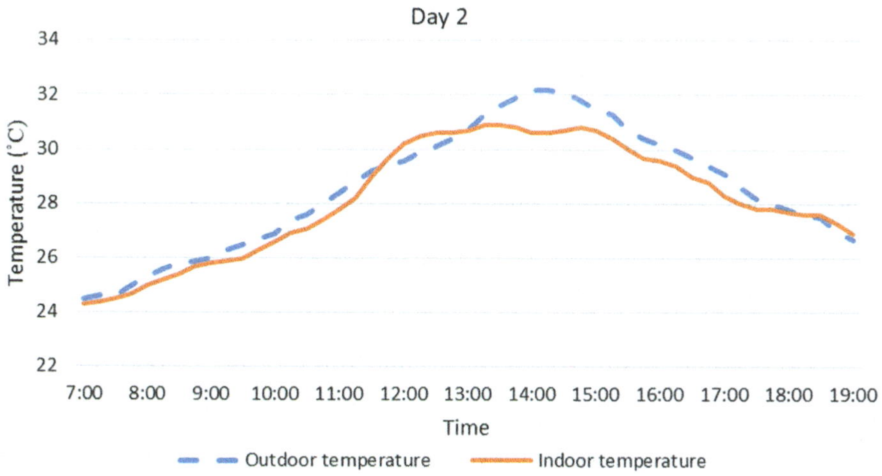

Fig. 7 Graph of Temperature vs Time on Day 2 (With PhotoVenti)

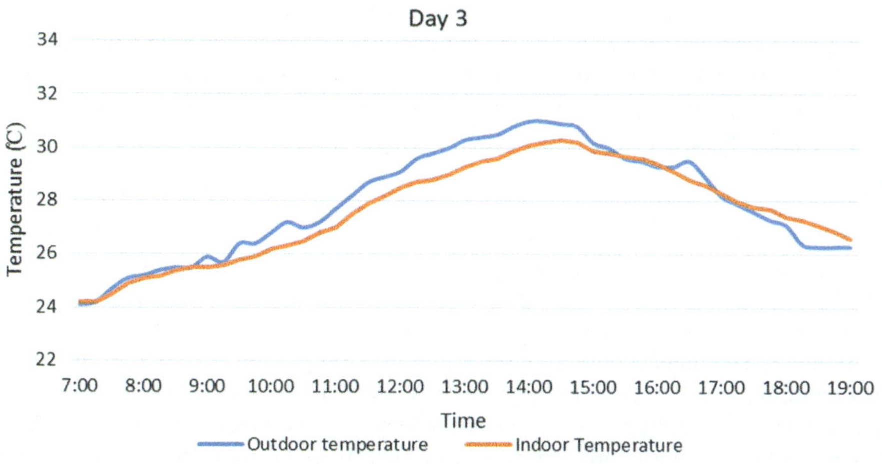

Fig. 8 Graph of Temperature vs Time on Day 3 (With PhotoVenti)

the sun started to pass overhead, the outdoor temperature increases at its highest and this adds more heat into the building. That explains why the indoor temperature continuously increasing above the outdoor temperature. Even though the outdoor temperature started to drop at the evening, the temperature keeps increasing and drops gradually. The heat in the building by diffusion through the building material and it takes a lot of time to lower the temperature. The average indoor temperature and the outdoor temperature is 28.8 °C and 28.2 °C respectively.

Day 2

The temperature pattern on Day 2 is shown in Fig. 7. On Day 2, the PhotoVenti is installed on the roof near the center of the building. The data from Day 2 shows the indoor and outdoor temperature increases gradually but the indoor temperature is slightly lower than the outdoor temperature. At 11:45, the indoor temperature started to increase rapidly as it exceeds the outdoor temperature. Then the indoor temperature remains between 30.0 and 30.9 °C with a slight change before the temperature drops while the outdoor temperature reaches its peak at 32.2 °C and started to drop gradually as well. However, the indoor temperature stays below the outdoor temperature for a period of time.

On Day 2, the indoor temperature remains slightly below the outdoor temperature before it started to increase higher than the outdoor temperature at 11:45. These changes shows the existence of heat build-up even though the Photovent has been installed. However, it can be seen that the temperature remains after that period while the outdoor temperature continues to increase to 31.5 °C. It proves that there is exchange of air or air flow in the building. The heat build-up could occur when there is no presence of wind to force the air out. Thus, the air relies on stack effect in order to escape from the room which also requires some time as it needs to create air pressure. As the sun started to set, the temperature remains to be below the outdoor temperature which differ from the Day 1 where the indoor temperature is higher than the outdoor temperature. The average outdoor and indoor temperature is 28.6 °C and 28.2 °C respectively.

Day 3

Referring to the Fig. 8 for the experimental data on Day 3, the indoor and outdoor temperature at 7:00 is about the same which is around 24.1 °C. Even though there is rapid change in outdoor temperature, but the indoor temperature gradually increases to its peak point which is 30.3 °Cand remains to be lower temperature than the outdoor until 15:45. The highest outdoor temperature is at 31 °C and started to drop. There is a rapid increase of outdoor temperature between 16:15 and 17:00 but the indoor temperature slowly drops. At 17:00 and afterwards, the outdoor temperature falls quickly but the indoor temperature decreases gently making the indoor temperature is higher than the outdoor temperature.

On Day 3, it can be seen that the outdoor temperature rapidly changes at some points but the indoor temperature changes little by little as heat is extracted out from time to time. Most of the time, the indoor temperature remains below the outdoor temperature. However, when it comes to the outdoor temperature drops quickly, the indoor temperatures require some time to drop to the same level. It also provides a good protection against temperature fluctuation. The average outdoor and indoor temperature for Day 3 are 28 and 27.6 °C.

Temperature Difference

Since the data of three different days cannot be compared directly due to the inconsistent weather, the comparison can be made through the analysis of temperature

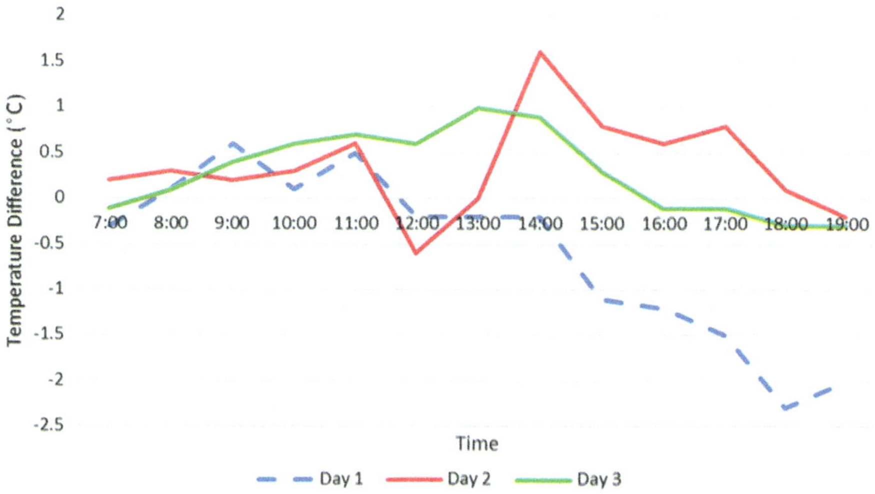

Fig. 9 Graph of Temperature Difference vs Time of Day 1, Day 2 & Day 3

difference. The outdoor temperature is subtracted with indoor temperature for each hour. Thus, the positive value represents the indoor temperature is lower than the outdoor temperature while the negative value represents otherwise. From the graph, we can conclude that, there is improvement in indoor air temperature as most of the time, the indoor temperature remains below the outdoor temperature. It is enough to prove that the device works for removing heat from a building. The temperature difference for the three days are illustrated in Fig. 9.

3.2 Daylighting Performance

The daylighting performance is measured only by considering the indoor illuminance. Then the performance of daylighting of Day 2 and Day 3 are compared to the Day 1 to analyze whether there is improvement after the device being installed. The data is illustrated as shown in Figs. 10, 11 and 12.

On Day 1, the graph shows there is inconsistency of the lux value. The highest illumination before the PhotoVenti is installed is only 9 lx and the minimum illumination is 0 lx. However, there is a huge difference when compared to Day 2 and Day 3 where the device was installed. On Day 2, the illumination increases gradually when the sun rises. The highest illumination is at 12:15 which is 876 lx then it drops to 567 lx before it rapidly increases to 612 lx. As the sun sets, the indoor illuminance slowly decreases as well. The lowest illuminance recorded on Day 2 is 7 lx. On Day 3, the indoor illuminance is very low from 7:00 to 8:00, then it quickly rises from 67 lx to 168 lx and continue to rise to 260 lx before it drops to 102 at 9:45. The

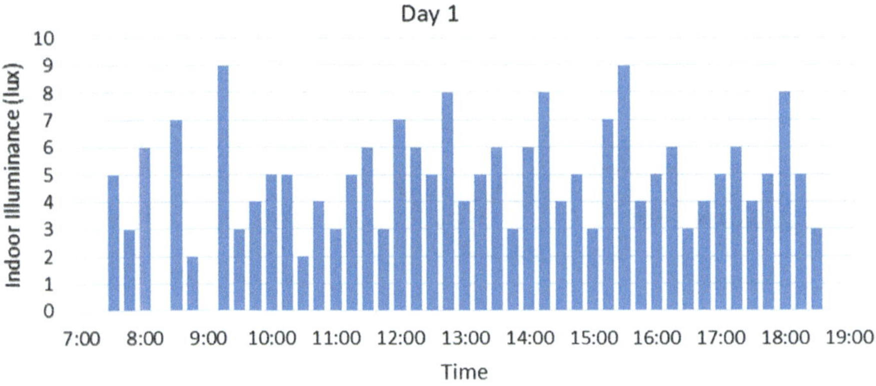

Fig. 10 Graph of Illuminance versus Time on Day 1 (Without PhotoVenti)

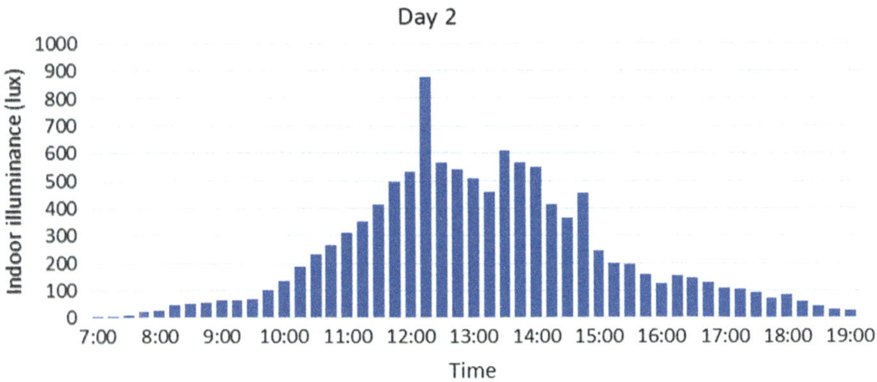

Fig. 11 Graph of Illuminance versus Time on Day 2 (With PhotoVenti)

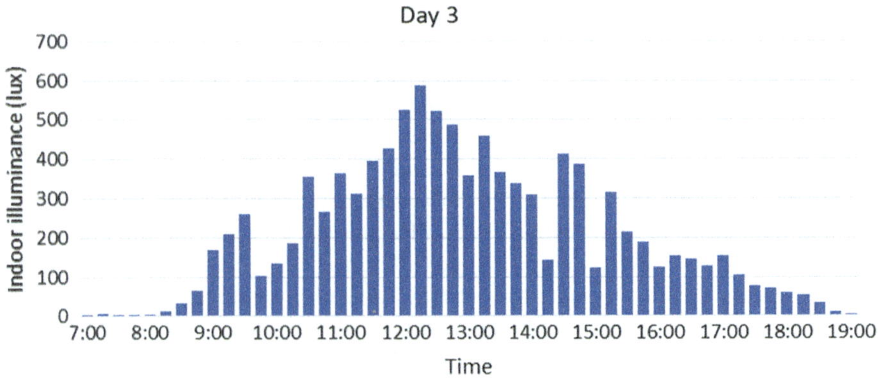

Fig. 12 Graph of Illuminance versus Time on Day 3 (With PhotoVenti)

highest illuminance for Day 3 is 589 lx at 12:15. Then, the illuminance started to drop with inconsistency throughout the day.

Daylighting Pattern

On Day 1, the indoor illuminance is low. The existing building only has one window and no other source for lighting such as lamp. Daylighting from the window is not enough to illuminate a large area. The highest illuminance it can achieve is 9 lx. On Day 2 and Day 3, there is a lot of improvements in term of daylighting. Malaysia's sky type is categorized as intermediate sky. The cloud formation causes frequent change of sky illuminance which may affect the daylighting. The illuminance for Day 2 seems more stable compared to Day 3. In addition to that, Malaysia experiencing monsoon season from October to March. Thus, the cloud cover could block the sunlight from passing through. That explains why Day 3 has inconsistent illuminance.

The average indoor illuminances for Day 2 and Day 3 are 232 lx and 208 lx respectively. The position of the device must consider the position of the sunrise and sunset to fully utilize the daylighting. Malaysia receives around 12 h of daylight every day, hence the usage of daylighting device is recommended to reduce carbon footprint.

3.3 Comparison with Standards

This Industry Code of Practice was approved by the Minister on 30 August 2010 and will replace the Code of Practice on Indoor Air Quality launched by the Minister on July 2005. A good quality of indoor air guarantees a healthy environment. Inability to keep the air quality remains good will cause short-term and long-term health problems. The most common health problems related to poor air quality includes eye irritation, respiratory problems and pneumonia. Thus, the Department of Occupational Safety and Health (DOSH) sets limits on various parameters related to Indoor Air Quality. The recommended temperature by DOSH is ranging from 23 to 26 °C. From the experimental data, the average indoor temperature two days are 27.6 and 28.3 °C. Even though the data does not meet the requirement, the device is proven to reduce indoor temperature in a building.

Room comfort condition is affected by various factors such as humidity, clothing, metabolism rate and air movement preference by the occupants A study [19] shows that the comfort temperature for tropical country is higher due to our ability of acclimation. However, a study [20] which conducted on terrace house in Malaysia proposed that the thermal comfort can be achieved below 28.69 °C as well. Another study proven that the suitable temperature for comfort can be between 23.6 and 28.6 °C [21].

The publication of MS 2680 and MS 1525 shows government's concern on energy efficiency and the use of renewable energy in buildings. MS 1525 and MS 2680 were published by the Department of Standards for non-residential and residential buildings respectively. The average indoor illuminances for Day 2 and Day 3 are

Table 2 Recommended lux for residential buildings

Area	Recommended Lux
Living room	200
Dining room & Kitchen	250
Bedroom	180
Bathroom	150

232 lx and 208 lx respectively. According to MS 2680, that level of illuminance is suitable for living room, bedroom and bathroom for residential buildings as shown in the Table 2.

MS 1525 is the Energy efficiency and use of renewable energy for non-residential buildings—Code of practice (Second Revision). This document outlines more details on recommended illumination level in non-residential buildings. Based on the experimental data, the illuminance level obtained is suitable for some area including the elevator, hotel bedroom, walkway, car park, restroom and lounge. These areas require the minimum illuminance of 100 lx.

4 Conclusion

The objective of this research is to develop a zero-energy device to improve ventilation and daylighting, measure its performance and compare with the standards. The device is an idea of combining the tubular skylight with roof ventilator in one unit and named PhotoVenti. Photoventi has been successfully built. Thus, the first objective has been achieved. It has the ability improve the ventilation and lighting in a building. Experimental data proves that this device can reduce the indoor temperature by exhausting air with two methods which are stack effect and wind energy. It also can transmit light into a building at the same time. The experiment was conducted on a house located in Panchor, Muar to meet the second objective. An average of 232 lx and 208 lx for indoor illuminance is obtained from Day 2 and Day 3 of experiment. The level of illuminance is enough for certain areas including living room, bedroom, lounge and walkway as recommend in the MS 1525 and MS 2680. The average for indoor temperature obtained are 27.6 and 28.3 °C while the recommended temperature by DOSH is between 23 and 26 °C. According to previous study, the comfort temperature may be higher for tropical countries. However, this device works great in removing excessive heat from a building. It is not limited to residential buildings but also applicable for other buildings as well especially at places where there is high heat gain from processes, machineries or human.

Acknowledgements This work would not have been possible without the financial support of the grant project Vot. H251 and "Geran Penyelidikan Pascasiswazah" (Vot. H595) by Universiti Tun Hussein Onn Malaysia (UTHM). We are grateful to all of those with whom we have had the pleasure to work during this and other related projects.

References

1. Bakhtyar B, Tarek K, Md AN (2017) A review on carbon emissions in Malaysian cement industry. Int J Energy Econ Policy 7:282–286
2. Shafie SM, Mahlia TMI, Masjuki HH, Andriyana A (2011) Current energy usage and sustainable energy in Malaysia: a review. Renew Sustain Energy Rev 15:4370–4377. https://doi.org/10.1016/j.rser.2011.07.113
3. Hassan JS, Zin RM, Abd Majid MZ, Balubaid S, Hainin MR (2014) Building energy consumption in Malaysia: an overview. Jurnal Teknologi 70:2180–3722. https://doi.org/10.11113/jt.v70.3574
4. Baglivo C, Bonomolo M, Beccali M, Congedo PM (2017) Sizing analysis of interior lighting using tubular daylighting devices. Energy Proc 126:179–186. https://doi.org/10.1016/j.egypro.2017.08.138
5. Soni GS (2015) Advantages of green technology. Soc Issues Environ Probl 3:1–5. https://doi.org/10.7753/IJSEA0609.1005
6. Khan N, Su Y, Riffat SB (2008) A review on wind driven ventilation techniques. Energy Build 40:1586–1604. https://doi.org/10.1016/j.enbuild.2008.02.015
7. Carter DJ, Al Marwaee M (2009) User attitudes toward tubular daylight guidance systems. Light Res Technol 41:71–88. https://doi.org/10.1177/1477153508096045
8. Kadir AA, Ismail LH, Kasim N, Kaamin M (2016) Potential of light pipes system in Malaysian climate. In: IOP conference series: materials science and engineering. https://doi.org/10.1088/1757-899x/160/1/012071
9. Zhang X, Muneer T (2000) Mathematical model for the performance of light pipes. Light Res Technol 32:141–146. https://doi.org/10.1177/096032710003200306
10. Yun GY, Hwang T, Kim JT (2009) Performance prediction by modelling of a light-pipe system used under the climate conditions of Korea. Indoor Built Environ 9:137–144. https://doi.org/10.1177/1420326X09358008
11. Lien STJ, Ahmed N (2011) Wind driven ventilation for enhance indoor air quality. University of New South Wales. https://doi.org/10.5772/17059
12. Ahmed S, Zain-Ahmed A, Abdul Rahman S, Sharif MH (2006) Predictive tools for evaluating daylighting performance of light pipes. Int J Low-Carbon Technol 1:315–328. https://doi.org/10.1093/ijlct/1.4.315
13. Kadir AA, Ismail LH, Kasim N (2004) Optimization of daylighting system by using light pipe system in a building. Sol Energy 78:772–780. https://doi.org/10.1016/j.solener.2004.09.002
14. Priyo AS, Triyogi Y (2018) The effect of inner fan blade angle to the ventilation rate of the turbine ventilator. Jurnal Rekayasa Mesin 9:227–233. https://doi.org/10.21776/ub.jrm.2018.009.03.10
15. Zain-Ahmed A, Sopian K, Zainol Abidin Z, Othman MYH (2002) The availability of daylight from tropical skies a case study of Malaysia. Renew Energy 25:21–30. https://doi.org/10.1016/S0960-1481(00)00209-3
16. Robertson AP, Hedges RC, Rideout NM (2009) Optimisation and design of ducted daylight systems. Light Res Technol 42:161–181. https://doi.org/10.1177/1477153509355500
17. Khan N, Su YH, Riffat S, Biggs C (2008) Performance testing and comparison of turbine ventilators. Renew Energy 33:2441–2447. https://doi.org/10.1016/j.renene.2008.01.016
18. Chi ML (2003) Experiments on the ventilation efficiency of turbine ventilators used for building and factory ventilation. Energy Build 35:927–932. https://doi.org/10.1016/S0378-7788(03)00024-0
19. Li B, Li W, Liu H, Yao R, Tan M, Ing S, Ma X (2010) Physiological expression of human thermal comfort to indoor operative temperature in the Non-HVAC environment. Indoor Built Environ 19:221–229. https://doi.org/10.1177/1420326X10365213
20. Zain Z, Taib M, Baki S (2007) Hot and humid climate: prospect for thermal comfort in residential building. Desalination 209:261–268. https://doi.org/10.1016/j.desal.2007.04.036

21. Sabarinah S, Steven VS (2007) The performance of a partially air conditioned apartment building in Kuala Lumpur. In: The 24th International conference on passive and low energy architecture, pp 608–614

Agile Project Management Software for Construction and Management Industries

Nur Hidayah Mohd Ali, Faridahanim Ahmad, Nur IzieAdiana Abidin, Saadatul Suhaili, Mohammad Ashraf Abdul Rahman, Hasnida Harun, Mariah Awang, Nor Hazren Abdul Hamid, Mohd Kamaruzaman Musa, Nuramidah Hamidon, Fatimah Mohamed Yusop, and M. M. Syafiq Syazwan

Abstract While Industry Revolution 4.0 concept is implementing, most of the working environment are automatized and interoperated with Internet of Things and Big Data being processed real time using cloud computing. The use of efficient project management software in the modern world is increasing in order to achieve organizational objectives. There are several software in market that being used by people in the idustry, but the satisfaction and opinion of the users need to be clarify. The objectives of this research is to identify the implementation of project management software PlanGrid and ZenTao, to analysed the advantages and disadvanatges of the two software. To develop a theoretical basis for this study, literature related to the experiences and lessons learned in the project management software and application was reviewed. A questionnaire survey was conducted to asses the level of implementation and the satisfaction of the software by the users. The findings revealed that the the people working in construction and management industry is not very aware and implement the project management software PlanGrid and ZenTao in their daily working life. Furthermore, users could be more happy if the interface come up with interactive display. The good thing about agile approach is that it has adaptability to changes over the life cycle of the project and to different projects in general. Including Kanban in project management software would be good since it can visualize the workflow. Thus, more effective and efficient work output can be produce via online and towards Industry Revolution 4.0.

Keywords Project management · Software · Kanban

F. Ahmad (✉) · M. A. A. Rahman · H. Harun · M. Awang · N. H. A. Hamid · M. K. Musa · N. Hamidon · F. M. Yusop · M. M. Syafiq Syazwan
Department of Civil Engineering Technology, Faculty of Engineering Technology, Universiti Tun Hussein Onn Malaysia, Pagoh, Muar 84600, Malaysia
e-mail: faridahanim@utm.my

N. H. M. Ali · N. I. Abidin · S. Suhaili
Department of Structure and Materials, School of Civil Engineering, Faculty of Engineering, Universiti Teknologi Malaysia, Johor Bahru, Johor 81310, Malaysia

M. A. A. Zaini et al. (eds.), *Proceedings of the 3rd International Conference on Separation Technology*, Lecture Notes in Mechanical Engineering,
https://doi.org/10.1007/978-981-16-0742-4_7

1 Introduction

The used of advanced technology in all industry are being applied and improved from day to day including in the field of construction industry due to the fast development of construction in the country. The technology used in construction industry is considered from the innovations of tools, machinery and software being used during construction phase to complete a project. There are a lot of software in market are being commercialized in order to achieve the satisfaction of the people in construction industry and to help in project management so that the project can be done within the targeted timing and budget.

Weak management capability of contractors can be one of the major challenges facing by the construction sector in developing countries [1]. Therefore, one of the key factors for improving contractor's capability in developed countries needs to be improving contractor's management ability. Moreover, according to previous study strongly believed that the project management skill of the people in construction industry is important part to be considered as it can affect the final result and product to be producing [2]. Good project management will help in smooth work progress and not to burst budget of the project which is can make profit to the company.

Kanban is a visual way to manage tasks and workflows, which utilizes a kanban board with columns and cards. The cards represent tasks, and the columns organize those tasks by their progress or current stage in development [3]. Thus, the overall situation of the project can be seen at a glance while the continuous flow of task cards indicates the progress of the project over the time [4]. Agile methodology is a type of project management process, mainly used for software development, where demands and solutions evolve through the collaborative effort of self-organizing and cross-functional teams and their customers [5].

There is several software's for project management in construction and management industry in market that usually being used by the construction industry players in order to help them to complete a project. However, there are many questions and uncertainties that need to be clarified such as the level of implementation and satisfaction of users towards the software. Based on the statistic record by The State of Project Management Survey in 2018, business dissatisfaction with project portfolio management is increasing from 45% to 50% and 56% for 2016, 2017 and 2018 respectively [6]. The application or software developed must achieve user's satisfaction in order to make the application successful. If the application does not reach user's expectation, users may unsubscribe or abandoning the software and applications.

The following objective has been conducted to complete this study:

1. To identify the implementation of PlanGrid and ZenTao software in construction industry; and
2. To analyse the advantages and disadvantages of both software from the assessment of people in this industry.

The research scope of this study is focused on literature review and the distribution of questionnaire through internet by Google form. The participant's background that involved answering the questions is developer, consultant and contractor. The experts or the workers that familiar with the project management software are needed in order to obtain different perspective about the implementation of PlanGrid and ZenTao in managing a certain project.

There has been variety research about project management effectiveness that being applied in any field in this country, but there are very few studies were found in project management software that support the revolution of industry 4.0 (IR 4.0) such as PlanGrid and ZenTao. Therefore, more comprehensive research is required to explore the connection between satisfaction of user and the effective of the project management software. The limitation of the study is limited cover are which is been done in Malaysia only. It is also limited by using only two project management software which is PlanGrid and ZenTao.

2 Literature Review

In short, we cannot address the issue of what project management is. Project management is so complex and cannot be defined adequately in only one way. To begin with, a definition of projects is necessary. Project is described as a job or operation intended to be performed over a period of time and directed at a particular goal [7]. Another definition, a project is a special, temporary attempt to achieve planned goals that can be specified in terms of benefits, outputs or outcomes [8]. A project is generally considered as successful if it meets its targets within the accepted timeline and budget, according to the approval requirements. The project concept is developed during initiation. Project solutions are tested, appraised and one selected, during planning and development. The project plan is carried out during implementation and the completed facility commissioned and handed over to the owner in the fourth phase [9].

Many writers and sources have described a project management, stressing its various aspects in different ways. One of them are from Project Management Institute that stated in order to fulfil the requirements of the project, the application of knowledge, skills, tools, and techniques implemented to the project activities is define as project management. Moreover, project management would use an ensemble of tools and techniques to guide the use of various resources to accomplish a specific, complex and one-time task within time, cost and quality constraints [10].

In many organizations, project management is generally used extensively in some form. No profession or industry has been identified where project management practices will not work [11]. Generally, by using project management would give a helps an organization to clarify targets and identify areas of problem and risk, isolating activities and track performance easily. Furthermore, by using project management can improves transparency as jobs can be separated and roles can be assigned. In addition, it helps to concentrate attention on a few unique and important tasks.

2.1 Kanban

In development of software, Kanban is increasingly become the central Lean practice [12]. In Japanese, Kanban is known as "signboard" or "visual card". Kanban's idea is to support the production system Just In Time (JIT) by implementing its pulling policy to attain its objectives. For JIT, the aim is to reduce the Work In Progress (WIP) inventory, eliminating waste to reduce costs and increase the quality, while Kanban's primary objective is to control the inventory of WIP. The concept is that in the production process, inventories control would reveal both the areas which need improvements and waste. Consequently, if corrective measures are applied, JIT will achieve its objectives [13]. Kanban's principles include of limiting WIP, visualize workflow, making explicative policies, recognizing opportunities for process improvements [14].

Despite the demand for visualization of the workflow there are no special guidelines about how to apply the content of the Kanban board. Kanban controls are usually applied in their basic form in production environments, with physical index cards commonly called tickets, going along with the content. These cards then function as tickets for flow control between the various work stations or processes. Every project task or card flows as it moves from one state to the next one from left to right. Thus, the overall situation of the project can be seen at a glance while the continuous flow of task cards indicates the progress of the project over the time [15].

2.2 PlanGrid

PlanGrid is a mobile application for construction, which helps professionals from the architecture, engineering and construction (AEC) industry to work together through a mobile device to develop their project plan, design specifications and photos on site and not just from the office. Up to 50 drawing sheets is available in the basic version of the software application. The user just need to upload the Portable Document Format drawings to PlanGrid.com and the service provider will automatically synchronize the drawings and make them communicable. All team-mates working on a specific project are permitted to share the updated sketches. In addition, marking tools can be used to change the drawings and send updates to anyone involved with the projects.

2.3 ZenTao

ZenTao is software developed by EasyCorp for open source project management. The key functions include the product management, project management, Test management, document management, bug tracking, CI management and to do management

[16]. It covers the main software development process and offers a qualified application life cycle management (ALM). Some of the features of this software are it is applying agile methodologies, client portal, cost-to-completion tracking, gantt charts, kanban board, time & expense tracking, collaboration tools, customizable templates, idea management, milestone tracking [17].

3 Methodology

This research consisted of three (3) key activities; exploratory study, research instrument used and the sampling size, and the analysis of the data and information collected. The below figure depicts the flow chart of the whole process flow for this research.

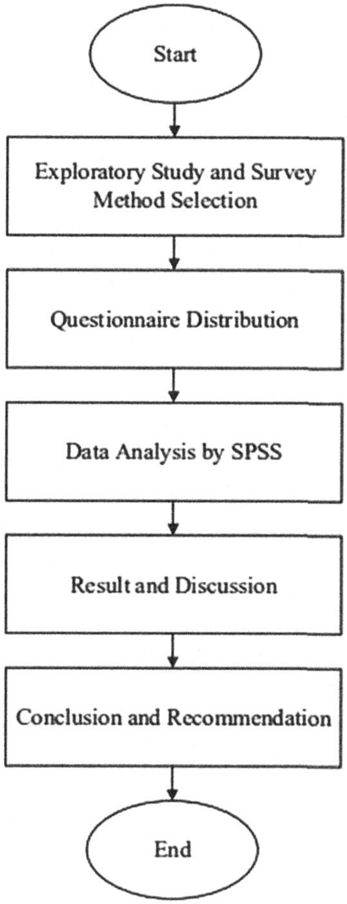

3.1 Exploratory Study

This research is an exploratory study. It aims to explore the opinion of software's user about the weakness and comparison of the both software mentioned in this research. Exploratory study is described as planning that researchers create to identify a problem, issue or phenomenon [18]. It is carried out when the topic of research is not well known to the public and information is limited [19].

3.2 Research Instrument and Sampling Size

This is a quantitative research, since questionnaires are adopted to collect answers effectively from a large sample group in which all respondents are asked to comment on the same collection of questions via Google form through Internet [18]. The questionnaire is use to gathers the samples and information relates to the respondent's attitudes, trends and opinions [19]. The target respondents for this research are the personnel working in Malaysian construction and management industry. This is to ensure the respondents have sufficient industrial knowledge and experience about the software to understand and able to answer the questionnaire. The study population consists of people working in construction and management industry. The study population for this research is 20. The number of respondents was referred to the Krejcie and Morgan table to represent the reliability of sample that being chosen. As for this study the population size is 20, meaning that the sample size is 19. So, the total questionnaires distributed were only 19.

3.3 Data Analysis

Data management and analysis were performed by using SPSS software. SPSS is short for Statistical Package for the Social Sciences, and it is used by various kinds of researchers for complex statistical data analysis. Once all the data are inserted in the SPSS, it will automatically provide the data that the users need such as mean rank, chi-square and Asymp Sig p-value. Therefore, the result and analysis of the evaluation of the two software; PlanGrid and ZenTao from the respondents can be obtained easily by using SPSS.

For Section B in the questionnaire survey, mean ranks of the statements are computed using Friedman test. The mean rank revealed the agreement of the respondents towards the statements in the questionnaire. The mean rank indicates the degree of acknowledgement on the statement from respondents [20].

4 Results and Discussion

The findings of this research are in line with the research objectives. The first result shows the attributes of respondents. The next section reports the results of data collected on the implementation of PlanGrid and ZenTao and the opinion and feedback of the software from the respondents. The evaluations of usability for both of the software are analyzed by SPSS software. The usability scores for each of the mobile applications presented in this chapter.

4.1 Respondents Background

The details of the respondents' attributes are summarized in Table 1 below. Majority of the respondent are from consultant firm (36.8%). Most of the respondents are male (57.9%) and the others are female (42.1%). Majority the respondents are in the age group of between 31 to 40 years old (42.1%) and there is only 21.1% of them have working experience more than 10 years and the rest are share the same frequency with 26.3%.

Table 1 Attributes of Respondents

General Information	Categories	Frequency	Percentage (%)
Gender	Male	11	57.9
	Female	8	42.1
Age	Below 25 years old	4	21.1
	25-30 years old	6	31.6
	31-40 years old	8	42.1
	Above 40 years old	1	5.3
Working Experience	Less than 2 years	5	26.3
	2-5 years	5	26.3
	5-10 years	5	26.3
	More than 10 years	4	21.1
Organization Business Activity	Developer	3	15.8
	Consultant	7	36.8
	Contractor	5	26.3
	Sub-contractor	–	–
	Supplier	2	10.5
	other	2	10.6

Table 2 Ranking for Evaluation about Plangrid

PlanGrid	Mean Rank	Chi-Square	Asymp. Sig
Frequently using the software.	4.05	45.679	.000
The software is easy to learn and use.	5.92		
The software is boring/not interesting.	3.45		
The various functions in this application were well integrated.	5.87		
The software is complex.	2.87		
Too much inconsistency in this application.	3.50		
Most people would learn to use this application very quickly.	5.42		
The application is better accumulation data.	4.92		

4.2 Evaluation Respondent About PlanGrid

The mean ranks of the evaluation from the respondents about PlanGrid software in construction and management industry are reported in Table 2. Friedman Test results revealed the differences of the mean rank are statistically significant ($\chi 2 = 45.679$, df = 7, N = 19, p = .000). The respondent ranked the software is easy to learn and use as the highest rank (5.92) and lowest rank (2.87) which is the software is complex.

Based on the answer from the questionnaire survey, the majority of the respondents strongly agree (42.1%) that they are prefer using PlanGrid software because it has the features that enable users to insert AutoCAD drawing as shown in Fig. 2. The result shows that 15.8% of the respondents answer for average and the rest share the same percentage which is 21.1% for do not use the software and agree that features to upload AutoCAD drawing is really helpful.

4.3 Evaluation Respondent About ZenTao

The mean ranks of the evaluation from the respondents about ZenTao software in construction and management industry are reported in Table 3. Friedman Test results revealed the differences of the mean rank are statistically significant ($\chi 2 = 24.447$, df = 7, N = 19, p = .001). The respondent ranked the software would be very quickly to learn as the highest rank (3.50) and lowest rank (3.32) which is too much inconsistency in this application as shown in Tables 4 and 5.

For the next question is about user's opinion about the features Kanban in ZenTao software is helpful and enable users to optimize their flow work. As shown in Fig. 3, most of the respondents rate this question as strongly agree with 47.4% and 26.3%

Table 3 Respondent's Vote about using PlanGrid software

Answer	Percentage (%)
Do not use	21.1
Strongly Disagree	0
Disagree	0
Average	15.8
Agree	21.1
Strongly Agree	42.1

Table 4 Ranking for Evaluation about ZenTao

PlanGrid	Mean Rank	Chi-Square	Asymp. Sig
Frequently using the software.	5.11	24.447	.001
The software is easy to learn and use.	5.13		
The software is boring/not interesting.	3.87		
The various functions in this application were well integrated.	4.03		
The software is complex.	3.63		
Too much inconsistency in this application.	3.32		
Most people would learn to use this application very quickly.	5.50		
The application is better accumulation data.	5.42		

Table 5 Respondent's Vote about the features Kanban in ZenTao software

Answer	Percentage (%)
Do not use	21.1
Strongly Disagree	0
Disagree	0
Average	5.3
Agree	26.3
Strongly Agree	47.4

agree with the statement and another 21.1% and 5.3% vote for do not use the software and average respectively.

5 Conclusion

The research findings revealed that, the people working in construction and management industries in Malaysia is not very aware and implement the project management software PlanGrid and ZenTao in their daily working life. Based on the data collected from the questionnaire survey, this study indicates that there are few things that can be improve to meet the user's satisfaction such as more colourful and interactive display, combine the feature of Kanban and allow the user to insert AutoCAD drawings. Therefore, the new interface designed with some improvement in order to meet the user's satisfaction.

Stratified random sampling is recommended as the sampling technique in future research. Stratified random sampling allows accurate representative sample groups that are meaningful and appropriate to the research topic by predetermined the number of respondents required in each stratum. Next, a more precise scope shall be defined in future research. A precise research scope helps researcher in focusing in the specific topic. A more concise questionnaire can be produced with the guide of the research scope.

In all areas, Internet of Things (IoT) intersects with project management, from teamwork to data collection. In a new age of dynamic planning and revolutionized project execution, monitoring real-time status via IoT can be expected. The IoT fundamentally changes the speed of project execution in term of project management. Organizations capitalizing on the IoT project will carry out their projects faster than others, and organizations not adapted to the IoT revolution are left behind hopelessly. Therefore, Industry 4.0 is leading to modern industrial automation that incorporates advanced technology with a view to rising efficiency, improving working conditions.

References

1. Long ND, Ogunlana S, Quang T, Lam KC (2004) Large construction projects in developing countries: A case study from Vietnam. Int J Project Manage 22(7):553–561. https://doi.org/10.1016/j.ijproman.2004.03.004
2. Dlungwana WS, Rwelamila PD (2004) Contractor development models that meet the challenges of globalization-a case for developing management capability of local contractors. Global Constr:347
3. The Ultimate Guide to Kanban. ProjectManager.Com. https://www.projectmanager.com/kanban
4. Ikonen M, Pirinen E, Fagerholm F, Kettunen P, Abrahamsson P (2011) On the impact of Kanban on software project work: An empirical case study investigation. In: 16th IEEE International conference on engineering of complex computer systems, ICECCS, Las Vegas, NV, pp 305–314. https://doi.org/10.1109/iceccs.2011.37
5. Terms for better understanding technical candidates. Glossarytech. https://glossarytech.com/terms/development_methodologies
6. The State of Project Management Survey. Wellingtone Limited. http://www.wellingtone.co.uk/wp-content/uploads/2018/05/The-State-of -Project-Management-Survey-2018-FINAL.pdf
7. Vale D, Mullaney S, Hartas L (1996) The cambridge dictionary. Cambridge University Press, Cambridge
8. Fahrenkrog SL, Bolles D, Blaine JD, Steuer C (2004). PMBOK® guide— third edition: an overview of the changes. In: Paper presented at PMI® Global Congress 2004—North America, Anaheim, CA. Newtown Square, PA, Project Management Institute

9. Muriithi N, Crawford L (2003) Approaches to project management in Africa: Implications for international development projects. Int J Project Manage 21(5):309–319. https://doi.org/10.1016/s0263-7863(02)00048-0
10. Atkinson R (1999) Project management: cost, time and quality, two best guesses and a phenomenon, it's time to accept other success criteria. Int J Project Manage 17(6):337–342. https://doi.org/10.1016/s0263-7863(98)00069-6
11. Cleland DI, Ireland LR (2007) Project management strategic design and implementation, 4th ed. The McGraw-Hill Companies
12. Anderson D, Concas G, Lunesu MI, Marchesi M (2011) Studying lean-kanban approach using software process simulation. In: International conference on agile software development, pp 12–26. https://doi.org/10.1007/978-3-642-20677-1_2
13. Lander E, Liker JK (2007) The Toyota production system and art: making highly customized and creative products the Toyota way. Int J Prod Res 45(16):3681–3698. https://doi.org/10.1080/00207540701223519
14. Anderson DJ (2010) Kanban: successful evolutionary change for your technology business. Blue Hole Press, Sequim, Washington
15. Ikonen M, Pirinen E, Fagerholm F, Kettunen P, Abrahamsson P (2011) On the impact of Kanban on software project work: an empirical case study investigation. In: 16th IEEE international conference on engineering of complex computer systems, ICECCS, Las Vegas, NV, pp 305–314. https://doi.org/10.1109/iceccs.2011.37
16. ZenTao Manual. https://www.zentao.pm/book/zentaomanual/free-open-source-project-management-software-zentao-8.html
17. SoftwareWorld. https://www.softwareworld.co/top-project-management-software/#ZenTao
18. Saunders M, Lewis P, Thornhill A (2016) Research methods for business students. Pearson Education, Harlow
19. Creswell JW (2009) Research design: qualitative, quantitative, and mixed methods approaches. SAGE Publications, Thousand Oaks, CA
20. Chiang I, Jhangiani R, Price P (2015) Constructing survey questionnaires. https://opentextbc.ca/researchmethods/chapter/constructing-survey-questionnaires/

Design Campus Map Using OpenStreetMap Digital Software

Mohamad Afi Asyraf Maidin, Faridahanim Ahmad, Nur IzieAdiana Abidin, Jauziah Suhaili, Mariah Awang, Mohammad Ashraf Abdul Rahman, Mohd Kamaruzaman Musa, Nuramidah Hamidon, Fatimah Mohamed Yusop, M. M. Syafiq Syazwan, Hasnida Harun, Nor Hazren Abdul Hamid, and Nur Atikah Kamil

Abstract Industry Revolution 4.0 could connect world of data, people, processes, facilities, devices, and the Internet of Things, enabling industrial assets to produce, exploit, and use actionable information. OpenStreetMap program is a software platform that is available on the Web. It is a community project that can help create the world's first editable, digital map. The problem facing this study is the lack of information about the location area that gives difficulties for the new student intake and parents to find the locations. This study aims to develop the digital mapping of College Campus Pagoh in OpenStreetMap. This study also improvises the Smart Campus Universiti Tun Hussein Onn Malaysia website content to make it more user-friendly. The website is developed using Joomla 3.8.13 with PHP version, 5.6.25. 34 locations were plotted using this link https://digitalcampus.uthm.edu.my/Index.php. Digital Campus based on OpenStreetMap can help students, admin, staff, lecturers, and visitors find location by using the digital platform. The methodology of the study was begun with identifying all the locations and the name of the building in the College Campus Pagoh. Then, the complete map of College Campus Pagoh in OpenStreetMap was embedded into the Smart Campus and finally improvises the interface website of Smart Campus. The map of the College Campus Pagoh on Smart Campus can be accessed by users to finding the location. Thus digital map on Campus is very useful for giving effective information about the location and is beneficial to all parties.

Keywords Digital map · Smart campus · OpenStreetMap

M. A. A. Maidin · M. Awang · M. A. A. Rahman · M. K. Musa · N. Hamidon · F. M. Yusop · M. M. Syafiq Syazwan · H. Harun · N. H. A. Hamid
Department of Civil Engineering Technology, Faculty of Engineering Technology, Universiti Tun Hussein Onn Malaysia, Pagoh, Muar 84600, Johor, Malaysia

F. Ahmad (✉) · N. I. Abidin · J. Suhaili · N. A. Kamil
Department of Structure and Materials, School of Civil Engineering, Faculty of Engineering, Universiti Teknologi Malaysia, Johor Bahru 81310, Johor, Malaysia
e-mail: faridahanim@utm.my

M. A. A. Zaini et al. (eds.), *Proceedings of the 3rd International Conference on Separation Technology*, Lecture Notes in Mechanical Engineering,
https://doi.org/10.1007/978-981-16-0742-4_8

1 Introduction

Industry Revolution 4.0 (IR4.0) relates to a new dominion that involves the interconnected between the Internet of Things (IoT) and Cyber-Physical System in such a way that the mixing of sensors, software, processors, and communication technologies plays a major role in producing 'things' that can provide information and eventually add value to production practices [1, 2]. IoT promotes human, computer, and equipment interaction and intelligent fabrication across the World Wide Web. OpenStreetMap (OSM) has been chosen and developed nowadays to edit the world map in contributors. The hub of the OpenStreetMap project is the main OSM website which helps to build a free, editable digital map for future updating of the world's mapping process.

The digital campus consists of a digital map with a responsive platform that uses flexible templates, flexible images, and cascading style sheet media queries to approach Web page development. The goal of responsive design is to build web pages that detect the visitor's screen size and orientation and change the layout accordingly. Science learning, design-based work provides a blueprint for developing and evaluating technologies for learning within the crucible of classrooms and other contexts. As a model that emphasizes iterative cycles of design and testing, design-based research is especially well suited to making evidence-based innovation improvements, in which feedback from implementation as well as tests guide changes that design teams make to learning innovations [3].

Implementation is an analysis which is the systematic study of software implementation. This provides systematic analyses of reliability and implementation variations, as well as analyzes the circumstances under which systems can be successfully implemented [4]. Implementation research is often conducted in larger outcome studies, to analyze and how much variation in implementation matters for effectiveness in innovation [5].

The objectives are to identify the geographic area of College Campus UTHM using current satellite data, furthermore to create the console admin for OpenStreetMap and inventory visual data map into OpenStreetMap. Lastly, to design a digital map by embedded OpenStreetMap into the Digital Campus Website. Using the satellite data, the geographic area created is more clear and detailed.

2 Literature Review

2.1 Responsive Web

Responsive design is a methodology used by designers to create an attractive visual interface irrespective of the size of the browser used and any limitations imposed on accessing the app. When it uses these three criteria, a concept is considered receptive which is a flexible grid, flexible media for images, and ultimately for media queries.

By using media queries, sensitive website designers can identify certain ranges of resolution as requirements for using certain meanings of Cascading Style Sheets (CSS) called fixed breakpoints. Thus, the designer will match which CSS concept will be applied to a given resolution to create a better visual experience for the users of the website. The information quality of a website can be observed through the accessibility, information design, and content dimensions of 3 of 12 mobile user experiences. The website's functionality should be told from its devices on its homepage. Therefore, later on in this research, we observed home functionality regarding the design aspect [6].

Web design responsive is the approach that suggests that design and development should respond to the behavior and environment of the user based on screen size, platform, and orientation. The technique is a combination of versatile grids and templates, images, and the good use of CSS media queries. As the user switches from their laptop to iPad, the website should switch automatically to accommodate capabilities for resolution, image size, and scripting. If they have a Virtual Private Network (VPN) for iOS on their iPad, for example, the website does not restrict the user's access to the page, one will also have to understand the settings on their computers. In other words, the website should have the technology to react automatically to the preferences of the user. This will remove the need for every gadget on the market to have a separate design and development phase [7].

2.2 *OpenStreetMap*

The wide availability of high-quality location information has made mass-market mapping possible based on inexpensive GPS units, home computers, and the Internet. Though a range of user-generated mapping-based projects has emerged, OpenStreetMap (OSM) is probably the largest and most effective project currently under development. We review the project in this article and provide an overview of the strategies and methodologies used within it [8].

As of May 2008, OSM had over 33,000 registered users (with around 3,500 currently active contributors), and the growth in data participation continues to increase steadily [8]. A large number of participants collaboratively edit the world map using the OSM technical infrastructure, and a core community, estimated at around 40 volunteers, dedicate their time to developing and enhancing the OSM infrastructure, including maintaining the server, writing the core software that manages the server transactions, and producing cartographic outputs. There is also an increasing community of software developers who are developing software tools to make OSM data available for further use across various application domains, software platforms, and hardware devices [8].

2.3 *Digital Campus*

Digital Campus is a campus network and an Internet built on the use of advanced information technology means and tools to achieve from the environment, resources, all-digital activities, centered on the traditional campus, to create a digital space, to enhance the productivity of the traditional campus, to extend the traditional campus role and, eventually, to inform the general educational process. In short, "digital campus" is set against the backdrop of teaching, studying, management, culture, a modern digital job, research, and living environment [9]. The digital campus is focused on the web and uses sophisticated information technology software and instruments to accomplish all the digital content for the environment, services, and activities. It can create a digital space centered on the traditional campus, expanding the role of the traditional campus and eventually introducing the comprehensive modernization of the education process [10]. Virtual reality is a newly developed computer technology sector, with integrated computer graphics technology, multimedia technology, sensor technology, machine-human interaction technology, network technology, stereo vision technology, and simulation technologies. Digital Campus is a significant application of virtual reality technology to make the topography of the school and the real scenery replicated on the screen and also can create a Digital Campus, a vibrant real campus atmosphere to provide users with a fully intuitive experience. Integrated with the database attributes, for a variety of related details, the system can satisfy interactive, real-time queries [11].

A final sketch has been conducted after all the data and information on the building and map of the college are were collected. This sketch consisted of all the important elements, such as building, parking, walkway, primary road, secondary road, restaurant, and others. This entire element is important to have to detail the map and to look realistic as the complete building today. When the entire place has been detailed as in the example, people will easily find their location, such as restaurant, parking nearby, rest and health and any other places that they need to go.

3 Methodology

Digital campus based on UTHM Pagoh Campus is a virtual reality to archive localization, map query, navigation guide, tracking, and other basic function. The software system is developed by using Content Management Jomlaa version 3.8.13 with PHP version 5.6.25. The digital campus system chooses Open Street Map application for embedded all block identified located mapped onto the digital by draw all block into Open Street Maps. The duration time of the project to be delivered is about one month started in September 2019. The system can be accessed either through the website via mobile phone, laptop, and desktop. The digital campus website enables users to search location by access to link https://digitalcampus.uthm.edu.my/index.php.

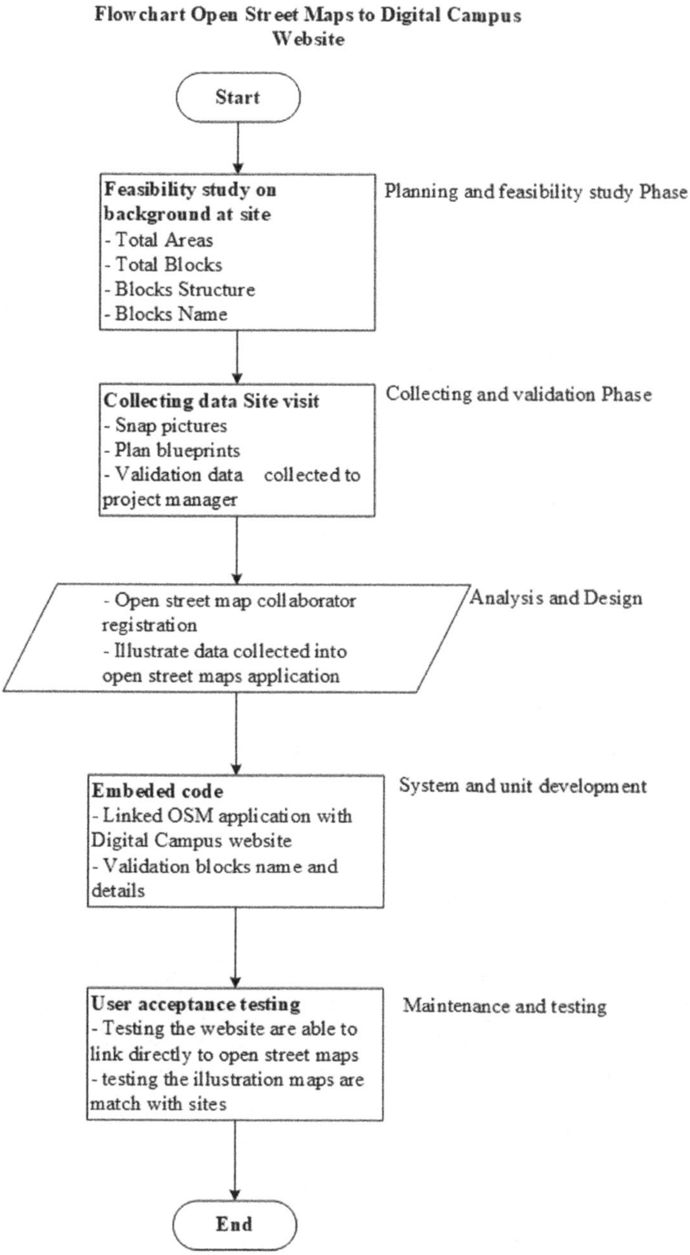

Fig. 1 Flow chart on processing using Open Street Map

Figure 1 shows the flow chart on processing using OpenStreetMap. In the first phase, the study on total areas, total blocks, blocks structure, and block name is done. Then, in the next phase, data such as pictures and blueprints are collected after the site visit. In the analysis and design phase, all the collections will be illustrated in OpenStreetMap application. After that, the OSM application will be linked to the Digital Campus website and every block's name and details will be validated. For the last phase, user acceptance testing is done to test the website and the illustration of the maps.

Figure 2 shows the flowchart of process digitalization mapping. The process will start with identifying all the locations and names of building in Campus Pagoh. Once all locations are identified, the system will illustrate the detail of every location and building identified before. After the illustration process is done, the name of every building will be verified. Before creating the digital mapping, the selection software of OpenStreetMap will be done to create the digital map. By using the OpenStreetMap software, the digital map can be created and inventory the complete digital into the Digital Campus. Lastly, embedded the interface website of Digital Campus to make sure it users friendly.

3.1 Administrator OpenStreetMap

Figure 3a shows that the login page of the administrator for OpenStreetMap, and Fig. 3b shows the username used to log in into the OpenStreetMap. After the login process is done, the digital map of College Campus Pagoh can be created.

3.2 Inventory Digital Mapping in OSM into Digital Campus

Figure 3c shows the process to construct the map in OpenStreetMap into the Digital Campus. After complete the sketch in the OSM, the map was saved and be uploaded into the digital campus. Since the digital campus was already linked with OSM, the map college of the UTHM will appear on the digital campus, but it will take several days to complete the data of the map. Within several days, the map of the college can be used to find the location.

3.3 Embedded the Interface Website of Smart Campus UTHM

The process is to improvise the interface of the website Smart Campus for administrators in accessing the website. The first step of the improvisation process is the login

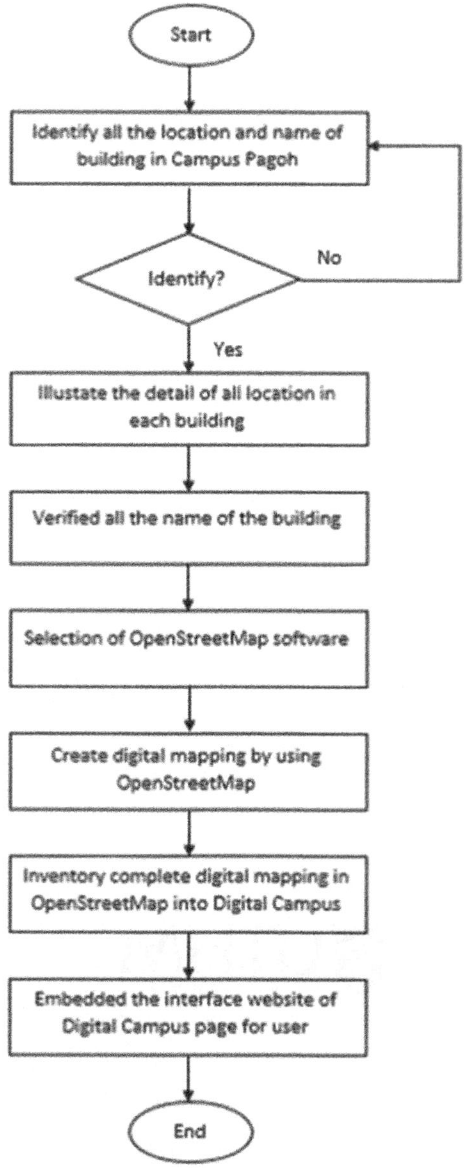

Fig. 2 The flowchart of process digitalization mapping

page for the administrator to access the website and the homepage after login successfully. This homepage includes the tools that can create any responsive website, such as Smart Campus UTHM. The homepage is the most important element and tool that can control all the information and design of any website that has been created included Smart Campus.

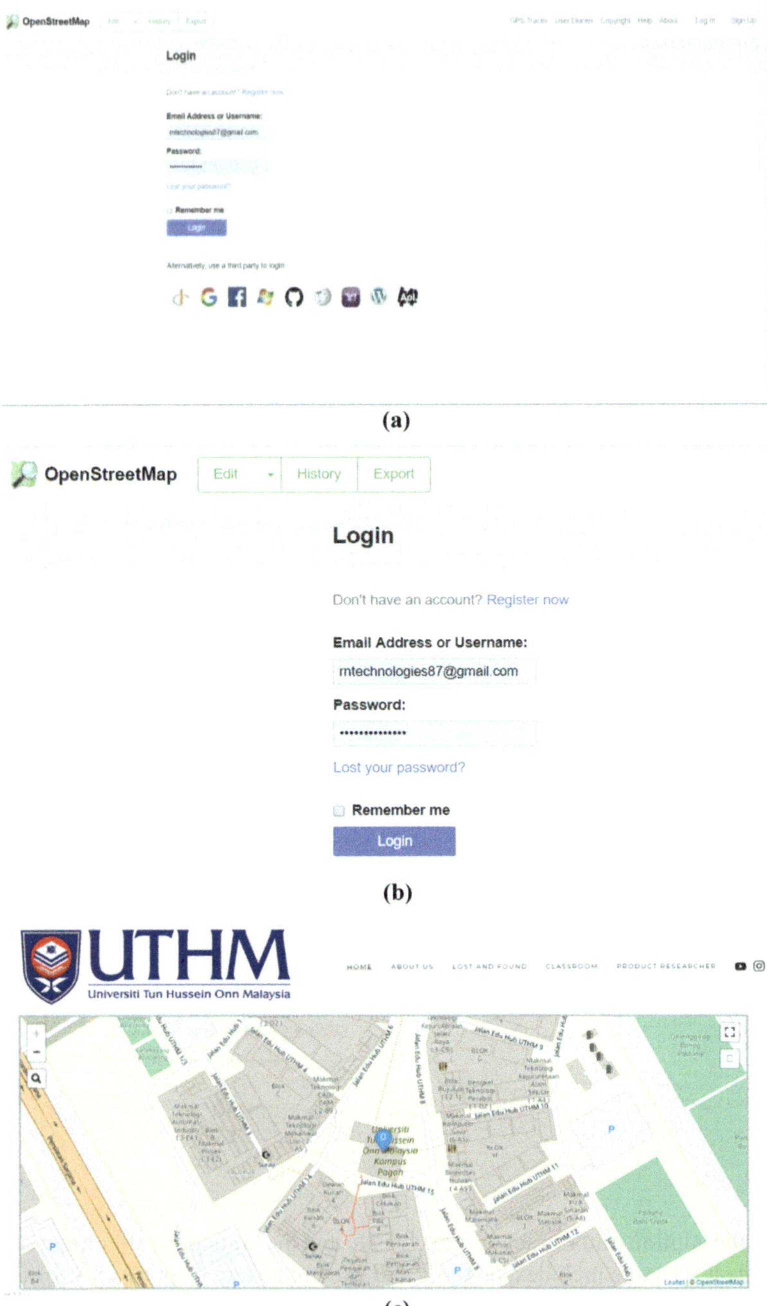

Fig. 3 **a** Administrator login page of OpenStreetMap. **b** Username login page. **c** OpenStreetMap in Digital Campus website

The next step is the Smart Campus web site editing department. The department was chosen for improvising the Smart Campus website. The Smart Campus editing process is done in the last step when the homepage is accessed. There is a step in this homepage which the administrator must follow during the editing process. The right resources need to be carefully selected and chosen, such as headline, text, image, map, contact information, and others.

In the editing process, the first method that needs to consider is the title of the page that needs to improvise. When the title has been chosen, the related picture that represents the page has been applied to the page. The picture acts as important information for users to analyze the page when they open it. Then, after all the pages were done in selecting the title and picture, the major information been applied into each page such as the headline that related to the title, videos, any information that can be used by users, contact (email, phone number, helpline), knowledge and products produce from UTHM. All of this method and element need to do by following the step until all the pages complete. Any mistake during the editing process might occur a problem such as missing data, the mistake in key in the data, and the worse problem that might occur is the loss of existing data.

3.4 Identify All the Location of Building in College Campus Pagoh

The initial sketch has been done on paper to identify the location of all buildings in college for the preparation before the final sketch up of the map in OpenStreetMap. The sketch is not only to identify the location of buildings but also to identify all the names of each building. After the entire sketch up completed, the major or specific room and its name in each building have been verified by the administration office of UTHM for further study.

The initial sketched has been done before all the processes started. The initial sketch is the process where the map has been sketched up first by manually on the A4 paper. This process has been done by referring to the layout drawing of UTHM in the administration office of Sime Darby, Pagoh. However, the layout was not completely the same as the reality of the complete building today, but there is still some similar area that needed to be referred to.

After the data collected from Sime Darby, the layout was being compared to the google map to detect the location of the building by referring to the road in the google map. When all the data collected, the final stage of the initial sketch is to manually visit each of the buildings in the college of UTHM and sketch it up on the paper. After all, the location of the building has been visited, the complete sketched came out. After the discussions, the final layout location needs to be verified by the Head of Department.

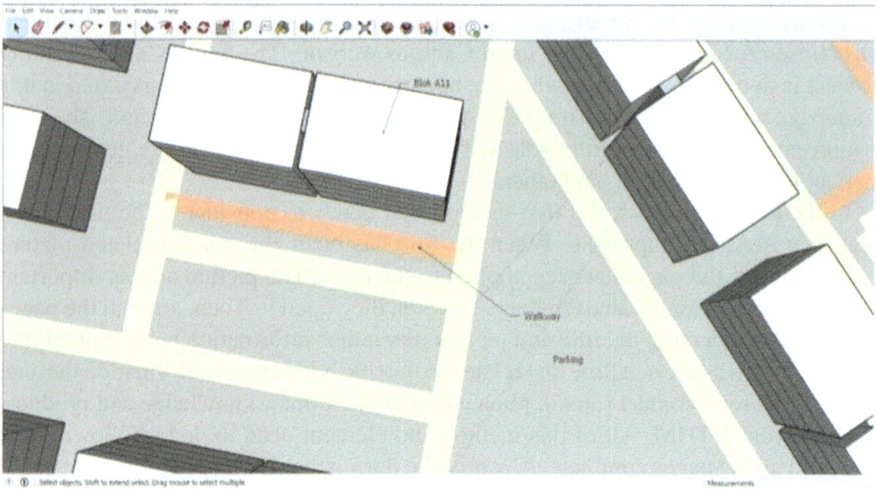

Fig. 4 The walkway is labeled as the yellow line

3.5 Drafting the College Campus Pagoh Map in Sketchup

For the beginning, before the map of college Campus UTHM sketch into the OSM, the drafting of the map has been done by using the Sketchup to indicate more clearly about the layout of the college as shown in Fig. 4. After every main building completed in drafting, the process of detailing the specific places in each building has been pointed to avoid the missing place when update into the OSM.

3.6 Consideration of Element in Mapping the Map

In the mapping process of OSM, other important elements need to be considered besides the building such as road, walkway, parking, and grasses. In this map of college, there is a walkaway beside Block A1 1 that is usually used by female students. Figure 4 also shows the walkway labeled as the orange line became the main walkway for them from their Block to the cafe area. Besides that, there is also a walkway beside the college hall, which continues beside the college cafe and ends at the college sure. Other than that, the next element is the car park area in the college. It is a very important element, as the majority of the students in college Campus Pagoh, brings their transportation, such as car, motorcycles, and bicycles. Besides that, the grass and the road element in the college also have been selected for the mapping process of the college Campus Pagoh.

4 Results and Discussion

4.1 Website

At the end of the study, Fig. 5 shows the map of the college Campus Pagoh has been successfully implemented in UTHM for students used in their daily life by using https://digitalcampus.uthm.edu.my/.

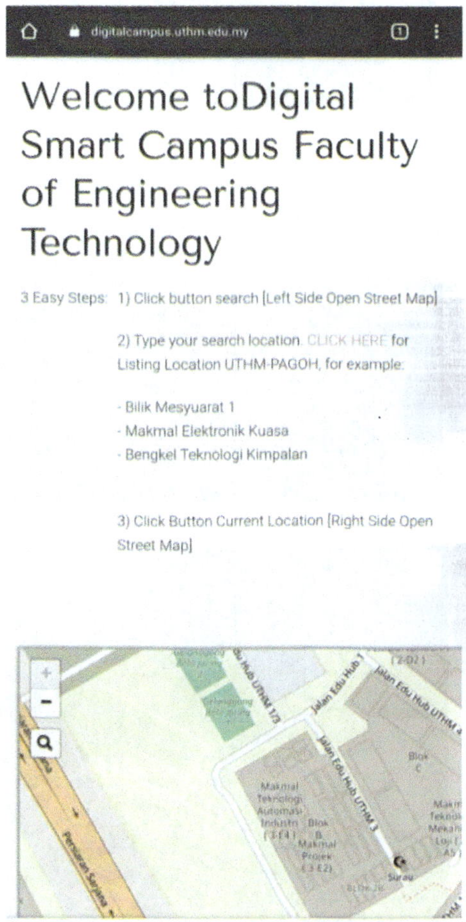

Fig. 5 Successfully implemented website in UTHM

(a)

(b)

(c)

Fig. 6 **a** Satellite view of College Campus Pagoh from Google map (2019). **b** The existing building in College Campus Pagoh. **c** Complete Map in OpenStreetMap

4.2 Geographic Area Study Using Satellite Data

Figure 6a shows the satellite data view of the College Campus Pagoh which is not updated on Google Map in 2019, whereas the existing facilities of road and building of the College Campus Pagoh area are shown in Fig. 6b. Therefore, the OpenStreetMap was chosen to create and update the digital map of College Campus Pagoh.

4.3 Complete Mapping of College Campus Pagoh in OpenStreetMap

The result from the drafting and sketch of the map was applied to the OpenStreetMap editor. After the sketched and mapping in OSM are completed, it will be updated to the home of OSM that can be used by all people to find their location as in Fig. 6c. The home page of OSM is used in Digital Campus as the new technology and facilities for students and people in UTHM in finding their locations.

4.4 Map of College Campus Pagoh in Smart Campus

Figure 7a shows a full view of the map of the college UTHM that was already done sketched in OpenStreetMap that was uploaded into the Digital Campus. The map took several days to fully update the new map before it ready to be used. After this stage, the map of the college was ready to use by the students and people in UTHM for their routine days in finding their target location. The map was uploading with its name of the building. It helps to assist people in finding the location because every single building has its name. The list can be review by students and people in the Digital Campus website before the process of finding a location by the Digital Campus.

4.5 Specific Location Point in OpenStreetMap

Figure 7b shows how the OSM function when students searching for the location which is has been detailed in one of the buildings. This specific place was the place that difficult to find manually by students. When the location has been identified, the blue zone will appear according to the target name of the building as Fig. 7c. This is because all of the specific place and the building have been pointed with their name according to the list name of the building during the mapping in OpenStreetMap. Any name listed in the name of the building can be searched in the OpenStreetMap with the correct spelling. Any spelling mistake in the name of the building might lead

(a)

(b)

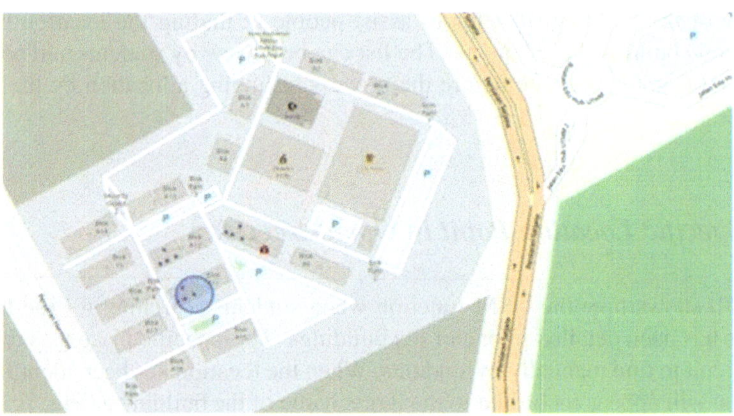

(c)

Fig. 7 **a** Full view of the map on Digital Campus. **b** Searching the Name of Location in OSM. **c** The blue zone shows the point of finding a location to unidentified data and the blue zone of pointed location will not appear to show the target location.

4.6 Addition Page of Smart Campus Website

In this study, the interface of Smart Campus also has been updated in content of page of Smart Campus website. In initial of the Smart Campus, there are only 2 pages which are consisting of Home Page and About Us. This website is not fully complete in editor and need to be improvised for other function that might be used by student. To improvise the Smart Campus website in order to help the website in more function not only in mapping, there is three pages has been added which is Lost And Found, Classroom and Product Researcher of UTHM.

4.6.1 Lost and Found

A lost and found is the tab in the website, where people can put their lost things in a bin or a tray and collect their things. Some lost and found offices will try to contact the owners of any lost items if there are any personal identifiers available. Practically all will sell, give or throw away items after a certain period has passed to clear their storage. In this website, student can update to the website their lost things and the characteristic of the things that they might lost. They also can leave their contact in order their lost thing has been found and kept at the offices. When the information has been updated, the offices that related such as security guard can inform and contact the student if there is any news about the lost things.

4.6.2 Classroom

Page of classroom which is consist the subject and knowledge which is it is the part of subject that the student of Faculty of Engineering Technology. This page been added in order to introduce the subject in the UTHM for the new students. The page consists of the video of Sustainable Construction Management and Computer Aided Building Services. With all the video, there is also some explanation about the video and the subject. All the video was linked to the Youtube which it is the online video that need to access with mobile data.

4.6.3 Product Researcher

The last page of the Smart Campus is the information about the product researcher of UTHM. In this page, the picture of the product can be seen when the page was opened which is the picture represented every single product researcher. The entire product actually produced by UTHM lecturer according to their flow of study and department. The information of the lecturer who was produces the product and the video about the product that has been produce in the right side. This page also introduces the new students about the product that produce from UTHM.

5 Conclusion

The mapping system for the campus to identifying and finding the location in this area by using the OpenStreetMap website is the good method and effort that has been applied in UTHM. This technology is the new facility that has been applied and proven to helps people in solving their problems regarding finding the location. As the Smart Campus website has been improved and updated, this website might be useful to the user in using the website. This study has contributed to students and universities to improve the system of the Digital Campus in solving the problem in finding the location.

As a recommendation for further study, OpenStreetMap should be improved in the content of the 3D layout. The OpenStreetMap shows the map in 2D layout, which is it is quite difficult to differentiate the level of the building. Since there is a majority building that is at a high level, the OpenStreetMap should appear in a 3D layout to differentiate the level of the building easily.

References

1. Schlechtendahl J, Keinert M, Kretschmer F, Lechler A, Verl A (2014) Making existing production systems Industry 4.0-ready. Prod Eng 9(1):143–148. https://doi.org/10.1007/s11740-014-0586-3
2. Industry 4.0 and manufacturing ecosystems in Deloitte. https://dupress.deloitte.com/dup-us-en/focus/industry-4-0/manufacturing-ecosystems-exploring-world-connectedenterprises.html
3. Cobb P, Confrey J, Disessa A, Lehrer R, Schauble L (2003) Design experiments in educational research. Educ Res 32(1):9–13. https://doi.org/10.3102/0013189x032001009
4. Penuel WR (2006) Implementation and effects of one-to-one computing initiatives. J Res Technol Educ 38(3):329–348. https://doi.org/10.1080/15391523.2006.10782463
5. Werner P (2018) Review of implementation of augmented reality into the georeferenced analogue and digital maps and images. Information 10(1):12. https://doi.org/10.3390/info10010012
6. Gupta A, Ahirwar M, Pandey R (2019) Creating website as a service using web components. SSRN Electron J. https://doi.org/10.2139/ssrn.3457787
7. Responsive Web Design—What It Is And How To Use It. https://www.smashingmagazine.com/2011/01/guidelines-for-responsive-web-design/
8. Haklay M, Weber P (2008) OpenStreetMap: User-generated street maps. IEEE Pervasive Comput 7(4):12–18. https://doi.org/10.1109/mprv.2008.80
9. Zhang Y, Cheng Y (2013) Design and implementation of a digital campus system based on the URP platform. In: Proceedings of the 2nd international conference on computer science and electronics engineering (ICCSEE 2013). https://doi.org/10.2991/iccsee.2013.793

10. Ronghuai H, Jinbao Z, Yongbin H, Junfeng Y (2012) Smart campus: the developing trends of digital campus. https://en.cnki.com.cn/Article_en/CJFDTOTAL-JFJJ201204004.htm
11. Xing JD, Yong GD, Long LN, Xin LQ (2007) Planning and practice of new generation digital campus in tsinghua university. J Xiamen Univ (Nat Sci) 46(Z2):173–178. http://en.cnki.com.cn/Article_en/CJFDTotal-XDZK2007S2047.htm

Multicriterial Approach for Traditional Shop Houses Façade Condition Assessment

M. A. A. Rahman, S. A. Rahman, S. H. Adnan, E. E. M. Safian, and N. Hashim

Abstract The facade of old shophouses' are the vital element on the timeline to preserve and conserve the value it carries in the history. Facades are the most important aspect of an early shop house because they can influence the exterior apperance and condition of the buildings considerably. The major problem we faced was the lack of equipment to measure the condition of the facade specificly to classify the current condition. Mainly, the façade elements of the shop house are cornice, upper level and lower level. Each element define building's unique architectural aesthetics. To ensure the facade's lifespan is always in good condition, timely performance assessment should be performed. Therefore, this paper to analyzed the performance assessment requirement of façade elements to establish a systematic method for classifying façade condition. An protocol interview with 20 panel of experts has been able to determine the weightage of main and sub elements of facade. The data being analyzed using Analytical Hierarchy Process (AHP).

Keywords Analytical hierarchy process (AHP) · Condition assessment · Façade · Heritage architecture · Shop house

1 Introduction

By definition, façade is a one exterior side of a building, usually the most prominent view of a building frontage [1]. According to [2, 3], façade of the traditional shop houses reflects the architecture of the days it was built and will not be the same if it was rebuilt in our days. Traditional shop house can be defining as a premise that has

M. A. A. Rahman · S. H. Adnan
Universiti Tun Hussein Onn Malaysia, 84600 Pagoh, Muar, Johor, Malaysia

S. A. Rahman (✉) · E. E. M. Safian
Zulfa Empire Trading, Taman Perindustrian Cemerlang, 81800 Ulu Tiram, Johor, Malaysia
e-mail: kefila89@gmail.com

N. Hashim
Universiti Teknologi Malaysia, 83100 Skudai, Johor, Malaysia

M. A. A. Zaini et al. (eds.), *Proceedings of the 3rd International Conference on Separation Technology*, Lecture Notes in Mechanical Engineering,
https://doi.org/10.1007/978-981-16-0742-4_9

been built before or on 31 January 1948 and subjected to the Rental Control Act 1966 [4]. Highly competitive and profound changes in the recent urban development are forcing local authorities to continuously improve and maintain the façade condition of traditional shop houses in order to ensure that the early heritage architecture remain relevant [5]. Among the priorities is through a facade condition assessment. The main objective of condition assessment is to evaluate the physical state of facade elements and to assess the maintenance needs of the facility [6]. To ensure the façade of traditional shop houses is safe and always in good condition, the appropriate periodic inspections need to be carried out in order to identify its current conditions and to prepare the remedial actions [7, 8]. In this study, an effort has been made for performance assessment of façade condition of early shop houses using Analytical Hierarchy Process (AHP). The assessment is done based upon different elements and their weightage is decided by the panel of experts.

1.1 Typical Elements of Building Facade

Facades of the shop houses according to [9, 10] can be divided into three key elements, namely:

(1) Lower level
(2) Upper level: and,
(3) Cornice level.

The illustration of the building façade is shown in Fig. 1.

Lower level or sometimes known as storefront usually consist of two pillars either in square or round in shape. The major elements in the lower level are between five-foot-way and the main entrance. The upper level is basically use as a residences

Fig. 1 Typical Traditional Shop House Building Façade

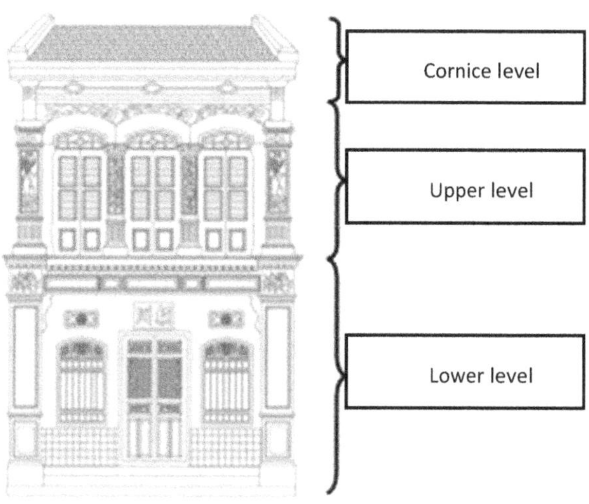

Table 1 Sub-key elements of each level

Level	Sub-key elements
Cornices (C)	C1. Cornice cap C2. Ornaments C3. Dentil
Upper (U)	UI. Window hood/lintel U2. Window sill U3. Upper wall U4. Upper window U5. Upper cornice
Lower (L)	L1. Transom L2. Display window L3. Piers/Pilasters L4. Bulkhead L5. Sign band L6. Recessed entry L7. Door head L8. Door frame L9. Plinth L10. Capital

and look more aesthetic in appearances since this level has stands out decoration compared to the lower level. A cornice is an ornamental moulding around the wall just below the ceiling [10]. Table 1 below explain sub-key elements of each level in detail.

2 Materials and Methods

The Analytic Hierarchy Process (AHP) is an effective tool for dealing with complex decision making [11]. AHP is a systematic tool for representing the elements of any problem hierarchically [12]. The pair wise comparison matrix is of size n x n, where n is the number of elements to be compared pair wise. The matrix will be filled up accordingly using following procedures:

(1) Each element compared with itself will get a value 1 = a(n,n) = 1
(2) When ith element is compared with jth element, it has got a value A(i,j), jth element being compared with ith element has got a value a(j,i) = 1/a(i,j). i.e. a(2,1) = 1/a(1,2), a(3,1) = 1/a(1,3), ... a(n,1) = 1/a(1,n).
(3) Relative weight (RW) = n$\sqrt{}$a(1,1) x a(2,1) x a(3,1) x a(4,1) x a(5,1)
(4) Normalized weight (NW) = RW/Σ RW
(5) Maximum Eigene Value (λMAX) = Σ column A x NW value row A + Σ column B x NW value row B +...... + Σ Column n x NW value row n
(6) Consistency Index (CI) = (λMAX−n)/(n−1)
(7) Random Index (RI) = 1.98x(n−2)/n
(8) Consistency Ratio (CR) = CI/RI, should be 10 percent.

(9) Composite Rank (COR) = NW Parameter 1 weightage x NW of that parameter + NW Parameter 2 weightage x NW of that parameter + NW Parameter n weightage x NW of that parameter.

3 Result and Findings

Set of questionnaires was developed and distributed to 20 expert panels. They were from industry practitioners and professionals with over 10 years of experience in the field of heritage architecture. Table 2 showed the list of expert panels. All panels were contacted and agreed to take part in answering the questionnaire given. Panels were ask to fill in their demographic background in Sect. 1 of the questionnaire and make pair wise comparisons of building's facade elements in Sect. 2 based on scaling matrices as in Table 3.

Table 2 Characteristic of expert committees

Characteristics	Expert committee (n = 20)
(a) Gender, n (%)	
• Male	16 (80)
• Female	4 (20)
(b) Age (years), median (q1,q3)	51 (44, 58)
(c) Experience (years), median (q1,q3)	22 (19, 29)
(d) Present professional setting, n (%)	
• Industrial/Private	12 (60)
• Institutional/Government	8 (40)
(e) Specialty, n (%)	
• Architect	1 (5)
• Surveyor	3 (15)
• Engineer	5 (25)
• Contractors	4 (20)
• Academician	7 (35)

Table 3 AHP scaling matrices

Level of importance	Scale
Equally	1
Slight more important	3
Obviously more important	5
Strong more important	7
Extremely more important	9
Intermediate	2,4,6,8

Table 4 Matrix for weightage to key element

Sub element	L	U	C	RW
L	1	1	1 1/3	1.100
U	1	1	1 1/3	1.100
C	3/4	3/4	1	0.825
Total	2.75	2.75	3.66	3.025

From Table 2, there are 20 expert committees involved in determining the weight of elements according to AHP scaling matrices. They were architects, surveyors, engineers, contractors and academicians. All of the experts indicated their best commitment in collaborating on this study.

Based on the scaling matrices in Table 3 above, the following judgment matrices for key elements were then established as shown in Table 4. The consistency ratio is acceptable when CR < 0.10.

RW of Lower level = $(^3\sqrt{}\ 1 \times 1 \times 1\ 1/3)\sqrt[3]{1x\ 1\ x\ 1\ 1/3)}\sqrt[3]{1x\ 1\ x\ 1\ 1/3)} = 1.100$

Similarly, RW for upper level and cornice were calculated.

$\lfloor_{MAX} = (2.75 \times 0.40) + (2.75 \times 0.40) + (3.66 \times 0.225) = 3.02$

CI = (3.02−3)/(3−1) = 0.02/2 = 0.01

RI = [1.98 x (n−2)]/n = 0.66

CR = 1.51% @ 0.015 (Acceptable)

Similarly, $\lfloor_{MAX}$, CI, RI and CR for upper level and cornice were calculated. The judgement matrices for sub elements were calculated and established as shown in Tables 5 and 6.

Table 5 Matrix for weightage to sub element cornice

Sub element	C1	C2	C3	RW
C1	1	1/2	1 1/9	0.82
C2	2	1	1 5/6	1.54
C3	1	5/9	1	0.82
Total	4	2.06	3.94	3.18

Table 6 Matrix for weightage to sub element upper level

Sub element	U1	U2	U3	U4	U5	RW
U1	1	2 1/5	1 5/6	3/4	4/7	1.12
U2	1/2	1	1	5/8	2/3	0.73
U3	5/9	1	1	1 1/7	1 2/3	1.01
U4	1 1/3	1 3/5	7/8	1	2 5/8	1.37
U5	1 5/7	1 3/7	3/5	3/8	1	0.89
Total	5	7	5	4	7	5.12

Table 7 below summarized the Relative Weights (RW) and Normalized Weights (NW) for facade key elements and sub elements.

Table 7 Criteria weightage for facade condition assessment

Level	Sub-key elements	RW	NW
Cornices (C)	C1. Cornice cap	0.82	4.45
	C2. Ornaments	1.54	8.36
	C3. Dentil	0.82	4.45
Upper (U)	UI. Window hood/lintel	1.12	6.08
	U2. Window sill	0.73	3.96
	U3. Upper wall	1.01	5.48
	U4. Upper window	1.37	7.43
	U5. Upper cornice	0.89	4.83
Lower (L)	L1. Transom	1.32	7.16
	L2. Display window	0.84	4.56
	L3. Piers/Pilasters	1.02	5.53
	L4. Bulkhead	0.92	4.99
	L5. Sign band	0.89	4.83
	L6. Recessed entry	1.13	6.13
	L7. Door head	1.01	5.48
	L8. Door frame	1.06	5.75
	L9. Plinth	0.93	5.05
	L10. Capital	1.01	5.48
Total (Σ)		18.43	100

4 Highlights

(1) The result of the AHP has successfully assigned individual weightage to all 18 sub-key facade elements of traditional shop houses.
(2) The AHP weightage found five sub-key elements that were ranked as the most important elements (NW > 6.0) which is ornaments (NW = 8.36), upper window (NW = 7.43), transom (NW = 7.16), recessed entry (NW = 6.13) and window hood/lintel (NW = 6.08).
(3) Out of all these 18 sub-key elements, window sill has recorded lowest NW values below 4.0 which is (NW = 3.96).
(4) The top ranking key element is lower level, followed by upper level and cornice level (Table 8).

4.1 Façade Condition Assessment Simulation Model

Based on the calculation of key elements and sub key elements weightage and the structure of the hierarchical assessment list of Table 1, a simulation model for

Table 8 Matrix for weightage to sub element lower level

Sub element	L1	L2	L3	L4	L5	L6	L7	L8	L9	L10	RW
L1	1	2.7	0.8	1.3	1.5	1.1	1.3	0.6	3	1.2	1.3
L2	0.3	1	1.2	1.4	1	1	1.3	0.6	0.6	0.6	0.8
L3	1.2	0.8	1	1	1	1	0.9	1.3	1.2	0.9	1.0
L4	0.8	0.6	1	1	1	0.5	1	1	1.2	1.6	1
L5	0.6	1	1	1	1	1.1	0.6	0.8	0.6	1.5	0.9
L6	0.9	1.1	1	1.8	1	1	1.3	1.3	0.8	1.4	1.1
L7	0.8	0.8	1.2	1	1.5	0.8	1	2.1	0.8	0.7	1.0
L8	1.5	1.6	0.8	1	1.3	0.8	0.5	1	2	0.8	1.1
L9	0.3	1.6	0.8	0.9	1.8	1.3	1.2	0.5	1	0.7	0.9
L10	0.8	1.4	1.2	0.6	0.6	0.7	1.4	1.3	1.3	1	1.0
Total	8	13	10	11	12	9	11	11	12	10	10

assessing façade condition was created then the condition of the building is determined using Condition Assessment Rating either in very good, good, fair, bad or in very bad condition. In the synthetic form, the model is shown in Fig. 2.

This simulation model has established the performance assessment elements that may affect some of the facade conditions of the traditional shop houses building by determining the relative importance of the sub-key elements of decision-making accuracy based on AHP. The result found that the ornament was a main factor affecting the condition of the shop house's facade. This is because the ornamental element is seen as a major contributor element that distinguishes between the traditional shop house and the modern shop house today. Although the results of this study refer to the decision factors of the facade elements of traditional shop houses in Malaysia, this model is still significant in comparison to other Asian countries such as Singapore and Thailand. Accordingly, these multi-factor priorities can serve as the basis for building owners and planners to understand the patterns of damage and subsequently plan to ensure that this heritage architecture remain protected.

5 Conclusion

The proposed method of performance assessment of façade condition using AHP helps in classified the existing condition of the shop houses. Hence, the same method can be used for various style of traditional shop house. AHP not only helps arrive at the best decision, but also provides a clear rational for the choice. The proposed method offers a more objective, simple and consistent performance assessment approach. This is very scientific approach and it gives correct results in very less time.

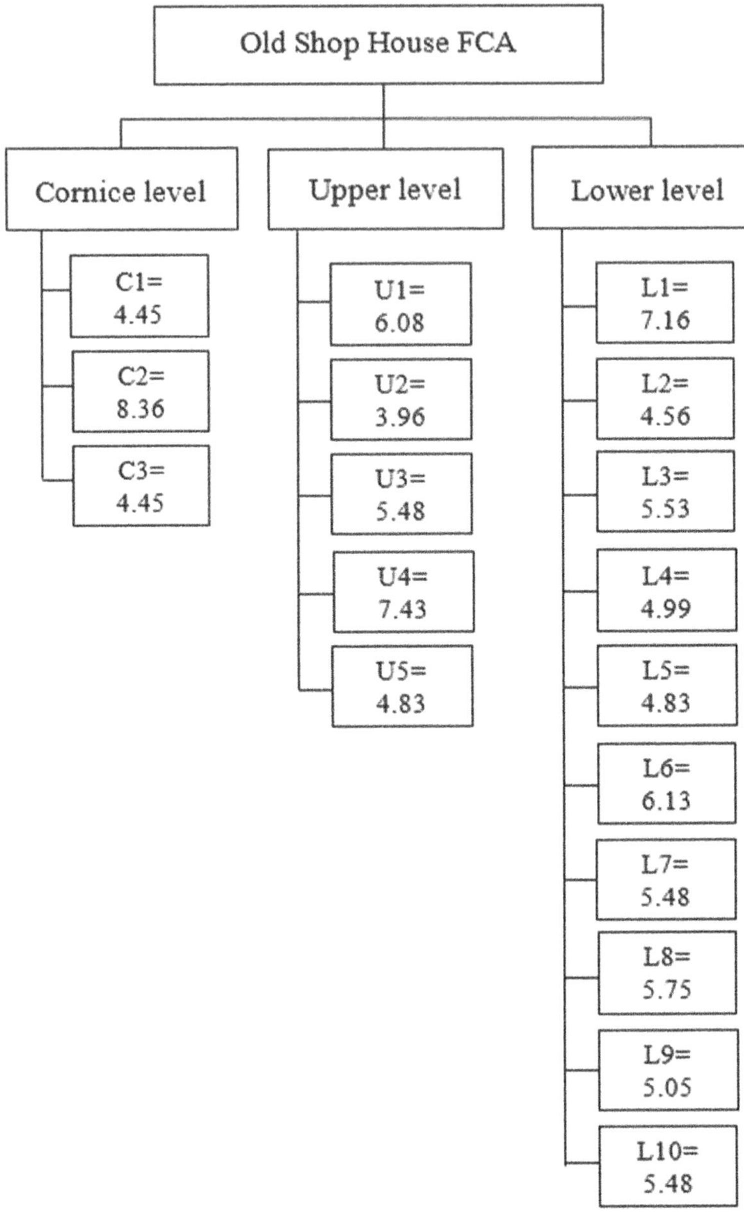

Fig. 2 Criteria weightage for facade condition assessment of old shop house

Acknowledgements The authors would like to express utmost gratitude to the Faculty of Engineering Technology of University Tun Hussein Onn Malaysia (UTHM) and to all colleagues for having provided support and valuable advice in carrying out the study.

References

1. Leon K (1984) Urban Components. Archit Des 54(7/8):43
2. Utami SA (2004) Permukiman Pengungsi Pasca Kerusuhan di Tentena Kecamatan Pamona Utara Kabupaten Poso. Master (S2) Thesis. Magister Perencanaan Kota dan Daerah. Gadjah Mada University. Yogyakarta
3. Masfaliza M (2007) Permasalah pembangunan semula premis sebelum perang kawasan kajian: Bandar Kluang. Universiti Teknologi Malaysia
4. Ab Haris RA (2015) Facade deterioration at shop houses in The Royal Town of Klang
5. Ismail WHW, Shamsuddin S (2005) The old shophouses as part of Malaysian urban heritage: the current dilemma. In: 8th International conference of the asian planning schools association, pp 11–14
6. Che-Ani AI, Ismail I, Johar S, Abd-Razak MZ, Hamzah N (2015) Condition survey protocol: a system for building condition assessment. Appl Mech Mater 747:347–350 (Trans Tech Publications)
7. Akasah ZA, Abdul RMA, Zuraidi SNF (2011) Maintenance management success factors for heritage building: a framework. WIT Trans Built Environ 118:653–658. https://doi.org/10.2495/STR110541
8. Mohamad SBHS, Akasah ZA, Rahman MAA (2015) A review of the maintenance performance factors for heritage buildings. In: InCIEC 2014. Springer Singapore, pp 177–187
9. Abd Rahman S, Abdul Rahman MA, Adnan SH (2017) An evaluation of important criteria for measuring the facade condition of old shophouses in Malaysia 3
10. Chang TC (2010) Bungalows, mansions and shophouses: encounters in architourism. Geoforum 41(6):963–971
11. Saaty TL (1990) How to make a decision: the analytic hierarchy process. Eur J Operat Res 48:9–26. https://doi.org/10.1016/0377-2217(90)90057-I
12. Zuraidi SNF, Rahman MAA, Akasah ZA (2018) A study of using AHP method to evaluate the criteria and attribute of defects in heritage building. In: E3S web of conferences, vol 65. EDP Sciences. https://doi.org/10.1051/e3sconf/20186501002

Thermal Comfort Evaluation at the Multipurpose Hall of an Academic Campus

M. K. Musa, C. M. Ting, M. A. A. Rahman, M. Awang, Nuramidah Hamidon, M. M. Syafiq Syazwan, Fatimah Yusop, and Faridahanim Ahmad

Abstract Most of the building design has a combination of natural ventilation and mechanical ventilation is used for indoor cooling purposes in many large enclosed spaces. However, this cooling method is insufficient to provide the occupants with a comfortable temperature level due to low wind velocity and the inability of a mechanical ventilator to remove warm air. The aim of this study is to evaluate the thermal comfort in a multipurpose hall. To assess the thermal perception of occupants in the multipurpose space, 179 occupants were randomly selected to participate in the survey. In addition, 4 indoor environmental variables were calculated using the Thermal Comfort Kit, including air temperature, air velocity, globe temperature and relative humidity. Data from questionnaire surveys and fieldwork measurements were obtained at the same time for 6 days between September and November 2019. All data collected were interpreted using the Predicted Mean Vote (PMV) and Predicted Percentage Dissatisfied (PPD) indices, which reflect the state of thermal comfort within the multipurpose hall. For this analysis, the PMV and PPD values for fieldwork measurements are 2.1 and 82.6%, while the thermal sensation voting values are 1.6 and 55%. Both PMV and PPD values for thermal environment measurement in the multipurpose hall and questionnaire survey exceed the respective upper limits as recommended in ASHRAE Standard 55 (2017) and the acceptability criterion in ISO 7730 (2005), indicating that the thermal comfort within the hall is warm.

Keywords Multipurpose hall · PMV-PPD model · Thermal comfort

M. K. Musa (✉) · C. M. Ting · M. A. A. Rahman · M. Awang · N. Hamidon · M. M. Syafiq Syazwan · F. Yusop
Faculty of Engineering Technology, Universiti Tun Hussein Onn Malaysia (Pagoh Campus), 84600 Pagoh, Muar, Johor, Malaysia
e-mail: kamaruz@uthm.edu.my; kamaruzaman.musa@gmail.com

F. Ahmad
Department of Civil Engineering, Universiti Teknologi Malaysia, Skudai, Johor, Malaysia

M. A. A. Zaini et al. (eds.), *Proceedings of the 3rd International Conference on Separation Technology*, Lecture Notes in Mechanical Engineering,
https://doi.org/10.1007/978-981-16-0742-4_10

1 Introduction

Thermal comfort is becoming increasingly of concern to academics and practitioners, while the living standards, economic development, urbanization, and climate change [1]. In the building environment, thermal comfort has an important role as people spend 80% of their time indoors, especially in summer conditions with extreme climatic conditions [2]. Estimated on global energy consumption, the Energy Information Administration (EIA) will continue to grow by 56% between 2014 and 2020. Malaysia's climate has the characteristics of a tropical hot-humid climate and may adversely affect indoor comfort of the occupants. In particular, at university, thermal comfort is a major aim for the built environment as it affects the satisfaction of occupants, health and productivity [3].

In the previous studies carried out by Seppänen et al. [4], the building's air temperature should rise between 20 and 23 °C whereas any rise above this range may result in a satisfactory negative productivity [5]. In the study conducted by [6], it was found that room air temperature appears to affect the occupant building by reducing the arousal, elevating symptoms of Sick Building Syndrome (SBS), and reducing manual dexterity.

The academic management has provided multipurpose hall that can accommodate 250 to 500 people at a time in providing facilities for student and staff comforts [7]. Multipurpose hall is the place where different activities are either formal, informal, or recreational. Malaysia's climate causes the temperature indoors to rise. The high temperature moving air transmits into the multipurpose hall which causes the temperature indoors to be much higher than the temperature in comfort. Only a few numbers of wall fans support for the entire ventilation system in the hall. Due to low wind velocity and the inability of wall fans to remove warm air, this cooling method is inadequate to provide a comfort temperature level for the occupants. It constitutes potential health hazard and impacts the ability of the occupant to function effectively when people are dissatisfied with their thermal environment.

The aim of this study is to identify the occupants' thermal perception through questionnaire survey in the multi-purpose hall, to measure the parameters of the indoor thermal environment in the multi-purpose hall, and to assess thermal comfort by implementing the Predicted Mean Vote (PMV) model and the Predicted Percentage Dissatisfied (PPD) index in the multi-purpose hall.

The scope of this study focuses mainly on the main hall of the multipurpose hall at the UTHM Pagoh Campus due to hot internal environmental problems. This study involves the distribution of a structured questionnaire on personal information, user thermal perception and clothing insulation. In addition, four thermal environmental parameters were measured using a thermal comfort kit consisting of air temperature, globe temperature, relative humidity and air velocity. Data from questionnaire surveys and fieldwork measurements were collected at the same time. The thermal comfort will be evaluated using the PMV model and the PPD index.

2 Thermal Comfort

The word "thermal comfort" refers to the human understanding of the thermal environment. Thermal comfort is described as a "state of mind, which exemplifies the satisfaction with the thermal environment"[8] in accordance with ASHRAE 55. Defining what a comfortable environment constitutes depends on the occupant's thermal perception and personal experience, dependent on numerous criteria [9]. Thermal comfort means that thermal balance is in equilibrium between the body and the environment [10]. For people who are in the same thermal environment and involved in various activities, thermal comfort is difficult to achieve [11]. The heat, ventilation, and air conditioning system of the building controls primarily thermal comfort. Ventilation can be used to decrease high room temperatures, increase air speed, and decrease thermal stress at high temperatures [12]. The architectural design of the building [13] also influences the thermal comfort.

2.1 Thermal Comfort Influences

Six primary factors can be grouped into two categories, indoor environmental variables (air temperature, air speed, relative humidity, and average radiant temperature) and personal variables (metabolic speed and insulation of clothing), affect thermal comfort directly. These six variable factors are important to maintain a healthy balance to maintain the satisfaction of the occupants in their environment.

Air Temperature. The air temperature is the human body's air temperature. In degree Fahrenheit or Celsius, air temperature is usually expressed. It is a common thermal comfort factor and can be quickly affected by passive and mechanical heating and refrigeration. Thermometer can be used to measure the air temperature. ISO 7726 (2001) requires a period of time in order to reach equilibrium in a given setting. There are 3 temperature sensor types, including expansion thermometers and electric thermometers.

Mean Radiant Temperature. Global temperature is determined by the black globe thermometer. Global temperature is a means of measuring the cumulative effects on human health of heat, air temperature and air velocity. The mean radiant temperature is the even temperature of an imaginary enclosure in which the radiant heat transfer is equivalent to that of the human body [14]. In the actual non-uniform enclosure. The average radiant temperature (MRT) can be calculated using the Eq. (1) based on the recorded indoor thermal environment parameters.

$$\mathrm{MRT} = \mathrm{Tg} + 2.37\sqrt{\mathrm{Vd}}\,(\mathrm{Tg} - \mathrm{Ta}) \tag{1}$$

where MRT is the mean radiant temperature(°C), T_g is the globe temperature, V_d is the air velocity(m/s), and Ta is the air temperature (°C).

Relative Humidity. Relative humidity e is the ratio between the partial pressure of water vapour pa, in humid air and the water vapour saturation pressure pas at the condition of same temperature and the same total pressure [14]. Humidity is the moisture content of the air, often expressed as a percentage in accordance with the following relationship. In the warm climate, sweat evaporation makes it harder to lose heat if the relative humidity increases [15].

Air Velocity. The air velocity is defined by direction and magnitude. In thermal environments, air speed is the quantity to be considered. The faster moving air will increase the heat loss by convection without changing air temperature in warm and moist conditions. Essential considerations are the calibration of the instrument, the reaction time of the sensor and the instrument, and the time measurement [14] in order to ensure accurate air velocity measurements.

Metabolic Rate. Human metabolism is an important element of prediction of thermal comfort. The heat generated by the human body by physical activity is affected. For persons living together in the same thermal climate, thermal comfort is difficult to achieve, particularly if they engage in several activities [16]. According to ISO 8996 (2004), person variation, work equipment variations, work speed, work technique and abilities are the key factors affecting the exactness of estimates [17]. 1 Met is described as the sedentary person's heat and is 58Wm-2 [18].

Clothing Insulation. Insulation is an inherent attribute clothes characteristic measured in the units of fabric (1 clo is equal to 0.155 (m2 K)/W) there are always air trapped in or around the fabric. A variable factor for thermal comfort is clothing isolation, and effective clothing modifications can extend the range of comfort and minimize building energy consumption [19]. This also reduces the heat lost through the skin and reduces the perceived warmth of the environment.

2.2 *Thermal Perception*

In addition to air temperature, significant wind speed, humidity and air density, thermal perception is the occupant's understanding of thermal environment. The thermal perception includes thermal sensations, thermal choice, personal acceptability and thermal tolerance, according to the subjective scales described in ISO 10551 (1995) [20]. In general, thermal perception offers accurate and comparative data on the subjective aspects of thermal comfort or thermal stress.

Thermal Sensation. Thermal sensation of 7 scaling points is a conscious sense that is generally graded. The conventional ASHRAE 7-point scale shown in Table 1 was the thermal sensation scale.

Thermal Preference. Thermal preference is often used to assess how climate change affects both the vulnerability to extinction and changes in activity. The thermal preference was the McIntyre 5-point preferred scale (prefer much warm, slightly warmer, no change and slightly cooler and much cool).

Thermal Tolerance. Thermal tolerance is the ability to tolerate subjection without adverse reaction to environmental conditions. The thermal tolerance scale would

Table 1 The 7-point scale of thermal sensation [28/khujuj]

Scale	Thermal sensation
+3	Hot
+2	Warm
+1	Slightly warm
0	Neutrality
−1	Slightly Cool
−2	Cool
−3	Cold

have a 5-degree unipolar range, including perfectly bearable, slightly difficult to bear, fairly difficult to bear, very difficult to bear, and unbearable to the thermal condition of building.

Thermal Acceptability. Thermal acceptability is used to determine an atmosphere which would consider thermally acceptable to a large majority of occupants. Thermal acceptability typically uses straightforward (acceptable, inacceptable) evaluation.

2.3 Predicted Mean Vote (PMV) and Predicted Percentage of Dissatisfied (PPD)

EN ISO 7730 (2005) is an International Standard which presents methods for predicting the general thermal sensation and degree of discomfort (thermal dissatisfaction) of people exposed to moderate thermal environments [21]. The PMV model of the Fanger can be measured using two sets of variables that are personal variables (clothing insulation and activity level) and indoor environmental variables (temperature, air velocity, mean radiant temperature, and relative humidity). The PMV model has been widely used by researches to evaluate the thermal comfort of human. According to Table 2, the PMV index suggested a number scale ranging

Table 2 Grading of PMV index thermal comfort [21]

Scale	Thermal comfort
+3	Hot
+2	Warm
+1	Slightly warm
0	Neutrality
−1	Slightly Cool
−2	Cool
−3	Cold

Table 3 Thermal acceptability criterion [21]

Class	A	B	C
PMV	$-0.2 < \text{PMV} < 0.2$	$-0.5 < \text{PMV} < 0.5$	$-0.7 < \text{PMV} < 0.7$
PPD (%)	<6	<10	<15

from −3 to +3 in which '0' represents the neutral, based on the human body's heat balance.

In addition, Fanger developed an empirical PMV relation with a "Predicted Percentage Dissatisfied (PPD)" that gives information about thermal unholding or thermal discomfort by predicting the percentage of people in a given environment who may feel uncomfortable, warm or cold. The PPD is available from the PMV. By using the equation below (2), a PDP index based on questionnaires can be calculated in any part of the building [15].

$$\text{PPD} = \frac{\text{The number of questionnaires having discomfortable}}{\text{Total number of questionnaires in any section}} \times 100 \quad (2)$$

Thermal environment is impossible satisfies 100% by occupant because individual thermal conception [10]. The criterion for thermal acceptance defines the spectrum of sensation of thermal in the PMV Equation as shown in Table 3, step by step with ISO7730 (2015) [21]. In class A, class B is the moderate satisfaction prerequisite and class C is the minimum test criterion in accordance with the criterion of thermal comfort.

2.4 Standard of Indoor Thermal Environment Parameters

Air temperature, humidity and air speed provide indoor environmental variables that decide or affect the thermal comfort of a person in this indoor environment. The quality of indoor environmental variables should be primarily considered for engineering design in order to provide the building's comfort. The suggested indoor design specifications have been provided in Table 4 [22, 23] according to MS 1525 (2014) and ICOP IAQ (2010) guidelines and standards.

Table 4 Thermal acceptability criterion [22, 23]

Parameter	MS 1525 (2014)	ICOP IAQ (2010)
Air temperature (°C)	24–26 °C	23–26 °C
Relative humidity (%)	50%–70%	40–70%
Air Velocity (m/s)	0.15–0.50 m/s	0.15–0.50 m/s

Fig. 1 The interior of multipurpose hall in hostel UTHM Pagoh

3 Methodology

3.1 Case Study Building

This study was carried out on the basis of thermally-perception of occupancy and field measurement at the multipurpose hall in the UTHM Pagoh hostel. The interior of the main hall is shown in Fig. 1. Multipurpose hall is a building used for many types of formal, informal, or recreational events or activities. It is (67850 × 34900 × 9150) mm in length. This hall uses natural ventilation methods and is supported by 10 mechanical fans. It is estimated that the main hall of the multi-purpose building can accommodate up to 320 people. The details of the hall are received from Koperasi UTHM Berhad as a drawing plan which included the specifications of the building design and ventilation system.

3.2 Data Collection

In order to evaluate the thermal comfort, 4 indoor environmental variables are determined in field measurement to assess thermal comfort. In addition, a structured questionnaire identifies the thermal perception of the occupants in the multipurpose hall. The thermal comfort survey questionnaire was conducted concurrently with the environmental measurement.

Thermal Comfort Survey Questionnaire. There are several tasks carried out in the multipurpose hall and this analysis only involves the occupants who conduct work at a met value of 1.1 (sedentary physical activity). The questionnaire is divided into 3 principal components: the first part deals with demographic data such as age, gender, weight, height and health; the second part with thermal sensation, thermal preference, thermal tolerance and thermal acceptance; and the third part is dedicated to clothing insulation. In line with the method [24], the number of representative samples was determined. Questionnaires were distributed face to face, including 179 students, lecturers, and staff. Each respondent took 5 to 10 min to react to the questionnaire. This questionnaire was analyzed using descriptive statistical methods involving frequency and percentage.

Thermal Environment Measurement. In main hall of the multipurpose hall, the air temperature, globe temperature, air velocity, and relative humidity are measured using the Thermal Comfort Kit. The height of measurement was 1.1 m, corresponding to the average height of the center of gravity for adults. Since this hall is optimally used for Saturday student events, data collection takes place every week for 6 weeks from 28th September 2019 to 2nd November 2019 on each of those days. The measurement was conducted in duration of 9 am to 5 pm. The time taken is about 15 and 20 min per measurement. In subsequent analysis the mean value of each measured variable was used. Depending on the area of the hall, thermal comfort parameters were measured by selecting sample points. The area in the multipurpose hall is a 425.82 m^2 drawing plan from Koperasi UTHM Berhad and five points were selected in the calculation of the thermal comfort parameters to obtain more precise details as illustrated in Fig. 2

The Thermal Comfort Kit Specifications are given in Table 5 and the location and sample are shown in Fig. 3.

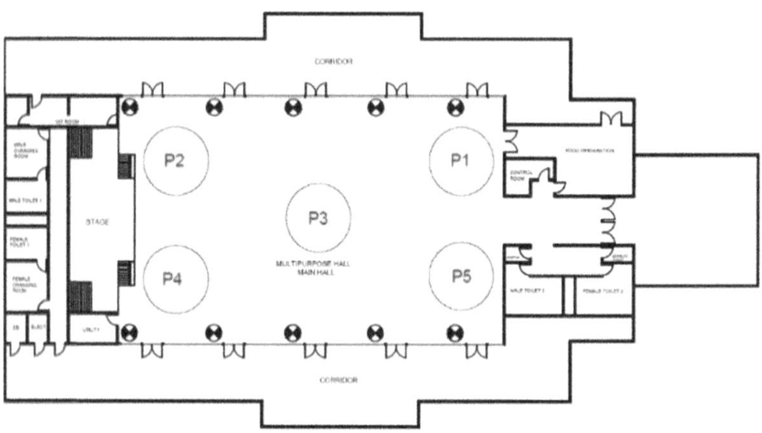

Fig. 2 Selected sampling point in multipurpose hall

Table 5 Thermal acceptability criterion [21, 29]

Trade/Brand	Parameter	Range	Accuracy
LSI LASTEM	Air temperature (°C)	−20°C −60°C	±0.8°C, ± 0.4°C (10−40°C)
	Mean radiant temperature (°C)	−20°C −60°C	±0.3°C
	Relative humidity (%)	0−100%	1.8%RH (10−90%)
	Air velocity (m/s)	0.01−20 m/s	±10 cm/s (0.5−1.5 m/s) 4% (>1.5 m/s)

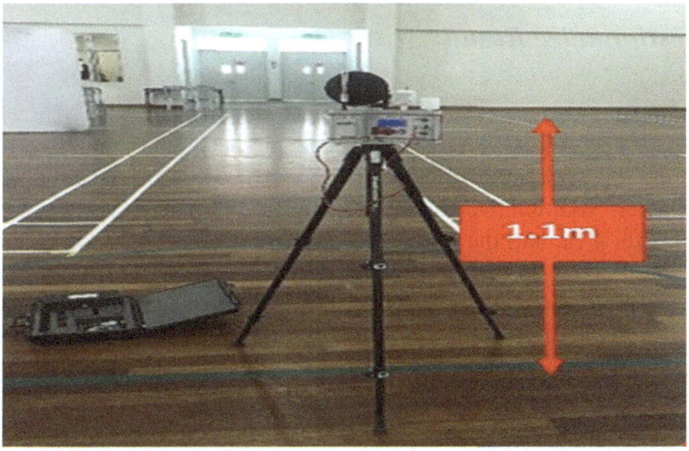

Fig. 3 Placement of thermal comfort kit and its probe

3.3 Data Analysis and Discussion

All the data collected in this study were first compiled into spread-sheets and then analyzed using Microsoft Excel software. The data gathered from the questionnaire were interpreted as pie chart and bar graph. To evaluate the average and graph out for study, the 4 indoor thermal environment parameters which are air temperature, air velocity, globe temperature, and relative humidity are measured. The data obtained from the calculation of the questionnaire and the fieldwork were interpreted using PMV model and PPD index. Calculated using CBE Thermal Comfort Tool, PMV and PPD. CBE Thermal Comfort Tool is a free online tool and develops a web-based graphical user interface according to ASHRAE Standard 55 for thermal comfort predictions.

4 Result and Discussions

Data from questionnaire survey and fieldwork measurement were gathered simultaneously from September 2019 to November 2019. The mechanical fans were opened to keep the building ventilated alongside questionnaire survey conducting and fieldwork measuring. All the collected data was interpreted using PMV model and PPD index.

4.1 Questionnaire Result

The total number of responses from survey questionnaires was 179, 70 of which were males and 109 were females. 96% of respondents between the ages of 15 and 25. 2% of respondents between the ages of 25–35 and 45–55. Consequently, most of the respondents were young university students due to the survey conducted within the university hostel area, and the rest included lecturers, university officers and staff. The highest proportion of respondents for the BMI category comes under the average weight group, which is 101 people (54%), while the lowest number of respondents (6%) or 11 people falls under the category of obesities. The total insulation of the clothing has been measured as 0.53clo. 4 Respondents who did not have a good health condition after conducting the questionnaire should be omitted because their interpretation would result in the study being unreliable.

Thermal Perception. Table 6 demonstrates the relationship between thermal sensation and demographic knowledge of the respondents. Because of the biological variability and perception from person to person [25], thermal comfort is difficult to please everyone in a room. Of the 179 respondents, a majority (40%) of respondents felt slightly warm and warm (36%) followed. From the survey results, males appear to feel warm as compared to females at the same temperature level. 20% of males felt the hall's indoor atmosphere was hot while 16% of females felt that way. In comparison, the older age group (45–55 years old) is more likely to feel moist and sticky about the climate, relative to the younger age group (15–25 years old). Age group respondents aged 15 to 25 are partially warm (37%) and partly warm (41%), while respondents aged 45 to 55 are more likely to feel hot (67%) and warm (33%) in the space. Of the respondents with obesity, the thermal sensation is the strongest, which is a complete 100% feeling slightly warm, warm, and hot. The greater the body mass index, then higher the thermal sensation [26].

Table 7 refers to the relationship between thermal preference and demographic information of the respondents. Thermal preference that included being much cool, slightly cooler, no change, slightly warmer, and much warm. Most respondents (65%) prefer a slightly cooler temperature while a minority of respondents (5%) prefer that the temperature at the multipurpose hall be no change. The thermal preference between male and female is very similar to each other, which is 27% of males and 31% of females who preferred a great deal of cool indoor climate. 100% of respondents

Table 6 Relationship between thermal sensation and respondents' demo-graphic information

Thermal sensations	Overall (%)	Gender		Age Group			BMI Category			
		M	F	15–25	25–35	45–55	Under weight	Normal Weight	Over weight	Obese
Cold	0	0	0	0	0	0	0	0	0	0
Cool	0	0	0	0	0	0	0	0	0	0
Slightly Cool	1	0	1	0	25	0	1	0	0	0
Neutral	6	1	8	5	25	0	6	5	5	0
Slightly Warm	40	42	39	41	25	0	43	22	22	40
Warm	36	37	36	37	25	33	33	53	53	50
Hot	17	20	16	17	0	67	17	20	20	10

Table 7 Relationship between thermal preference and respondents‘ demo-graphic information

Thermal Sensations	Overall (%)	Gender		Age Group			BMI Category			
		M	F	15–25	25–35	45–55	Under weight	Normal Weight	Over weight	Obese
Much Cool	30	27	31	28	50	100	23	31	33	27
Slightly Cooler	65	67	64	67	67	0	74	64	60	73
No Change	5	6	5	5	5	0	3	5	7	0
Slightly Warmer	0	0	0	0	0	0	0	0	0	0
Much Warmer	0	0	0	0	0	0	0	0	0	0

between the ages of 45–55 enjoyed a lot of coolness during multipurpose hall events. This is because the person in an early adulthood community can respond more rapidly to the thermal environment as opposed to those already in the late adulthood stage [25]. Of the respondents with obesity the thermal preference for the thermal tolerance is the maximum, which is a total of 100% favored much cooler and slightly cooler. In the hot climate the obese people were found to be less comfortable [27].

Table 8 shows the relation between thermal tolerance and demographic details of the respondents. Most respondents (43%) felt difficult to handle while the minority (3%) felt completely bearable and unbearable to the thermal situation. As a result, the females were more sensitive to the hot thermal atmosphere and less satisfied. In the multipurpose area, the age group between the ages of 15 and 25 is more likely to have difficulties in bearing with the heat inside the multipurpose hall. From 45 to 55 years of age, the older age group has the highest vote (33%) to give zero tolerance to the thermal condition in the multipurpose space. And the higher the age level, the lower the thermal resistance. 90% of the respondents are subject to obesity in a difficult thermal response and 10% more are experiencing a very difficult thermal bearing situation.

Table 9 describes the thermal acceptability of the respondent for multipurpose hall thermal condition. When respondents were asked to recognize the multipurpose hall thermal condition, 45% of them were acceptable, while 55% were unacceptable. There were no major gaps between males and females. In total 57% of male respondents and 54% of female respondents are inappropriate to the hall's indoor climate. The thermal tolerance for respondents with an age group between the ages of 45 and 55 is the maximum, which is a total of 100% who cannot tolerate the multipurpose hall thermal environment. The percentage of respondents selecting unacceptable rises with the growing category of BMI, which is 54% unaccepted for average weight, 60% unaccepted for overweight and 70% for obesity, respectively. Overall, elderly people are less suited to warm or hot conditions than younger people, and the lower the body mass index showed higher thermal tolerance.

4.2 Thermal Environment Measurement Result

Table 10 summarizes the findings of the measurement of the thermal environment at 5 sampling points in multipurpose hall. The thermal ambient parameters including air temperature, globe temperature, air velocity and relative humidity were measured using a thermal comfort kit. The main multipurpose hall area is 425.82 m^2 and 5 points were selected while calculating the thermal comfort parameters. During the survey time the outdoor average temperature was 31.1 °C with a minimum average temperature of 29.8 °C and a maximum average temperature of 35.5 °C for a total of 6 days of observation. At point 3 the highest air temperature was at 31.10 °C whereas at point 2 the lowest temperature was at 30.58 °C. Point 3 was the middle point of the multi-purpose hall and the wall fans were insufficient to expel warm air from the multi-purpose hall in the center. The results show that the average radiant

Table 8 Relationship between thermal tolerance and respondents' demographic information

Thermal Sensations	Overall (%)	Gender		Age Group			BMI Category			
		M	F	15–25	25–35	45–55	Under weight	Normal Weight	Over weight	Obese
Perfectly Bearable	3	4	2	3	25	0	1	1	10	0
Slightly difficult to bear	32	29	34	33	0	60	34	34	12	0
Fairly difficult to bear	43	43	44	43	75	17	49	49	40	90
Very difficult to bear	19	24	16	19	0	17	16	16	30	10
Unbearable	3	0	4	2	0	6	0	0	8	0

Table 9 Relationship between thermal acceptability and respondents' demographic information

Thermal Sensations	Overall (%)	Gender		Age Group			BMI Category			
		M	F	15–25	25–35	45–55	Under weight	Normal Weight	Over weight	Obese
Acceptable	45	43	46	45	50	0	50	46	40	30
Unacceptable	55	57	54	55	50	100	50	54	60	70

Table 10 Summary of the thermal environment measurement result according 5 sampling points

Average Outdoor Temperature (°C)	Point	Air Temperature °C	Mean Radiant Temperature °C	Air Velocity (m/s)	Relative Humidity (%)
31.1	1	30.83	31.20	0.08	65.55
	2	30.58	30.97	0.12	66.08
	3	31.10	31.50	0.05	64.77
	4	30.68	31.07	0.12	65.15
	5	30.97	31.33	0.07	65.13

temperature was 3 (31.50 °C) while the lowest temperature was 2 (30.97 °C). It is because the heat from the sun aimed at the building's roof center which will cause the highest point 3 temperature. The presence of the awning roofs that surround the building allows the radiant temperature to be marginally lower at the points near the ground.

The air velocity at each point in the multipurpose hall was nearly 0 m/s, because only mechanical ventilation was used during the survey period. At point 3 the lowest air velocity was at 0.05 m/s. The mechanical fans were located around of hall; thus, the mechanical fans are unable to provide wind velocity at center point of the multipurpose hall. The highest relative humidity was at point 2 (66.08%) while the lowest relative humidity was at point 3 (64.77%). On the contrary, point 3 air temperature was the highest as compared with other points. Relative humidity is inversely related to the temperature of the air. As the temperature is increased, the relative humidity should decrease.

4.3 Thermal Comfort Evaluation

Using the CBE Thermal Comfort Tool, the PMV and PPD were determined for each respondent based on the four measured environmental parameters including air temperature, mean radiant temperature, relative humidity, air velocity and the two personal parameters included clothing level and metabolic rate. Table 11 shows data inputs and the effects of PMV and PPD from the CBE Thermal Com-fort Tool. The PMV result is 2.1 and the PPD is 82.6%, suggesting a warm and extremely dissatisfied thermal condition

In the case of environments to assess the PMV and PPD, thermal sensation vote (7-degree scale) shall be applied from the questionnaire survey. Thermally dissatisfied persons are those who vote on the 7-point thermal sensation for hot, warm, cool or cold [14]. Table 12 displays the thermal sensation vote via questionnaire survey and the PPD index was determined using the equation below (3) using the following expression.

Table 11 Summary of the thermal environment measurement result according 5 sampling points

Parameter	Average Data
Air temperature, (°C)	30.8
Mean Radiant Temperature, (°C)	31.4
Relative Humidity, (%)	65.3
Air Velocity, (m/s)	0.1
Metabolic Rate, (met)	1.1
Clothing Insulation, (clo)	0.53
PMV	2.1
PPD, (%)	82.6
Air temperature, (°C)	30.8

Table 12 Thermal sensation vote (TSV) through questionnaire survey

Cold	Cool	Slightly cool	Neutral	Slightly warm	Warm	Hot
0	2	1	10	73	67	32
Dissatisfied = 2		Satisfied = 84			Dissatisfied = 99	

By using Eq. (3)

$$PPD = \frac{99 + 2}{185} \times 100$$
$$PPD = 55\% \tag{3}$$

Figure 4 displays PPD graph versus PMV. There are 55% of occupants who are dissatisfied with the multi-purpose hall thermal environment. The PMV value is + 1.6 and lies slightly warm and warm between the thermal condition.

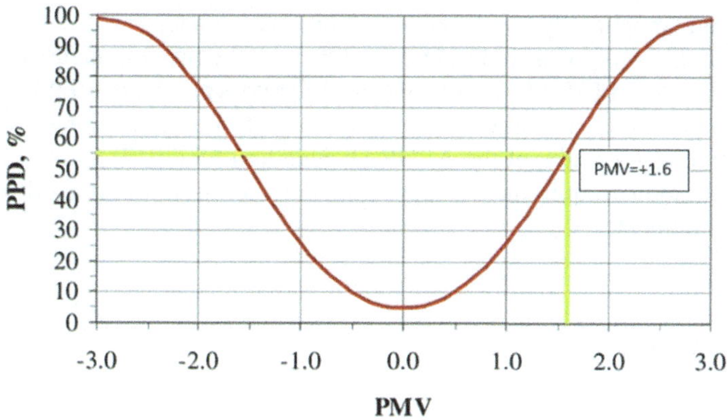

Fig. 4 PPD versus PMV graph

Table 13 The comparison between the measured results with standard

Parameter	Data collected at multipurpose hall	MS1525, (2014)	ICOP IAQ (2010)
Air temperature (°C)	30.83 °C	24–26°C	°C
Relative hunidity (%)	65.34%	50%–70%	40% –70%
Air velocity (m/s)	0.08 m/s	0.15–0.50 m/s	0.15–0.50 m/s

Table 14 The comparison between the calculated PMV and PPD with 2 international standards, which are ASHRAE Standard 55 (2017) and ISO 7730 (2005)

Data	Data calculated from thermal environment measurement	ASHRAE Standard 55 (2017)	ISO 7730 (2005)
PMV	2.1	1.6	−0.5 < PMV <+0.5
PPD (%)	82.6%	55%	<20%

The field-based measurements of PMV and PPD values are 2.1 and 82.6%, while thermal sensation values are 1.6% and 55%. It may conclude that the PMV model tends to classify the thermal environment as warmer than the user's actual thermal sensation. Reported by [9], there are wide variations in physiological and psychological aspects from person to person.

4.4 Compare Data with Standard

The relation of calculated results with standard is shown in Table 13. The thermal measurement outcome of the air temperature was, according to Table 13, superior to the guidelines for MS 1525 (2014) and the ICOP IAQ (2010) standard. Therefore, air velocity has not reached the standard. The standard was complied with only by relative humidity. These three parameters should be considered and maintained in order to build an optimally comfortable indoor environment.

The comparison of measured PMV and PPD with the two international standards of ASHRAE Standard 55 (2017) and ISO 7730 (2005) is provided in Table 14. All the PMV and PPD values from multipurpose hall and the questionnaire survey exceed the respective upper limits of the ASHRAE Standard 55 (2017) and the ISO 7730 (2005) acceptability requirements, suggesting that thermal comfort inside the hall is a wall measurement.

5 Conclusion

All the objectives accomplished in this study. One of the aims of this study is to determine the thermal perception of the occupants by means of a multipurpose hall questionnaire. In general, there is a 'slightly' warm and 'warm' feel in occupants 'while a 'slightly colder' thermal choice. The thermal tolerance of occupants is toward 'fairly difficult to bear' and their thermal acceptability is 'unacceptable' to the thermal environment of multipurpose hall. The occupants may have a somewhat significant effect on the thermal sensation, thermal comfort, thermal tolerance, and thermal acceptability of their gender, age, and body mass index.

The Second aim is to measure the thermal indoor parameters in the multipurpose hall. Data obtained from 28th September and 2nd November 2019 are 4 indoor thermal environment parameters. The building's air temperature and air velocity do not comply with the ICOP IAQ standard (2010). Only the relative humidity is complying with the standard. Temperature between 23 and 26 °C and relative humidity between 40 and 70% must be maintained to create an optimally comfortable indoor environment.

In addition, the third objective of this study is to assess the thermal comfort of the hostel multipurpose hall in UTHM Pagoh by using the Predicted Mean Vote (PMV) and Predicted Percentage Dissatisfied (PPD). The PMV and PPD results exceed the respective upper limits as recommended in the ASHRAE Standard 55 (2017) and the acceptability criterion of ISO 7730 (2005). In conclusion, the thermal comfort of multipurpose hall is uncomfortable and in warm condition.

Acknowledgements Universiti Tun Tun Hussein Onn Malaysia (UTHM) funded this work under Tier 1 grant (H228). The authors are completely appreciative of UTHM and UTM team members who have provided input and experience to support research.

References

1. Yao J, Yang F, Zhuang Z, Shao Y, Yuan PF (2018) The effect of personal and microclimatic variables on outdoor thermal comfort: a field study in a cold season in Lujiazui CBD, Shanghai. Sustain Cities Soc 39:181–188.https://doi.org/10.1016/j.scs.2018.02.025
2. Sari KAM, Mastaza KFA, Rahman MAA, Saji N, Muslim R, Mustafa MSS, Ghing TY (2019) Assessment of indoor air quality parameters at Ambulatory Care Centre XYZ, Malaysia. In: IOP conference series: earth and environmental science, vol 373, no 1, p 012013. IOP Publishing
3. Kim J, Schiavon S, Brager G (2018) Personal comfort models-A new paradigm in thermal comfort for occupant-centric environmental control. Build Environ 132:114–124. https://doi.org/10.1016/j.buildenv.2018.01.023
4. Seppänen OA, Fisk W (2006) Some quantitative relations between indoor environmental quality and work performance or health. Hvac&R Res 12(4):957–973. https://doi.org/10.1080/10789669.2006.10391446
5. Rahman MAA et al (2014) The review on significant adverse impact of poor indoor air quality on employees health. Adv Mater Res 931–932:749–753. 10.4028/www.scientific.net/AMR.931-932.749
6. Wyon DP, Wargocki P (2006) Room temperature effects on office work. In: Creating the productive workplace, pp 181–192. Taylor & Francis, London, New York
7. Awang M, Tham CS, Ruddin NMB, Rahman MAA, Hamidon N, Ahmad F, Rahman MSA (2020) Assessment of energy saving potential and lighting system in teaching building. J Adv Res Fluid Mech Thermal Sci 65:159–169

8. ASHRAE, ANSI/ASHRAE Standard 55-2017 (2017) Thermal environmental conditions for human occupancy. American Society of Heating, Refrigerating and Air Conditioning Engineers, Inc., Atlanta. https://ashrae.iwrapper.com/ViewOnline/Standard_55-2017
9. Cardoso V, Ramos NM, Almeida RM, Barreira E, Martins JP, Simões ML, Ribeiro B (2017) Thermal comfort evaluation in cruise terminals. Build Environ 126:276–287. https://doi.org/10.1016/j.buildenv.2017.10.008
10. Khan MH, Pao W (2015) Thermal comfort analysis of PMV model prediction in air conditioned and naturally ventilated buildings. Energy Proc 75:1373–1379; 183:484–499. https://doi.org/10.1016/j.egypro.2015.07.218
11. Sari KAM, Mastaza KFA, Rahman MAA, Musa MK, Awang M, Saji N (2018) Indoor air quality assessment at care centre ABC, Malaysia. In: Financial and economic tools used in the world hospitality industry: Proceedings of the 5th international conference on management and technology in knowledge, service, tourism and hospitality, SERVE 2017, pp 39–42. CRC Press, Balkema. https://doi.org/10.1201/9781315148762-8
12. Santamouris M, Wouters P (2006) Building ventilation: the state of the art. UK and USA. Earthscan. https://doi.org/10.1016/j.egypro.2015.07.218
13. Robert MP Eng (2006) Fundamentals of HVAC IP book: chapter 3 thermal comfort. Copyright © 2006 Elsevier Ltd. https://doi.org/10.1016/B978-012372497-7/50018-X
14. ISO (2001) Ergonomics of the thermal environment Instruments for measuring physical quantities. BS EN ISO 7726:2001
15. Mustafa MSS, Yusop F, Abdullah MA, Rahman MAA, Sari KAM, Fahmi AR, Hariri A (2019) Humidity control strategies in operation theatre Malaysia. In: IOP Conference series: earth and environmental science, vol 373, no 1, p 012016. IOP Publishing. https://doi.org/10.1088/1755-1315/373/1/012016
16. Khalid W, Zaki SA, Rijal HB, Yakub F (2019) Investigation of comfort temperature and thermal adaptation for patients and visitors in Malaysian hospitals. Energy Build. https://doi.org/10.1016/j.enbuild.2018.11.019
17. ISO B (2004) 8996: 2004 Ergonomics of the thermal environment-Determination of metabolic rate. BSI, London
18. Hall MR (2010) Materials for energy efficiency and thermal comfort in buildings. Woodhead publishing limited
19. Wu Z, Li N, Wargocki P, Peng J, Li J, Cui H (2019) Adaptive thermal comfort in naturally ventilated dormitory buildings in Changsha, China. Energy Build. https://doi.org/10.1016/j.enbuild.2019.01.029
20. International Organization for Standardization (1995) ISO 10551: 1995. Ergonomics of the thermal environment-assessment of the influence of the thermal environment using subjective judgement scales
21. ISO standard 7730, 3rd ed. (En), ISO (2005) Ergonomics of the thermal environment. Analytical determination and interpretation of thermal comfort using calculation of the PMV and PPD indices and local thermal comfort criteria
22. Department of Standard Malaysia (2014) Malaysian standard: code of practice on energy efficiency and use of renewable energy for non-residential buildings. MS1525:2014
23. Malaysia DOSH (2010) Industry code of practice on indoor air quality 2010, JKKP DP (S) 127/379/4-39. Minist Hum Resour Dep Occup Saf Heal, 1–50
24. Krejcie RV, Morgan DW (1970) Determining sample size for research activities. Educ Psychol Measur 30(3):607–610. https://doi.org/10.1177/001316447003000308
25. Khamidun MH, Abdul Rahman MA (2017) Analysis of mass transfer resistance for adsorption of phosphate onto industrial waste materials in plug-flow column. In: MATEC web of conferences, vol 103. EDP Sciences. https://doi.org/10.1051/matecconf/201710306004
26. Shaharon MN, Jalaludin J (2012) Thermal comfort assessment-a study toward workers' satisfaction in a low energy office building. Am J Appl Sci 9(7):1037. https://doi.org/10.3844/ajassp.2012.1037.1045
27. Maykot JK, Rupp RF, Ghisi E (2018) A field study about gender and thermal comfort temperatures in office buildings. Energy Build 178:254–264. https://doi.org/10.1016/j.enbuild.2018.08.033

28. Gilbert AL, Miles DB (2017) Natural selection on thermal preference, critical thermal maxima and locomotor performance. Proc R Soc B Biol Sci 284(1860):20170536. https://doi.org/10.1098/rspb.2017.0536
29. Pourshaghaghy A, Omidvari M (2012) Examination of thermal comfort in a hospital using PMV-PPD model. Appl Ergon 43(6):1089–1095. https://doi.org/10.1016/j.apergo.2012.03.010

Materials in Engineering

Thermodynamic Analysis of Light Hydrocarbon Production from Bio-oil Model Compound Through Co-cracking

Lim Jian Liang, Mahadhir Mohamed, Norzita Ngadi, Mazura Jusoh, and Zaki Yamani Zakaria

Abstract Thermodynamic equilibrium analysis of bio-oil model compound to light olefins was carried out by employing the total Gibbs free energy minimization method. Thermodynamic equilibrium compositions for acetic acid co-cracking were decided based on four operating parameter range, which were temperature of 300–1200 °C; AMR (acetic acid/methanol ratio) and AER (acetic acid/ethanol ratio) of 1:12, 1:6, 1:3, 1:1, 2:1 and pressure of 1 bar. Detailed analysis of stipulated feasible reactions showed hydrogen emerged as the primary product while carbon monoxide and ethylene as other important products. Equilibrium analysis hinted that ethylene formation does not occur spontaneously and ethylene is minutely produced compared to hydrogen and carbon monoxide. Optimum condition for ethylene production are AMR 2:1, T = 700 °C, P = 1 bar and AER 1:12, T = 1100 °C, P = 1 bar. From thermodynamic equilibrium analysis, ethylene formation was found to be not thermodynamically feasible at equilibrium but study involving experimental work involving catalyst can enhance the formation of light hydrocarbons.

Keywords Thermodynamic modeling · Acetic acid co-cracking · Equilibrium constant · Light hydrocarbons production · Gibbs free energy

1 Introduction

Many environmental issues have been caused by fossil fuels burning, for example, air pollution, damage to land surface, acidification, greenhouse gas accumulation, water pollution and ground-level ozone. Presently, 80% of the prime demand for petroleum derived refined liquid fuels belongs to the transport sector [1]. From this

L. J. Liang · M. Mohamed · N. Ngadi · M. Jusoh · Z. Y. Zakaria (✉)
School of Chemical Engineering, Teknology University of Malaysia, 81310 Skudai, Malaysia
e-mail: zakiyamani@utm.my

Z. Y. Zakaria
Centre for Engineering Education, Teknology University of Malaysia, 81310 Skudai, Malaysia

M. A. A. Zaini et al. (eds.), *Proceedings of the 3rd International Conference on Separation Technology*, Lecture Notes in Mechanical Engineering,
https://doi.org/10.1007/978-981-16-0742-4_11

liquid fuel, fossil diesel is identified as significant contributor to numerous environmental concern mentioned earlier [2]. If the situation is not prevented, environmental catastrophic disaster can take place and the natural stable ecosystem of planet earth would be jeopardized.

In view of this situation, alternatives for fossil fuel that is environmentally friendly and economically viable must be put forward. The worsening environmental situation has urged researchers to commence investigation on possible options for this purpose since the past three decades. Biofuels such as biodiesel and bio-gasoline are deemed as having spectacular potential as the replacement of fossil fuel in order to reduce or minimize the impacts toward environment [3]. As a result from extensive research and development, biodiesel has been successfully commercialized and the consumption is gaining massive attention. The percentage of biodiesel blended with fossil diesel kept increasing with years, thanks to the government support and policy from numerous countries around the globe. On the other hand, bio-gasoline is also having the right devotion from researchers and could soon follow suit the breakthrough made by biodiesel.

However, sole dependency on biodiesel and bio-gasoline would not be sufficient to cope for the massive world energy demand. Additional renewable and green energy source is still required to support the space and gap for clean fuel or bio-oil. One of the abundantly available natural resources is biomass.

Biomass is widely available and is always regarded as undesired, waste and product of no value. Because of this, it is best to utilize the availability of this feedstock. Various researches have been conducted on the transformation of biomass to fuel (bio-oil) and value added chemicals. Chemical reactions such as cracking, gasification and pyrolysis, and biochemical approaches such as fermentation are among the methods to extract useful compound from biomass. Unfortunately, the chemical and physical properties are too diverse that it is difficult to experimentally predict the outcome products from biomass reaction.

Bio-oil obtained from biomass fast pyrolysis process is a complex compound and its application is limited due to the drawbacks of crude bio-oil such as high moisture content, high acidity, high oxygen content, high viscosity and low heating value [4].

Other bio-oil compounds have different cracking behaviours and most bio-oil compound possesses the capability to be transformed into light hydrocarbon or liquid fuels. Unfortunately, the catalyst deactivation problem has to be fixed in order to obtain better quality fuels [5]. Most crude bio-oil has low effective hydrogen-carbon ratio which is not ideal for fuel source. Ethanol could be added into bio-oil and can serve as a solvent for bio-oil that can improve the effective hydrogen-carbon ratio.

The examples mentioned above proves that conducting experiment several times could be very costly and consumes a lot of time and energy, and the end result is not always aligned with the hypothesis. Such trial and errors are something that researchers would like to avoid. One way to overcome this situation is by performing thermodynamic analysis of the reaction. Such thermodynamic study will skip and eliminate all the troubles from extensive experimental work. It can provide important general idea on the type and amount of products that can be produced from specific biomass product, provided that the components in the biomass is known.

The thermodynamic study can be basically adopted to numerous chemical reaction applications. Saimon et al. [6] had performed thermodynamic analysis of hydrogen production from methanol-ethanol-glycerol mixture through dry reforming and discovered the trend of hydrogen, CO, CO_2, CH_4, C_2H_4, C_2H_6, H_2O formation trend. Zakaria and co-worker [7] investigated the thermodynamic analysis of a type of biomass, which was glycerol for steam reforming to ethylene. In both cases light hydrocarbon was obtained but appeared to be very minute, around 0.003–0.018 kmol, but it provided indication on the possibility of obtaining light hydrocarbon [8]. The amount of light hydrocarbon increased when suitable heterogeneous acid catalysts were adopted in the glycerol biomass steam reforming [9].

As a starting point, it is important to know the components of biomass and use it as a model compound that can represent the actual biomass. The usual components of biomass are acetic acid, aldehyde, ketone, glucose and others. Subsequently, one or more of the components can be chosen to be a model compound of biomass to be thermodynamically analyzed for its transformation into fuel or useful chemicals. Chosen model compound can be subjected to chemical process such as cracking to extract desired profitable products such as hydrogen, light olefins and other precious chemicals. Similar cracking reaction could also lead to the formation of undesired compound such as CO, CO_2 and coke [6, 7]. In view of this situation, it is imperative to suppress the formation of undesired compounds. Cracking reaction could be enhanced by introducing new reactant to stimulate favourable products, and this process is called co-cracking. Co-cracking has been proven to be able to improve the effective hydrogen-carbon ratio, depending on what the co-reactants are. Such study is not widely available when it comes to the thermodynamic analysis of biomass co-cracking to bio-oil, light olefin—particularly ethylene and other value added chemicals.

In this research, thermodynamic modeling was used to analyze co-cracking process of bio-oil model compound by varying the parameters: feed ratio and temperature. The thermodynamic investigation on acetic acid (representing bio-oil model compound) conversion to hydrogen and light hydrocarbon, specifically ethylene, is still not available in the literature. Hence, the prime goal of this study is to thermodynamically analyze the equilibrium reactions of producing hydrogen and ethylene from acetic acid by employing minimization of the total Gibbs energy.

2 Methodology

Thermodynamic study of bio-oil model compound to light hydrocarbons was achieved using HSC Chemistry software. The analysis proceeded by the application of total Gibbs free energy minimization method which is one of the main feature in the software. Equilibrium constants for every possible reaction mechanism were included in this study. The species considered were acetic acid, acetic acid with methanol and acetic acid with ethanol as feedstock. Reaction products for this study include hydrogen, ethylene, carbon monoxide and coke.

In this investigation, only ethylene was considered to represent light hydrocarbon due to the complex product compositions if other light hydrocarbon is to be included. A 2 kmol feed was fixed as the total input of reactants. The processing temperature range was 300–1200 °C while the AMR (acetic acid/methanol ratio) and AER (acetic acid/ethanol ratio) were 1:12, 1:6, 1:3, 1:1 and 2:1. Throughout the process the pressure was locked at 1 bar. Thermodynamic analysis revealed that complete conversion (100%) of bio-oil model compound and encouraging product yields trend were noticed in all cases, implying the feasibility of acetic acid co-cracking process.

3 Results and Analysis

Probable reactions that may occur in acetic acid co-cracking process and its computed heat of enthalpy are tabulated in Table 1. The equilibrium constants of all reactions in Table 1 are exhibited as a function of temperature in Fig. 1. According to the thermodynamic principals, when the Gibbs free energy change of reaction (ΔGr) is negative,

Table 1 Possible reaction in co-cracking of acetic acid [10–12]

R	Reaction type	Chemical reaction	ΔH(kJ/mol)
1	Thermal decomposition	$CH_3COOH \Longleftrightarrow 2H_2 + 2CO$	+ 213.70
2	Thermal decomposition	$CH_3COOH \Longleftrightarrow CH_4 + CO_2$	−33.50
3	Water gas shift reaction	$CO + H_2O \Longleftrightarrow H_2 + CO_2$	−41.14
4	Methanation	$CO + 3H_2 \Longleftrightarrow CH_4 + H_2O$	−206.10
5	Methanation	$CO_2 + 4H_2 \Longleftrightarrow CH_4 + 2H_2O$	−165.10
6	Methanation	$2CO + 2H_2 \Longleftrightarrow CH_4 + CO_2$	−247.30
7	Steam reforming	$CH_3COOH + 2H_2O \Longleftrightarrow 4H_2 + 2CO_2$	+131.40
8	Oxidative coupling of methane	$2CH_4 + CO_2 \Longleftrightarrow C_2H_6 + CO + H_2O$	+106.00
9	Oxidative coupling of methane	$2CH_4 + 2CO_2 \Longleftrightarrow C_2H_4 + 2CO + 2H_2O$	+284.00
10	Dehydrogenation of ethane	$C_2H_6 \Longleftrightarrow C_2H_4 + H_2$	+136.33
11	Methane decomposition	$CH_4 \Longleftrightarrow 2H_2 + C$	+74.52
12	Disproportionation	$2CO \Longleftrightarrow CO_2 + C$	−172.44
13	Hydrogenation of CO_2	$CO_2 + 2H_2 \Longleftrightarrow 2H_2O + C$	−90.16
14	Hydrogenation of CO	$H_2 + CO \Longleftrightarrow H_2O + C$	−131.30
15	Ketonization	$CH_3COOH \Longleftrightarrow CH_2CO + H_2O$	+144.40
16	Ketene decomposition	$2CH_2CO \rightarrow C_2H_4 + 2CO$	−76.90
17	Allene formation	$2CH_2CO \rightarrow C_3H_4 + CO_2$	−110.70
18	Hydrocarboxylation of ethylene	$C_2H_4 + H_2O + CO \rightarrow CH_3CH_2COOH$	−171.59

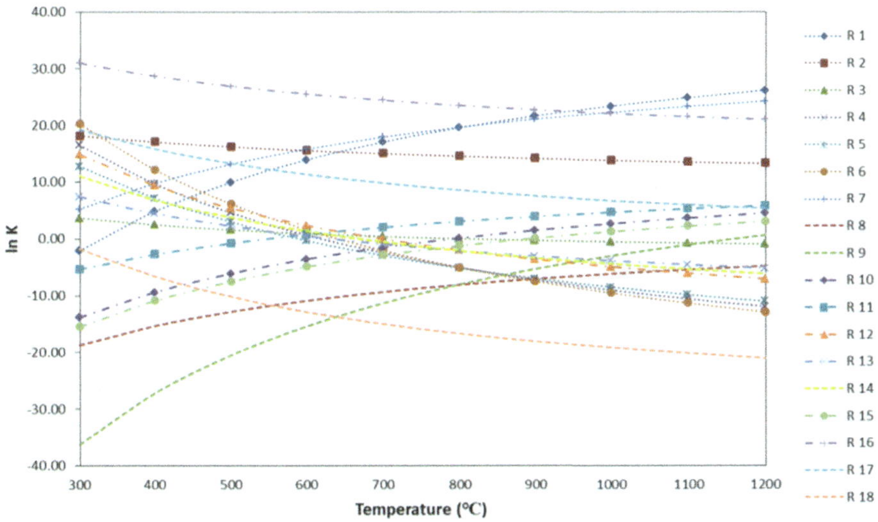

Fig. 1 Equilibrium constants of reactions involving co-cracking of acetic acid at atmospheric pressure

a larger ln K indicates a spontaneous reaction is more feasible to occur. Conversely, for positive ΔGr, the reaction is thermodynamically limited. The equilibrium constant (K) determines the extent to which the reaction occurs.

Figures 2a and 3b illustrate the effect of AMR and AER on the hydrogen production by acetic acid co-cracking at predetermined operating temperatures at one bar pressure. Hydrogen production was observed to steadily increase with temperature but slightly decreased once reaching the peak. The AMR did not have significant effect on the hydrogen production. Meanwhile, higher AER tended to produce less amount of hydrogen and this may be due to the fact that some hydrogen gas were from ethanol solvent. The rise of hydrogen was initiated from reaction R1 and R7. Methanation reactions (R4, R5 and R6) were not likely to absorb hydrogen since they were only possible at low temperature. In this study hydrogen appeared to gradually drop with carbon dioxide at high temperatures. Simultaneously, the amount of carbon monoxide and water gently increased. This scenario can be elucidated by the reverse water–gas shift reaction (R3).

Figures 2b and 3c exhibit the amount of moles of carbon monoxide formed from co-cracking of acetic acid with AMR and AER at 1 bar pressure at the designated temperature range. In all cases, carbon monoxide yield increased with temperature. As AMR and AER increased at 1 bar pressure, the carbon monoxide yield increased. The sudden increase of carbon monoxide that began from 600 °C was related to the thermal decomposition of acetic acid (R1) and decomposition of ketene (R16). On the other hand, low carbon monoxide at temperature <600°C is ascribed to the methanation reaction (R4, R5 and R6) that rigorously consumed carbon monoxide.

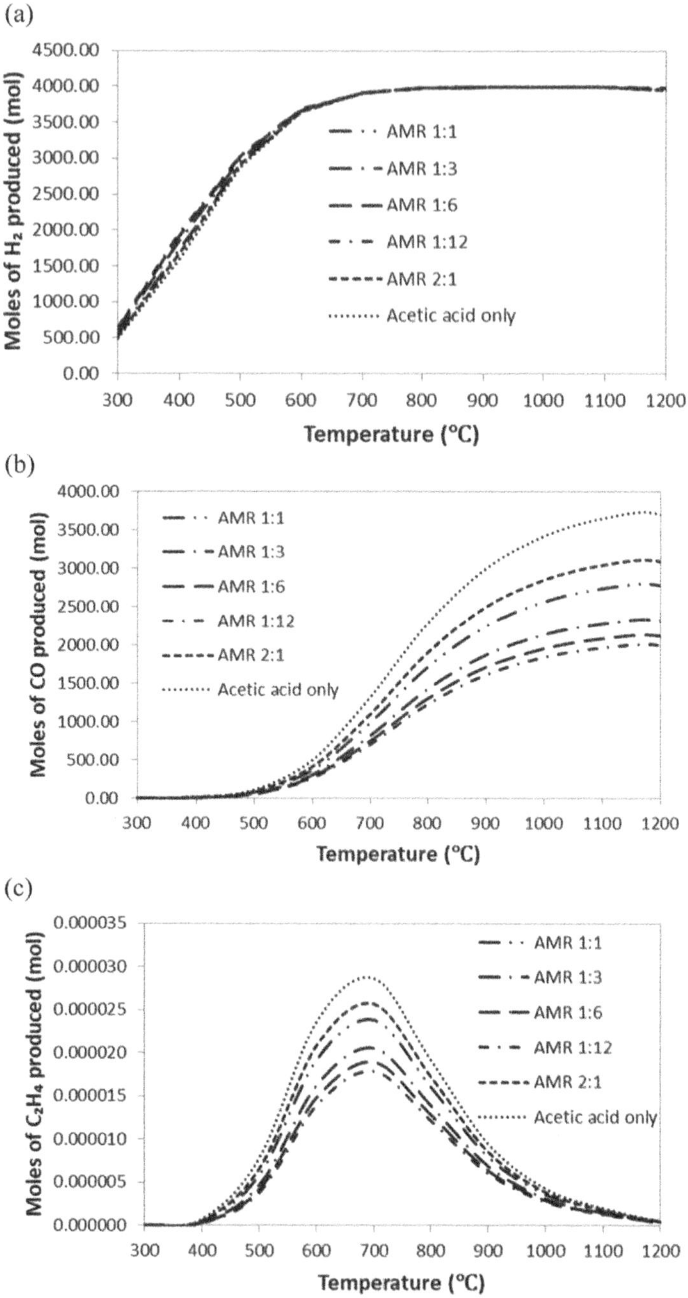

Fig. 2 Product yield: **a** H_2, **b** CO and **c** C_2H_4 for different AMR at 1 bar

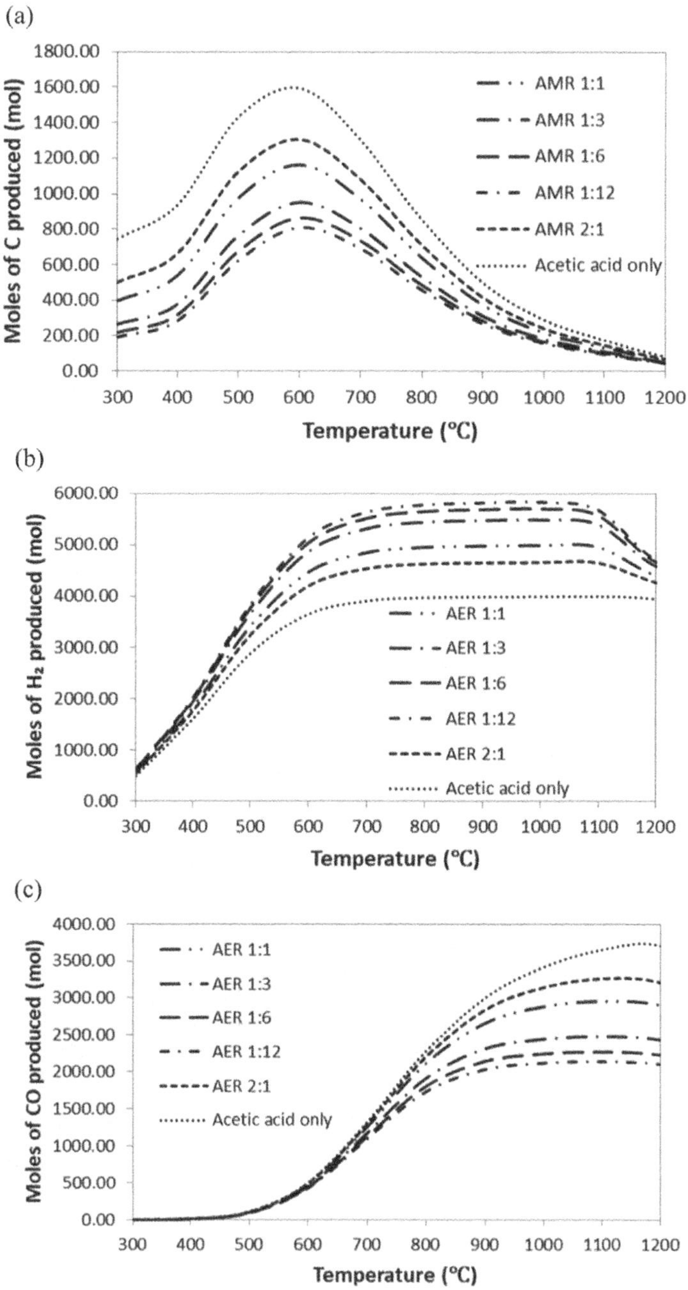

Fig. 3 Yield of **a** C for different AER at 1 bar, **b** H_2 and **c** CO for different AER at 1 bar

The number of moles of ethylene produced due to the effect of AMR and AER with different temperatures at constant pressure are demonstrated in Figs. 2c and 4a. Higher AMR produced more ethylene while higher AER produced less ethylene.

This phenomenon occured because ethylene was largely boosted by R10 at temperature more than 700 °C compared to R9 that was more constrained to equilibrium limitations. Thermodynamically, based on R10, the production of ethylene was unlikely to occur at temperature below 800 °C. The enormous variances amid

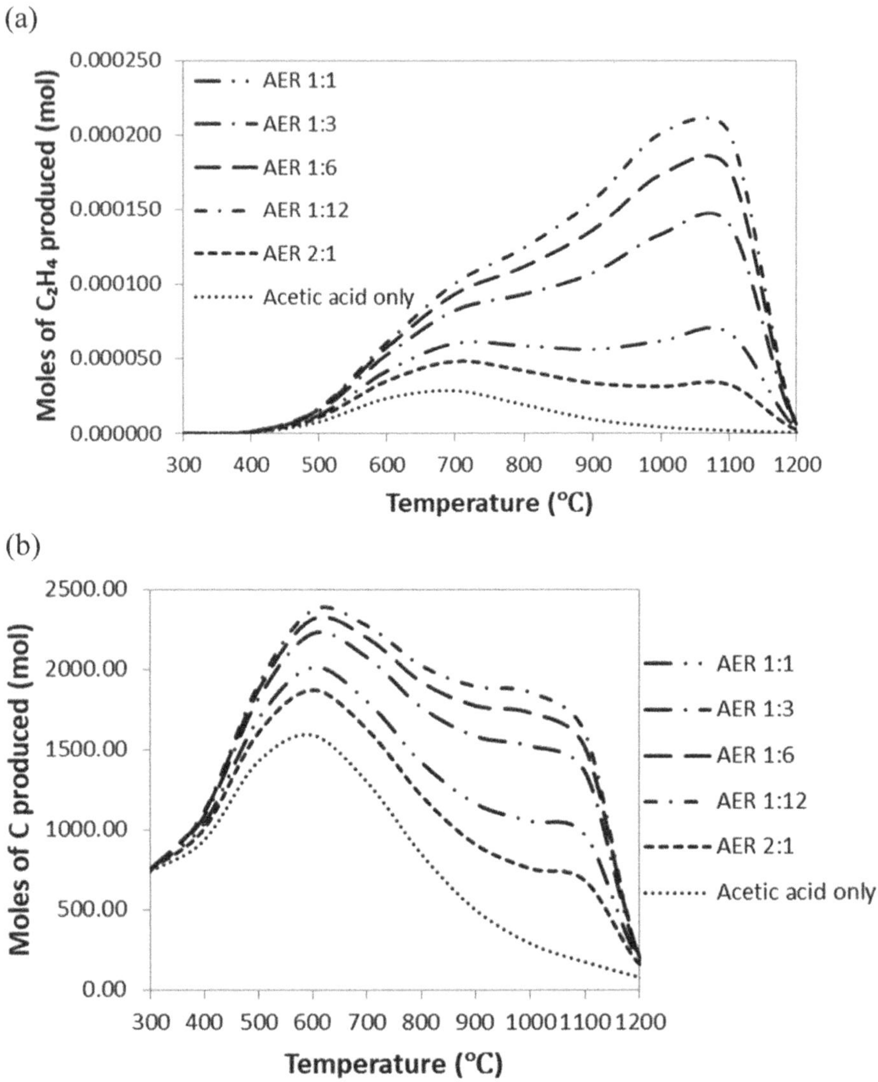

Fig. 4 Yield of **a** C_2H_4 and **b** C for different AER at 1 bar

thermodynamic modeling and experimental investigation appear mainly from catalyst synergistic effect which originated from chemical and physical properties in supporting continuous co-cracking of acetic acid, thus promoting R8 and R9.

During reaction, catalyst increases the reaction rate and this will lower the energy barrier (lowering the activation energy) allowing the production of ethylene. In addition, reactions that involve methanol and ethanol may contribute in the production of light hydrocarbons but these reactions are not within the scope of this study.

Carbon (coke) is an unwanted product in the acetic acid co-cracking reaction as it poisons (deactivates) solid heterogeneous acid catalyst and negatively trigger undesired pressure drop occurring in reactor. Coke formation at designated temperatures with AMR and AER are illustrated in Figs. 3a and 4b. The coke formation was more rapid for large AMR because more carbon atoms were available while when AER was low, the carbon formation was more significant as the reactions of ethanol may contribute more in carbon formation. It is worthy to note that coke formation was originated from R11, R12, R13 and R14. These reactions were easily influenced by the processing parameters since the equilibrium constants of the reactions are relatively low. It could be seen that disproportionation of carbon dioxide (R12), also identified as Boudard reaction is significantly important at temperatures below than 800 °C. The enthalpy of formation of carbon dioxide is greater than carbon monoxide but its entropy formation is subjacent in this reaction. Consequently, the carbon dioxide total free energy change of formation by oxidation is almost constant disregarding the temperature. This suggests that at smaller temperatures, catalyst could easily be poisoned as the equilibrium compliments exothermic carbon dioxide and coke formation. The production of coke from R11, R13 and R14 are mostly implausible since the reactions are negatively influenced by the equilibrium limitation.

4 Conclusion

Thermodynamic modeling and equilibrium analysis revealed unpromising yield for ethylene production, but not for hydrogen and syngas production. It can be observed that optimum condition for acetic acid with methanol as feedstock to maximize the production of ethylene was at $T = 700$ °C and $P = 1$ bar with AMR 2:1. Meanwhile, for acetic acid with ethanol as feedstock, the optimum condition was at $T = 1100$ °C and $P = 1$ bar with AER 1:12. When the current study was compared with almost similar previous experimental studies, the formation of ethylene can be inexorably enhanced in the existence of catalyst. Henceforth, appropriate shape selective and acidic catalysts are required to improve the production of light olefins and restrain coking at the optimal reaction conditions.

Acknowledgements This research is supported by Research University Grant Scheme (Vote No: 21H37 and 17J16). The authors totally acknowledged Universiti Teknologi Malaysia (UTM) and Ministry of Higher Education (MOHE), Malaysia for the approved fund.

References

1. Neamhom T, Polprasert C, Englande AJ (2016) Ways that sugarcane industry can help reduce carbon emissions in Thailand. J Cleaner Prod 131:561–571
2. Dayaratne SP, Gunawardana KD (2015) Carbon footprint reduction: a critical study of rubber production in small and medium scale enterprises in Sri Lanka. J Cleaner Prod 103:87–103
3. Ong YK, Bhatia S (2010) The current status and perspectives of biofuel production via catalytic cracking of edible and non-edible oils. Energy 35:111–119
4. Wang S, Guo X, Liang T, Zhou Y, Luo Z (2012) Mechanism research on cellulose pyrolysis by Py-GC/MS and subsequent density functional theory studies. Bioresour Tech 104:722–728
5. Gayubo A, Aguayo A, Atutxa A, Aguado R, Bilbao J (2004) Transformation of oxygenate components of biomass pyrolysis oil on a HZSM-5 Zeolite. I. alcohols and phenols. Ind Eng Chem Resear 43:2610–2618
6. Saimon NN, Jusoh M, Kamarudin MJ, Arsad A, Zakaria ZY (2017) Thermodynamic analysis of hydrogen production from methanol-ethanol-glycerol mixture through dry reforming. Chem Eng Transact 56:967–972
7. Zakaria ZY, Johari JM, A, Tuan Abdullah TA, Hassim MH, Kidam K, Kamaruddin MJ, Wan Sulaiman WR, (2015) Thermodynamic analysis of hydrogen production from ethanol-glycerol mixture through steam and dry reforming. Proc Manufact 2:92–96
8. Zakaria ZY, Linekoski J, Amin NAS (2014) Thermodynamic analysis of glycerol steam reforming to ethylene. Jurn Tek 67:109–115
9. Zakaria ZY, Linnekoski J, Amin NAS (2012) Catalyst screening for conversion of glycerol to light olefins. Chem Eng J 207:803–813
10. Goicoechea S, Ehrich H, Arias PL, Kockmann N (2015) Thermodynamic analysis of acetic acid steam reforming for hydrogen production. J Power Sources 279:312–322
11. Zakaria ZY, Amin NAS, Linnekoski J (2014) Thermodynamic analysis of glycerol conversion to olefins. Energy Procedia 61:2489–2492
12. Satyro M, Yaws CL (2018) The Yaws handbook of thermodynamic properties for hydrocarbons and chemicals. Elsevier Science.

Ultrasonic Assisted Synthesis of Bimetal Composite Strontium Oxide/Iron(III) Oxide for the Adsorption Isotherm Analysis of CO_2 Capture

Azizul Hakim Lahuri, Mohd. Ambar Yarmo, and Maratun Najiha Abu Tahari

Abstract Carbon dioxide (CO_2) capture by the bimetal composite of strontium oxide impregnated on iron(III) oxide (SrO/Fe_2O_3) was studied. The adsorbent was calcined at 200–600 °C and their adsorption capacities including adsorption and desorption were investigated. The adsorbents were characterized using X-ray diffraction (XRD), N_2 adsorption–desorption isotherm, CO_2 adsorption isotherm at 25 °C, 1 atm and temperature programmed desorption for CO_2 (CO_2-TPD). The SrO/Fe_2O_3-200 was observed to be the most efficient with adsorption capacity for physisorption and chemisorption of 5.08 and 100.03 mg CO_2/g adsorbent respectively. It exhibited a lower maximum desorption temperature of 523 °C compared to SrO and Fe_2O_3 only. The order of best fit linearized adsorption isotherm models for CO_2 adsorption isotherm is Freundlich > Dubinin-Radushkevich > Temkin > Langmuir. The Freundlich isotherm possesses Freundlich constant, $n = 1.196$ corresponded to favorable adsorption, highest R^2 value (0.98) and the calculated adsorption quantity closest to the experimental value. Adsorption process governed by CO_2 molecules forms a successive multilayer on the SrO/Fe_2O_3-200 surfaces and mainly exhibits a physisorption process. The heat of adsorption ($b_T = 5.238$ J/mol) from the Temkin isotherm suggests the adsorption is an exothermic reaction and mean free energy ($E = 2886.8$ J/mol) from Dubinin-Radushkevich isotherm indicates the CO_2 adsorption is physical in nature.

Keywords Adsorption isotherm · Bimetal composite · CO_2 capture · Iron(III) oxide · Strontium oxide

A. H. Lahuri (✉)
Department of Science and Technology, Universiti Putra Malaysia Bintulu Campus, Nyabau Road, P.O Box 396, 97008 Bintulu, Sarawak, Malaysia
e-mail: azizulhakim@upm.edu.my

Mohd. A. Yarmo · M. N. A. Tahari
Catalysis Research Group, School of Chemical Sciences and Food Technology, Faculty of Science and Technology, Universiti Kebangsaan Malaysia, 43600 Bangi, Selangor, Malaysia

M. A. A. Zaini et al. (eds.), *Proceedings of the 3rd International Conference on Separation Technology*, Lecture Notes in Mechanical Engineering,
https://doi.org/10.1007/978-981-16-0742-4_12

1 Introduction

Anthropogenic CO_2 gas is widely emitted to the environment and ecologically harmful due to its natural ability in trapping heat. It is the largest greenhouse gas compared to methane, nitrous oxide and fluorinated gases [1]. Human activities and industrial evolution are important sources of CO_2 emissions to the atmosphere. This resulting average annual CO_2 growth rate from 2018 to 2019 was increased at 2.58 ppm/year, reported by National Oceanic and Atmospheric Administration (NOAA). The CO_2 level in atmosphere in April 2020 was reported at 413.95 ppm which is higher than that of CO_2 safety level in atmosphere of 350 ppm [2]. In Southeast Asia, The World Bank Data [3] recorded the CO_2 emissions from 1970 to 2014, largely contributed by Indonesia compared to other countries in that region and followed by Thailand and Malaysia (Fig. 1). The inset diagram in Fig. 1 shows Brunei, Cambodia, Myanmar and Laos contributed less to the CO_2 emissions.

Based on the statistics, it is closely related to climate change of global warming with rise in global mean surface temperature and rise in global sea level. The estimation of average annual rate sea level rise of ~0.43 cm/year during the year ranging from 2000–2010 which means approximately ~17 cm sea level rise ranging from 2010–2040 [4]. Surprisingly, the estimation of sea level rise will be far less about ~5.7 cm for the same duration provided without 90 carbon producers. Therefore, a necessary technology that is economically feasible is needed to mitigate the CO_2 emissions and to reduce the risks associated to its highly presence in atmosphere.

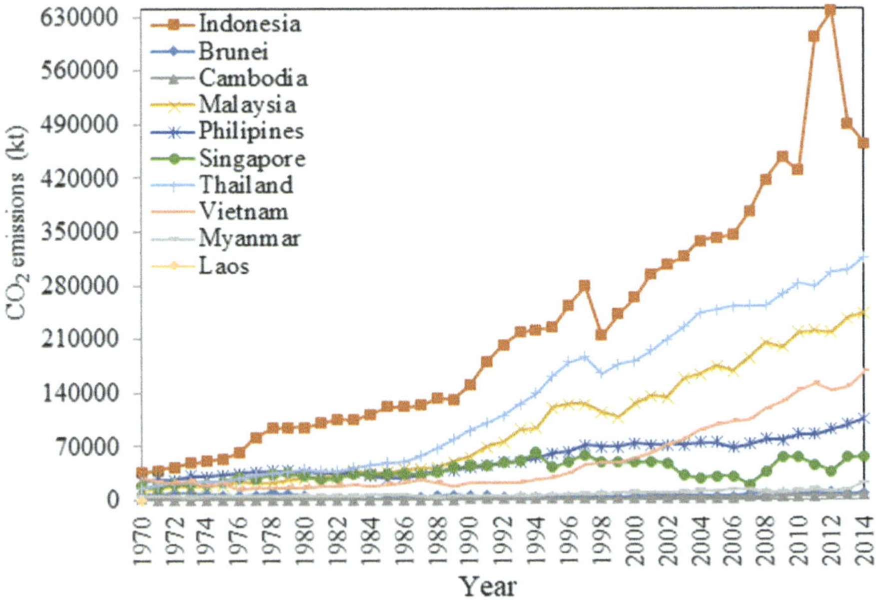

Fig. 1 CO_2 emissions (kilotonnes, kt) from 1970 to 2014 in South East Asia countries

An appropriate CO_2 reduction technique is necessary based on the type of combustion which is the main contribution in CO_2 formation. The available CO_2 capture technologies in the market are generally expensive which constitute 70–80% of the total cost of a complete Carbon Capture and Storage system involving capture, transport and storage [5, 6]. Hence, intensive efforts on research and development emphasized on the economically feasible reaction could reduce the operating cost. The main three CO_2 capture systems associated to different combustion processes are pre-combustion, oxyfuel combustion, and post-combustion. Usually, oxyfuel combustion and post-combustion are available to both gas fired plants and coal, meanwhile pre-combustion is mainly applied to coal-gasification plants. The most mature combustion process for CO_2 capture is post-combustion [7, 8].

Gas–solid adsorption and gas–liquid absorption are the main CO_2 capture techniques to isolate the CO_2 from the industrial flue gas and fuel gas stream. Other advanced technologies that have been developed are wet scrubber, dry regenerable sorbents, membranes, cryogenics, pressure and temperature swing adsorption [9]. Among these technologies, the most applicable technique for CO_2 capture is chemical absorption using aqueous alkanolamine solutions [10, 11]. Nonetheless, the alkanolamine aqueous solutions cause high equipment corrosion rate, high energy consumption in regeneration, and a large absorber volume required [12]. As for, ammonia functionalised on support material faces similar problem of dissociation of ammonia during CO_2 desorption [13, 14]. Consequently, other adsorbents are suggested and favourable to avoid those inherent drawbacks in amine based capture system.

Solid surfaces possess strong affinity toward gas molecules which results in binding or trapping of molecules during adsorption [15, 16]. The nature of interactions for adsorption is composed of physisorption and chemisorption. Physisorption occurs on adsorbates that are physically adsorbed by physical forces such as dipole–dipole, electrostatic, apolar, hydrophobic associations or Van der Waals, whereas chemical adsorption related to the adsorbates is chemically adsorbed by chemical bond covalent, ionic or metallic [16, 17]. Furthermore, the bond energy for physical adsorption and chemical adsorption are 8 to 41 kcal mol^{-1} and 60 to 418 kcal mol^{-1} respectively.

CO_2 regeneration for the physisorption and chemisorption phenomena are different by simple degassing and thermal desorption methods respectively. The CO_2 regeneration of solid matrices for chemisorption is very challenging ascribed from the adsorbates which are strongly bonded and trapped. For instance, CaO and soda lime will interact chemically with acid gases of CO_2, SO_2, NO_2 [17, 18]. Surface modification by improving the basicity of the adsorbents could enhance CO_2 capture due to the acidic in nature of CO_2. A commonly reported calcium oxide based adsorbents are ascribed from abundance and relatively cheap material. However, it requires extremely high desorption temperature around 800–950 °C [19, 20]. A bimetal system was reported by using calcium oxide based but still facing similar problem [21]. High regeneration temperature will cause sintering effect on the adsorbent which defect of the active site to attract CO_2 and inefficient operating cost due to high energy consumption.

Several studies reported on CO_2 capture by using SrO sorbent and SrO-based sorbents. Computational methods of density functional theory and lattice phonon dynamics enable this method to obtain general overview for CO_2 capture screening on alkali earth metal oxides of SrO [22]. Furthermore, $SrCO_3$ can be regenerated to $Sr(OH)_2$ in the presence of water. The adsorption/desorption temperatures are based on Gibss free energy ($\Delta G^{o}_{rxn} = 0$ kJ/mol) in which equilibrium temperature for exothermic adsorption and endothermic desorption is at 1175 °C [23]. Thus, most of the previous studies were conducted at carbonation temperature above 1000 °C [24] to obtain optimum adsorption capacity which highly consume energy in the capture process. Specifically, adsorption temperature for SrO at 1150 °C and $SrCO_3$ desorption temperature was conducted above equilibrium temperature at 1235 °C [25]. Hence, this work attempts to discover the lower desorption by modification using SrO.

Previous study discovered that Fe_2O_3 can be an alternative efficient adsorbent [26–29] for capturing CO_2 due to its basicity properties and stability compared to FeO and Fe_3O_4 [27]. Therefore, the idea of modification through bimetal composite anticipates the enhancement in adsorption properties. Bimetal of Cu-Fe loaded on zeolite showed a promising adsorption capacity and recyclability [30]. The computational method by using density functional theory showed the capability of bimetallic system of Cu-Ni nanoparticles in CO_2 chemisorbs and physisorbs depending on the decorated formation of Cu-Ni [31]. Furthermore, bimetallic system using CaO/Fe_2O_3 is able to provide chemical shift in CO_2 desorption temperature due to the indirect bonding of CO_2 to the monometallic system [32].

This study attempts to synthesize bimetal composite via wet impregnation, determine the CO_2 adsorption capacity, identify the CO_2 desorption temperature of bimetal composite and understand the adsorption phenomena using adsorption isotherm models. The surface modification of the Fe_2O_3 by addition of alkali earth metal, SrO presumes the enhancement of its basicity properties and reduce CO_2 regeneration temperature. The adsorption phenomena was studied by plotting the experimental CO_2 adsorption data using linearized Langmuir, Freundlich, Temkin and Dubinin-Radushkevich (D-R) isotherm models. Thus, it can further understand the nature of adsorption from isotherm parameters.

2 Experimental

2.1 Materials

The chemicals for Fe_2O_3, $Sr(NO_3)_2$ and ethanol were purchased from Sigma Aldrich (USA), Hopkin & Williams (UK) and Systerm® ChemAR® (Malaysia) respectively.

2.2 Synthesis of SrO/Fe_2O_3

In this research, bimetal composites SrO/Fe_2O_3 were synthesized by conventional impregnation method. Fe_2O_3 was pre-treated by thermal treatment under N_2 flow at 150 °C. The adsorbent was synthesized using wet impregnation method. Initially, 20 wt% of $Sr(NO_3)_2$ from the total weight of Fe_2O_3 (4 g) was dissolved in disperse medium and it was stirred at ambient temperature for 2 h. The disperse medium consisted of ethanol: distilled water with the ratio of 1:4. The mixture was ultrasonicated for 30 min for improving the homogeneous mixture. It was stirred continuously until dry at the temperature of 80–100 °C and placed in the oven for 24 h at 110 °C. The adsorbents were ground to powdery form and calcined under ambient air at 200–600 °C. The adsorbents denoted as SrO/Fe_2O_3-x with 'x' indicating calcination temperature of 200–600 °C. The phase composition was analyzed by using Bruker AXS D8 Advance type with X-ray radiation source of Cu Kα (40 kV, 40 mA). The crystal structures were verified by recording 2θ diffraction angle from 10 to 90° at wavelength ($\lambda = 0.154$ nm). The crystalline phase was interpreted by matching the x-ray diffractograms with standard diffraction data (JCPDS) file. Nitrogen (N_2) adsorption–desorption isotherms and textural properties measurement was conducted using a static volumetric technique instrument (gas sorption analyzer, Micromeritics ASAP 2020). Prior to analysis, the adsorbents (500 mg) were outgassed for 6 h at 200 °C under vacuum in order to remove humidity content. Liquid N_2 circulating bath was used during analysis to control the analysis temperature of 77 K.

2.3 CO_2 Capture

CO_2 physisorption at 25 °C and 1 atm was measured using aforementioned static volumetric technique with water circulating bath to control the adsorption temperature at 25 °C. The purity of CO_2 as adsorbate is 99.9%. CO2 chemisorption was analyzed by using temperature programmed desorption (CO_2-TPD) from a chemisorption analyzer model Micrometrics 2920 Chemisorb. The adsorbent (50 mg) was cleaned with He flow at the temperature of 150 °C for 30 min to remove traces impurity gases. The probe molecule for adsorption was composed of synthetic gas mixture 5% (v/v) CO_2 in He. CO_2 adsorption with saturated flow was performed at 30 °C. The excess weakly adsorbed CO_2 was removed by He flow (30 mL/min) at 50 °C for 30 min and followed by TPD from 30 to 900 °C with ramp rate of 10 °C/min. The CO_2 effluent gas stream during desorption was detected quantitatively by thermal conductivity detector (TCD).

2.4 Adsorption Models

Langmuir equation was derived theoretically from a kinetic approach by assuming composed of many homogeneous adsorption mechanism on a monolayer. Later, Freundlich and Brunauer-Emmet-Teller (BET) equation arise from the Langmuir expression [33]. The adsorption isotherm was recognized by Freundlich [34], describing heterogeneous surfaces considered to be a composite surface. Subsequently, the derivation of BET equation is for multilayer adsorption [35] by adopting the Langmuir mechanism, but introducing a number of simplifying assumptions. The CO_2 adsorption data were analyzed with established adsorption isotherms of Langmuir, Freundlich, Temkin and Dubinin-Radushkevich (D-R) models.

The Langmuir model is derived by assuming constant energies of adsorption onto surface without migrating of adsorbate in the plane of the surface [36]. Adsorption is proportional to the empty adsorbent surfaces whereas desorption is directly proportional to the part of the adsorbent surface that is monolayer covered [37]. According to the assumptions, Langmuir can be expressed as follow:

$$Q_e = \frac{Q_o K_L P_e}{1 + K_L P_e} \tag{1}$$

The linear form of Langmuir adsorption parameters was represented the following Eq. 2.

$$\frac{P_e}{Q_e} = \frac{1}{Q_o K_L} + \frac{P_e}{Q_o} \tag{2}$$

where P_e is the equilibrium pressure of adsorbate (bar), Q_e is the amount of adsorbate adsorbed per gram of the adsorbent at equilibrium (cm^3/g), $Q_{\circ}$ is maximum monolayer coverage capacity (cm^3/g) and K_L is Langmuir isotherm constant that is related to the energy of adsorption. The linear plot of P_e/Q_e againts P_e gives $1/Q_{\circ}$ is the slope and $1/Q_{\circ}K_L$ is the intercept. The equilibrium parameter or separation factor, R_L is essential feature for Langmuir isotherm which is refers to dimensionless constant [38].

$$R_L = \frac{1}{1 + (1 + K_L C_o)} \tag{3}$$

where $C_{\circ}$ is initial concentration of the adsorbate.

The Freundlich adsorption isotherm is usually used to describe the adsorption properties on heterogeneous surface [39] by fitting with the empirical equation proposed by Freundlich:

$$Q_e = K_F P_e^{\frac{1}{n}} \tag{4}$$

where K_F is Freundlich isotherm constant (cm^3/g·bar$^{1/n}$), n is adsorption intensity, P_e represents the equilibrium pressure of adsorbate (bar) and Q_e is the amount of adsorbate adsorbed per gram of the adsorbent at equilibrium (cm^3/g). The linearizing of Freundlich isotherm is expressed as:

$$\log Q_e = \frac{1}{n}\log P_e + \log K_F \tag{5}$$

The K_F constant is an indicator for approximate adsorption capacity whereas $1/n$ is a function for the adsorption strength in adsorption [40]. The linear plot of log Q_e versus log P_e gives $1/n$ is the slope and log K_F is the intercept.

The Temkin model assumes a uniform distribution of binding energies (up to maximum binding energy) as presented in non-linear Temkin isotherm Eq. 6 [41]. The isotherm possesses a factor that strictly concerns adsorbent-adsorbate interactions. The temperature as a function for heat of adsorption applying to all molecules in the layer would decrease linearly rather than logarithmic with coverage provided neglecting the extremely low and large value of concentrations [41]. Thus, Temkin isotherm is restricted for an intermediate range of ion concentrations [42].

$$Q_e = \frac{RT}{b_T}\ln(A_T P_e) \tag{6}$$

$$Q_e = \frac{RT}{b_T} ln A_T + \frac{RT}{b_T} ln P_e \tag{7}$$

$$B = \frac{RT}{b_T} \tag{8}$$

$$Q_e = B ln A_T + B ln P_e \tag{9}$$

where by A_T is Temkin isotherm equilibrium binding constant (L/g); b_T is Temkin isotherm constant for heat of sorption (J/mol); R is universal gas constant of 8.314 J/mol.K; T is absolute temperature at 298 K and B is constant related to heat of sorption. The linearized form of Temkin model Eq. 9 was plotted adsorption quantity Q_e against $ln\ P_e$ with the constants were determined from the slope and intercept.

The D-R isotherm is generally applied to multi-layer adsorption and expresses the adsorption mechanism with a Gaussian energy distribution onto a heterogeneous surface [43]. The model has been found to successfully fit the data of high adsorbate activities and intermediate range of concentrations [44]. The linearized isotherm (Eq. 11) was plotted logarithm of amount adsorbed ($ln\ Q_e$) against the square of Polanyi sorption potential (ε^2) equation which is expressed in Eq. 12 [45].

$$Q_e = Q_s e^{-K_{ad}\varepsilon^2} \tag{10}$$

$$\ln Q_e = \ln Q_s - K_{ad}\varepsilon^2 \tag{11}$$

$$\varepsilon = RT\ln[1 + \frac{1}{P_e}] \tag{12}$$

where Q_e is amount of adsorbate in the adsorbent at equilibrium (cm^3/g); Q_s is theoretical isotherm saturation capacity (cm^3/g); ε (J/mol) is Polanyi sorption potential; R is universal gas constant of 8.314 J/ mol.K; T is absolute temperature at 298 K; P_e is adsorbate equilibrium pressure (bar) and K_{ad} is D-R isotherm constant (mol^2/J^2) related to the mean energy of sorption per mole of adsorbate. Physical and chemical adsorption of metal ions with its mean free energy, E per molecule of adsorbate can be distinguished by applying this approach. This energy E is determined by the Eq. 13 [46].

$$E = \frac{1}{\sqrt{2K_{ad}}} \tag{13}$$

The type of adsorption can be estimated by the magnitude of adsorption energy, E. Physical adsorption will be determined if E value is less than 8 kJ/mol, ion exchange will be determined if the value is between 8 and 16 kJ/mol, and the strong chemical adsorption will be determined when the value is above 16 kJ/mol [47].

3 Results and Discussion

3.1 *Characterization of Adsorbents*

X-ray diffraction was used to obtain the composition information about the changes of the Fe_2O_3 structure upon impregnation of SrO. The X-ray diffractograms for all adsorbents are displayed in Fig. 2. Calcination temperature influenced the intensity of the observed peak. Peak intensity indication of the crystallinity was observed as increasing with higher calcination temperature of bimetal composite SrO/Fe_2O_3 series as shown in the selected 2θ of 35–37° in Fig. 2. Most of the peaks broadened at higher calcination temperature describe smaller crystallite size [48] generated upon annealing at 600 °C. The peak positions for bimetal composite SrO/Fe_2O_3-300 to SrO/Fe_2O_3-600 shifted to the right compared to Fe_2O_3 only, which attributed to the decrease of the lattice parameters. All of the peaks corresponded to rhombohedral phase of hematite (α-Fe_2O_3) with lattice parameters of a = b = 5.0356 and c = 13.7489 (matched with Joint Committee on Powder Diffraction Standard, JCPDS, file number 33–0664). Surface modification by addition of SrO was not detected due to it well dispersed on Fe_2O_3.

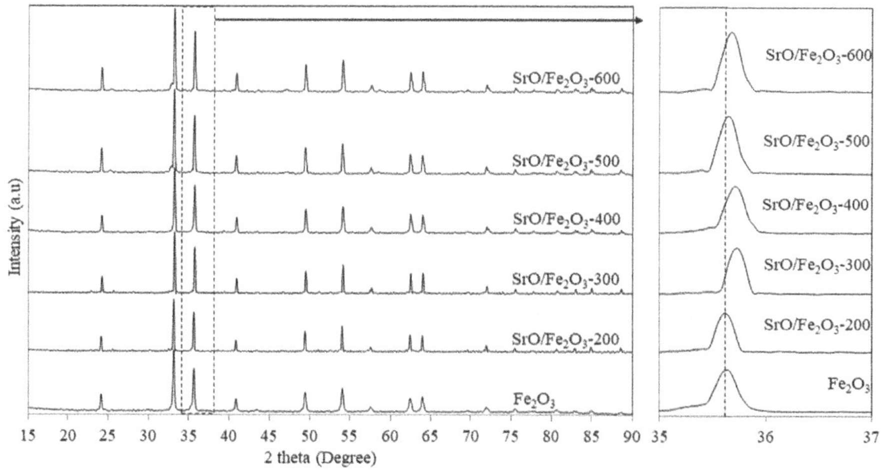

Fig. 2 XRD patterns of the adsorbents

N_2 adsorption–desorption isotherms of the adsorbents are described in Fig. 3 and the textural properties can be explained through its parameters. The parameters of BET surface area, mesopore surface area, total pore volume and pore diameter are shown in Table 1. The shape of the N_2 adsorption–desorption isotherm shows typical sloping adsorption which obeys the type III profile according to The International Union of Pure and Applied Chemistry (IUPAC). Further adsorption continues when increasing the interaction between adsorbate layers rather than the interaction with the adsorbent surface. The hysteresis was determined from the space in between the adsorption and desorption curves. It follows H3 type hysteresis with desorption branches covering a large range of relative pressure (*P/Po*). Although the interaction of adsorbent—adsorbate is weak but it is possible for porous adsorbents [49]. The hysteresis indicates plate-like particles giving rise to slit-shaped pores with large ranges of pore diameters [50].

The BET surface area of bimetal SrO/Fe_2O_3 increases (10.10 m^2/g) with higher calcination temperature up to 500 °C for SrO/Fe_2O_3-500 due to nucleation growth of particles and well-developed morphology. However, SrO/Fe_2O_3-600 shows a decrease in BET surface area (6.88 m^2/g) which suggests agglomeration of the particles or sintering effect upon exposure to a calcination temperature of 600 °C. Pores generated on the SrO/Fe_2O_3-500 surfaces are smaller with an average pore diameter of 14.2 nm. Pores are generated is higher with the surface modification by adding SrO. Thus, the total pore volume for SrO/Fe_2O_3-500 (0.0359 cm^3/g) was enhanced and ascribed from the bimetal. These features assume that the high BET surface area and pore volume enable the adsorbate to occupy pores and trapped on the adsorbent [51]. It was observed that SrO/Fe_2O_3-200 exhibits large pore diameter correlated to the pore size distribution in Fig. 4 due to initiation of morphology development.

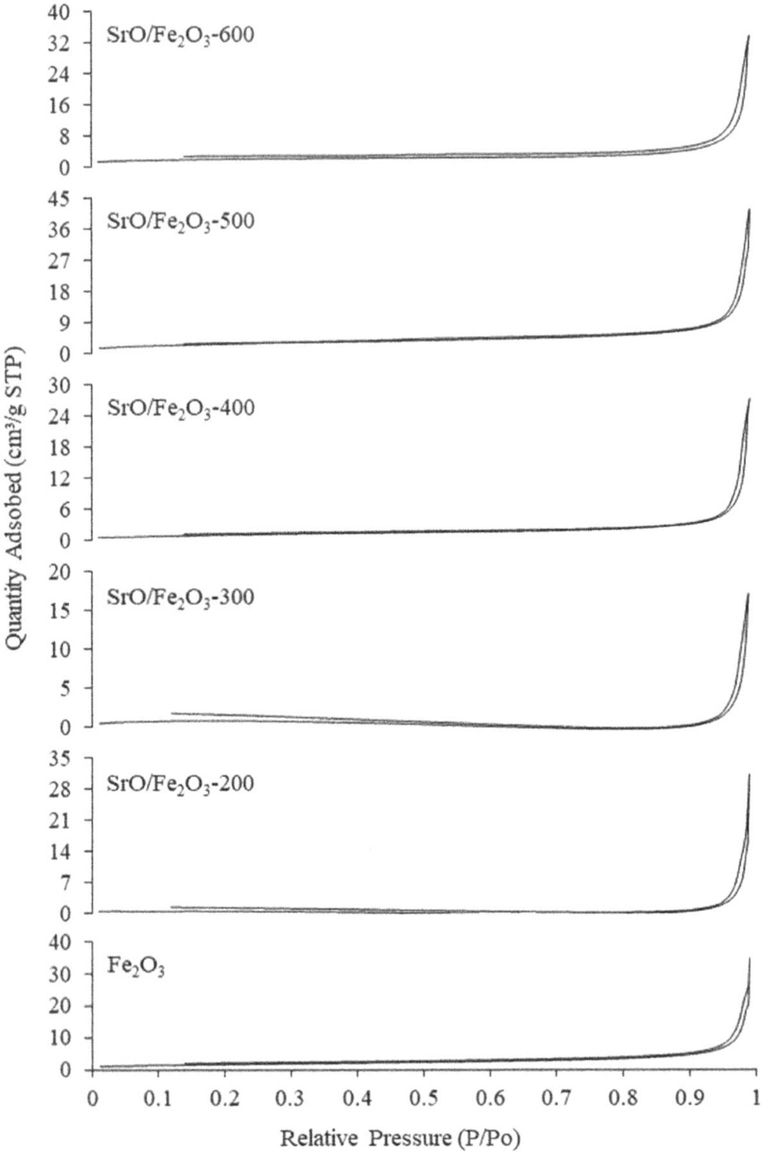

Fig. 3 N_2 adsorption–desorption isotherm for the adsorbents

Pore size distribution curve described in Fig. 4 was computed using the Barrett-Joyner-Halenda (BJH) method for accurate mesopore and macropore filling mechanism. Based on the pore size distribution curve as shown in the inset Fig. 4, high BET surface area and total pore volume for SrO/Fe_2O_3-500 were attributed by the mesopore structure generated around 2.0–4.3 nm (20–43 Å). It was believed that the

Table 1 Textural parameters of the adsorbents obtained from the N_2 adsorption–desorption isotherms

Adsorbents	BET surface area (m^2/g)	Total pore volume (cm^3/g)	Pore diameter (nm)
Fe_2O_3	5.84	0.0283	19.4
SrO/Fe_2O_3-200	1.49	0.0179	48.0
SrO/Fe_2O_3-300	2.66	0.0135	20.3
SrO/Fe_2O_3-400	4.07	0.0244	23.9
SrO/Fe_2O_3-500	10.10	0.0359	14.2
SrO/Fe_2O_3-600	6.88	0.0291	16.9

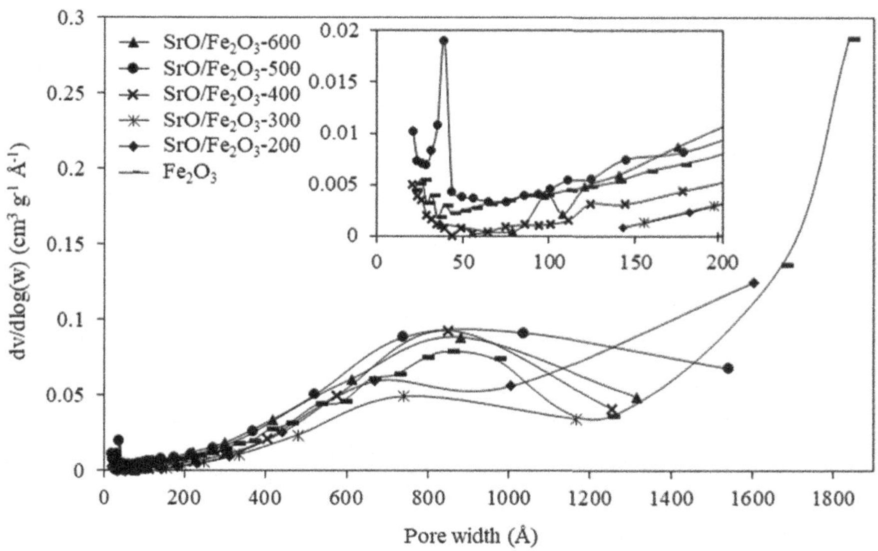

Fig. 4 Pore size distribution curves for the adsorbents

impregnation of SrO on Fe_2O_3 occurs by deposition on the Fe_2O_3 surfaces. Hence, SrO particles could generate pores enhancing the total existing pore volume by Fe_2O_3 only. By referring to the same region as shown in the inset of Fig. 4, textural properties for SrO/Fe_2O_3-600 degraded due to the thermal surface defect and sintering effect on particles. Subsequently, lower distribution at larger mesopore region around 7.5–11.1 nm (750–1110 Å).

3.2 CO_2 Capture Activities

CO_2 adsorption isotherms at 25 °C and 1 atm describes CO_2 physisorption for bimetal composite SrO/Fe_2O_3 are shown in Fig. 5 and the adsorption capacity was tabulated in Table 2. The bimetal composites SrO/Fe_2O_3 enhance the adsorption capacity by exhibiting higher CO_2 physical adsorption compared to Fe_2O_3 only. SrO/Fe_2O_3-500 shows the highest adsorption capacity (8.03 mg CO_2/g adsorbent) followed by SrO/Fe_2O_3-200 (5.08 mg CO_2/g adsorbent). It is noteworthy that the SrO/Fe_2O_3-200 has a higher adsorption capacity than SrO/Fe_2O_3-600, which physisorption does not follow its textural properties. This suggests that active sites responsible to attract CO_2 might have defected at the calcination temperature of 600 °C. During CO_2 adsorption, traces amount of moisture from the gas feed may affect the formation of hydroxyl (OH) group that acts as the active site. Accidental adsorption of traces

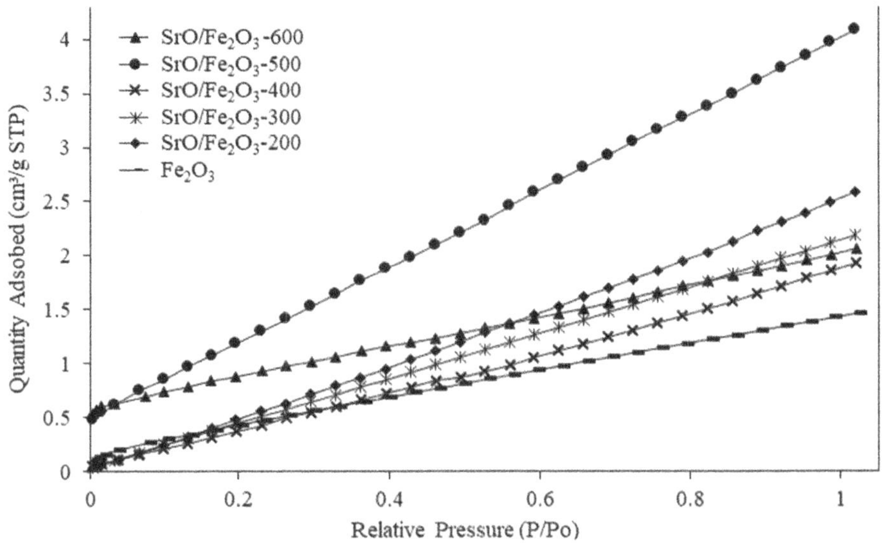

Fig. 5 CO_2 adsorption isotherms at 25 °C and 1 atm

Table 2 Adsorption capacity for physisorption

Adsorbents	CO_2 adsorption at 25 °C (cm^3/g)	Adsorption capacity at 25 °C (mg/g)
Fe_2O_3	1.46	2.87
SrO/Fe_2O_3-200	2.59	5.08
SrO/Fe_2O_3-300	2.19	4.31
SrO/Fe_2O_3-400	1.93	3.79
SrO/Fe_2O_3-500	4.09	8.03
SrO/Fe_2O_3-600	2.06	4.05

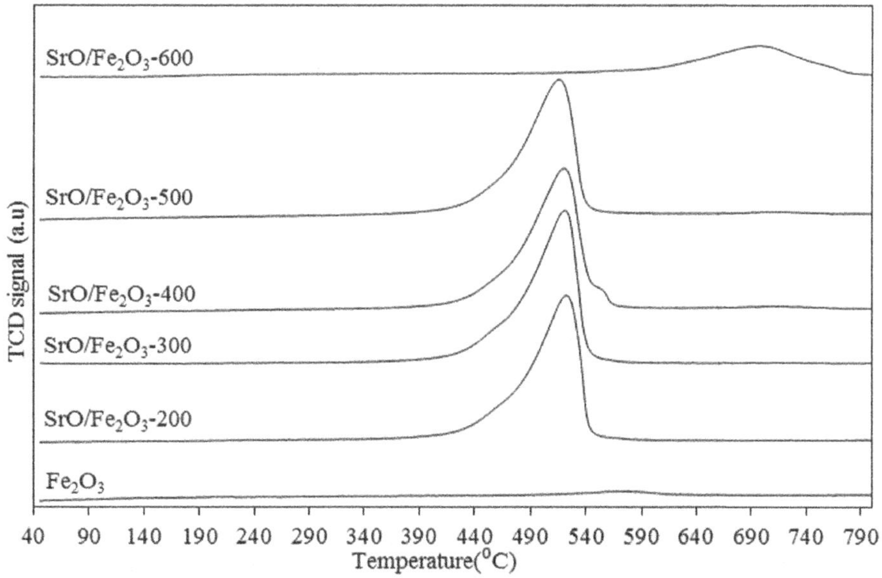

Fig. 6 TPD-CO_2 curves for bimetal composite SrO/Fe_2O_3 at different calcination temperature

amounts of water as OH group is difficult to avoid and difficult to detect even at UHV [52]. The presence of traces amount of OH group on SrO/Fe_2O_3-200 surfaces is chiefly responsible for the bicarbonate (HCO_3^-) formation.

CO_2 regeneration was conducted by using TPD-CO_2 analysis (Fig. 6) which shows the chemical shift of desorption temperature range (range$_{DT}$) to lower temperature (377–570 °C) compared to Fe_2O_3 only. It is noteworthy to take into account that modifying Fe_2O_3 surfaces with SrO generates lower desorption temperature. The desorption temperature was increased around 485–800 °C for SrO/Fe_2O_3-600. It might be due to the formation of different phase composition of alloy which can be formed at calcination temperature of 600 °C [53], as compared to the work reported by Bagherisereshki et al. [25] which requires high adsorption (1150 °C) and desorption (1235 °C) temperature for forward and reverse reaction in Eq. 14.

$$SrO\ (s) + CO_2(g) \nu SrCO_3 \tag{14}$$

Bimetal of SrO/Fe2O3 displays the feature of a surface alloy affecting the binding energies of adsorbates due to the kinetic site blocking effect. As a result, CO_2 binding energy on the alloy surfaces is reduced as compared to the single metal oxide [54]. The surface modification of Fe_2O_3 with addition of SrO is believed to be responsible for this phenomena. The interactions of bimetal for strontium iron oxide are proven by the XRD creating alternative active sites for CO_2 molecules, consequently, lowering its binding energy with CO_2 that attributed to the lower desorption temperature. Nevertheless, the range$_{DT}$ is observed as increasing (485–800 °C) with higher

Table 3 CO_2 chemisorption for the adsorbents

Adsorbents	Chemical adsorption			Reference
	Desorption temperature range, $Range_{DT}$, (°C)	Maximum desorption temperature, Max_{DT} (°C)	Adsorption capacity (mg/g)	
Fe_2O_3	180–231 464–644	208 575	0.12 3.83	This work
SrO	n.a	1235	400	22
SrO/Fe_2O_3-200	377–570	523	100.03	This work
SrO/Fe_2O_3-300	350–490	522	100.19	This work
SrO/Fe_2O_3-400	387–575	520	97.59	This work
SrO/Fe_2O_3-500	464–638	516	92.73	This work
SrO/Fe_2O_3-600	485–800	696	38.36	This work

n.a – not available

calcination temperature of 600 °C for SrO/Fe_2O_3-600. The increment presumed that it might be different active sites generated that causes the only formation of carbonate species with stronger binding energy. The active site of traces OH group was diminished around 300–400 °C [55] which leads to lower desorption temperature of HCO_3^- species [56]. The formation mechanism of the HCO_3^- species is generated by the chemical reaction between surface OH groups and CO_2, whereas between surface oxygen of the adsorbent and CO_2 is considered to produce strongly bonded of monodentate carbonate, bidentate carbonate and polydentate carbonate species [57]. This information supports adsorption capacity (Table 3) for SrO/Fe_2O_3-600 substantially decreasing (38.36 mg CO_2/g adsorbent) as compared to SrO/Fe_2O_3-300 (100.19 mg CO_2/g adsorbent) which is the highest among bimetal composite series. Although SrO was reported high adsorption capacity at 400 mg CO_2/g adsorbent, it is inefficient due to high energy consumption during adsorption and desorption [25]. Throughout the CO_2 physisorption and chemisorption, the most efficient adsorbent is SrO/Fe_2O_3-200 with both total adsorption capacities of 105.11 mg CO_2/g adsorbent and low calcination temperature.

3.3 Adsorption Isotherms Analysis

The significance of adsorption isotherm is anticipating and comparing the adsorption properties, describing the adsorption mechanism, and optimizing the adsorption system [58]. Thus, the CO_2 adsorption performance of the bimetal composite SrO/Fe_2O_3 can be described through the adsorption isotherm. The CO_2 uptake for SrO/Fe_2O_3-200 from CO_2 adsorption isotherm at 25 °C and 1 atm was analyzed using several models including Langmuir, Freundlich, D-R, and Temkin isotherm

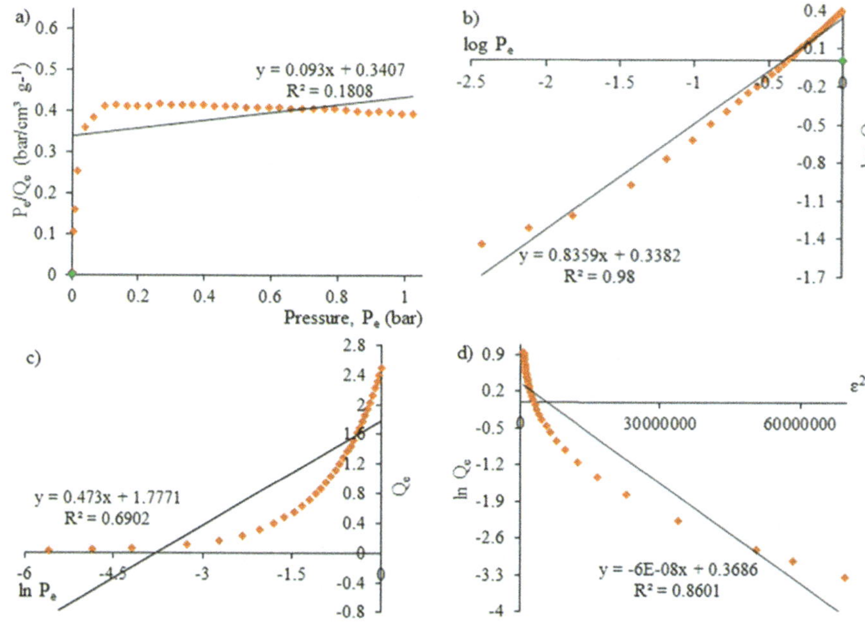

Fig. 7 Linearized form adsorption isotherms by **a** Langmuir, **b** Freundlich, **c** Temkin and **d** D-R models using CO_2 adsorption isotherms at 25 °C and 1 atm on the SrO/Fe_2O_3-200

model. Linearized form of the isotherms is shown in Fig. 7 and the parameters are tabulated in Table 4.

Langmuir adsorption is mainly created to best imagine the gas–solid phase adsorption and is also used to compare with approximately to determine the adsorptive capacity [59]. It is best describing the surface adsorption by taking into account dynamic equilibrium including relative rates of adsorption and desorption. The data is not well fitted with the Langmuir isotherm with a minimum value of $R^2 = 0.1808$. The values of theoretical adsorption quantity, Q_o and adsorption energy, K_L were obtained from the slope and intercepts of the plot. These values explain maximum adsorption referring to a saturated monolayer of adsorbate molecules on adsorbent surfaces without transmission of adsorbate in the plane of the adsorbent surface. The value of Q_o (10.753) deviates from the experimental Q_e value (2.59). The CO_2 adsorption is favorable based on the dimensionless separation factor, R_L value of 2.94×10^{-6} which is in the range of 0–1. This supporting Freundlich constant, n that lies between 1 and 2 as shown in Table 4.

The assumption of the Freundlich model is that multilayer adsorption occurs onto heterogeneous surfaces. The surface heterogeneity is depending on the physicochemical properties of active sites and adsorbed molecules [60]. As the number of molecules bound to adsorption sites increases, exponentially higher energy requires another molecule bind at the same adsorption sites which implies smaller chances of molecules bind at the same sites [61]. The high correlation coefficient ($R_2 = 0.98$)

Table 4 Adsorption isotherm parameters for CO_2 adsorption at 25 °C and 1 atm

Isotherm	Parameters	Values	$Q_{e\ exp.}$ (cm^3/g)
Langmuir	$Q_{\circ\ cal.}$ (cm^3/g)	10.753	2.59
$P_e/Q_e = 1/Q_\circ K_L + P_e/Q_\circ$	K_L	0.273	
P_e/Q_e versus P_e	R_L	2.94×10^{-6}	
	R^2	0.1808	
Freundlich	K_F ($cm^3/g \cdot bar^{1/n}$)	2.179	
$\log Q_e = 1/n \log P_e + \log K_F$	$1/n$, (n)	0.8359 (1.196)	
$\log Q_e$ versus $\log P_e$	R^2	0.9800	
Temkin	A_T (L/g)	42.82	
$Q_e = B \ln A_T + B \ln P_e$	b_T (J/mol)	5238	
Q_e versus $\ln P_e$	B	0.473	
	R^2	0.6902	
Dubinin-Radushkevich	Q_s (cm^3/g)	1.446	
$\ln Q_e = \ln Q_s - K_{ad}\varepsilon^2$	K_{ad} (mol^2/J^2)	6×10^{-8}	
$\ln Q_e$ vs. ε^2	E (J/mol)	2886.8	
	R^2	0.8601	

for the Freundlich isotherm suggests the best fit towards the experimental data over the ambient temperature and pressure. Freundlich intensity parameter of $1/n$ with the value of 0.8359 shows normal adsorption representing greater surface heterogeneity. The adsorption strength referring to the value of $1/n < 1$ indicates that the smaller $1/n$ is, greater the heterogeneity is expected, while $1/n > 1$ corresponds to driving force or cooperative adsorption [60, 62] which an adsorbate molecule adsorbed to adsorption site will facilitate next molecule to be adsorbed. The value of the Freundlich constant, n (1.196) is greater than 1 suggesting that the adsorption process is favorable. The constant K_f has a value of 2.179 (cm^3/g) which is closest to experimental adsorption data among other models, an indication of the adsorption affinity between CO_2 and SrO/Fe_2O_3-200.

Temkin isotherm model considers the effects of indirect interactions between the adsorbates in the adsorption. The assumption of the model derives the relation of surface coverage which is linearly indirect proportional with the heat of adsorption (ΔH_{ads}) for all molecules in the layer [63]. The bond constant (A_T) corresponds to the maximum binding energy, and heat of adsorption (b_T) are 42.82 L/g and 5238 J/mol respectively. According to the D-R isotherm, the D-R isotherm constant, K_{ad} is related to the mean energy of sorption as it is transferred from the adsorbate to the surface of the solid is 6×10^{-8} mol^2/J^2. The approach is usually applied to obtain mean free energy, E (2886.8 J/mol). The b_T and E values from Temkin and D-R isotherms respectively are below 8 kJ/mol which corresponds to the adsorption that is physical in nature [64]. The D-R isotherm parameter shows theoretical isotherm saturation capacity, Q_s (1.446 cm^3/g) deviate from measured adsorption quantity, Q_e

with R^2 value of 0.8601 which fitted well to experiment adsorption data rather than Temkin isotherm ($R^2 = 0.6902$).

By referring to R^2 value, the Freundlich isotherm model is higher and more suitable to describe the adsorption process than the Langmuir isotherm model. Therefore, it implies that a multilayer CO_2 adsorption occurs on heterogeneous surfaces of SrO/Fe_2O_3-200 and does not reflect a chemisorption reaction by monolayer formation. The Temkin adsorption quantity deviates from measured adsorption quantity and R^2 value shows the model did not fitted well. Though D-R and Freundlich isotherms show a good correlation coefficient, Freundlich adsorption quantity was closest compared to D–R. It can be concluded that Freundlich isotherm is the best fit among other isotherms.

4 Conclusion

Bimetal composite of SrO/Fe_2O_3 was successfully synthesized by using wet impregnation method. Physicochemical characterization of the adsorbents was performed to identify the metal composition and textural properties. The most efficient adsorbent of SrO/Fe_2O_3-200 for CO_2 capture was determined by considering the physisorption and chemisorption capacities of 5.08 and 100.03 mg CO_2/g adsorbent. It was found that the regeneration of CO_2 discovers the chemical shift to lower desorption temperature at around 370–570 °C. The experimental adsorption data fitted to various isotherm models such as Langmuir, Freundlich, D-R, and Temkin isotherms. It is noteworthy to find that Freundlich isotherm is to be the best fit with the highest regression value and closest theoretical adsorption quantity. This implies the CO_2 adsorption on the heterogeneity surface of SrO/Fe_2O_3-200 by multilayer adsorption. Consequently, the bimetal composite of SrO/Fe_2O_3-200 is a potential sorbent for CO_2 reduction and anthropogenic greenhouse gas remediation.

Acknowledgements The authors wish to express gratitude for the research work supported by Universiti Putra Malaysia (UPM) under GP-IPM/2018/9657200, GP-IPM/2020/9683100 and GP-IPB/2019/9671302.

References

1. Kroeze G, Pulles T (2015) The importance of non-CO_2 greenhouse gases. J Integr Environ Sci 12:1–4. https://doi.org/10.1080/1943815X.2015.1118131
2. National Oceanic and Atmospheric Administration, Earth System Research Laboratory, Global Monitoring Division, USA. https://www.esrl.noaa.gov/gmd/ccgg/trends/global.html
3. CO_2 emissions, The World Bank Data. https://data.worldbank.org/indicator/EN.ATM.CO2E.PC
4. Ekwurzel B, Boneham J, Dalton MW, Heede R, Mera RJ, Allen MR, Frumhoff PC (2017) The rise in global atmospheric CO_2, surface temperature, and sea level from emissions traced

to major carbon producers. Clim Change 144:579–590. https://doi.org/10.1007/s10584-017-1978-0
5. Blomen E, Hendriks C, Neele F (2009) Capture technologies: improvements and promising developments. Energy Procedia 1:1505–1512. https://doi.org/10.1016/j.egypro.2009.01.197
6. Irlam L (2017) Global costs of carbon capture and storage. Global CCS Institute
7. Zero Emissions Resource Organization. https://www.zeroCO2.no
8. De Guido G, Compagnoni M, Pellegrini LA, Rossetti I (2018) Mature *versus* emerging technologies for CO_2 capture in power plants: key open issues in post-combustion amine scrubbing and in chemical looping combustion. Front Chem Sci Eng 12:315–325. https://doi.org/10.1007/s11705-017-1698-z
9. Leung DYC, Caramanna G, Maroto-Valer MM (2014) An overview of current status of carbon dioxide capture and storage technologies. Renew Sustain Energy Rev 39:426–443. https://doi.org/10.1016/j.rser.2014.07.093
10. Rochelle GT (2009) Amine scrubbing for CO_2 capture. Science 325:1652–1654. https://doi.org/10.1126/science.1176731
11. Barzagli F, Mani F, Peruzzini M (2016) A Comparative study of the CO_2 absorption in some solvent-free alkanolamines and in aqueous monoethanolamine (MEA). Environ Sci Technol 50:7239–7246. https://doi.org/10.1021/acs.est.6b00150
12. Zhu J, He B, Huang J, Li C, Ren T (2018) Effect of immobilization methods and the pore structure on CO_2 separation performance in silica-supported ionic liquids. Micropor Mesopor Mater 260:190–200. https://doi.org/10.1016/j.micromeso.2017.10.035
13. Ahn CK, Lee HW, Chang YS, Han KW, Kim JY, Rhee CH, Chun HD, Lee MW, Park JM (2011) Characterization of ammonia-based CO_2 capture process using ion speciation. Int J Greenh Gas Con 5:1606–1613. https://doi.org/10.1016/j.ijggc.2011.09.007
14. Abu Tahari MN, Hakim A, Tengku Azmi TSM, Wan Isahak WNR, Hisham MWM, Yarmo MA (2016) Studies on adsorption-desorption of CO_2 by long chain fatty amine supported on SiO_2. Mater Sci Forum 840:343–347. https://doi.org/10.4028/www.scientific.net/MSF.840.343
15. Lahuri AH, Adnan R, Mansor MH, Tajudeen NFW, Nordin N (2020) Adsorption kinetics for carbon dioxide capture using bismuth(III) oxide impregnated on activated carbon. Malays J Chem 22:33–46
16. Lahuri AH, Michael Ling NK, Rahim AA, Nordin N (2020) Adsorption kinetics for CO_2 capture using cerium oxide impregnated on activated carbon. Acta Chim Slov 67:570–580. https://doi.org/10.17344/acsi.2019.5572
17. Ho TH, Howes T, Bhandari BR (2014) Encapsulation of gases in powder solid matrices and their applications: a review. Powder Technol 259:87–108. https://doi.org/10.1016/j.powtec.2014.03.054
18. Ishibe T, Sato T, Hayashi T, Kato N, Hata T (1995) Absorption of nitrogen dioxide and nitric oxide by soda lime. Br J Anaesth 75:330–333. https://doi.org/10.1093/bja/75.3.330
19. Florin N, Fennell P (2011) Synthetic CaO-based sorbent for CO_2 capture. Energy Procedia 4:830–838. https://doi.org/10.1016/j.egypro.2011.01.126
20. Berstad D, Anantharaman R, Jordal K (2012) Post-combustion CO_2 capture from a natural gas combined cycle by $CaO/CaCO_3$ looping. Int J Greenh Gas Con 11:25–33. https://doi.org/10.1016/j.ijggc.2012.07.021
21. Naeem MA, Armutlulu A, Imtiaz Q, Muller CR (2017) CaO-based CO_2 sorbents effectively stabilized by metal oxides. ChemPhysChem 18:3280–3285. https://doi.org/10.1002/cphc.201700695
22. Duan YH, Sorescu DC (2010) CO_2 capture properties of alkaline earth metal oxides and hydroxides: a combined density functional theory and lattice phonon dynamics study. J Chem Phys 133:074508. https://doi.org/10.1063/1.3473043
23. Arlt W, Wasserscheid P (2011) Storing and transporting energy, comprises e.g. using one-component system of A and B in a chemical reaction under heat absorption and removal of carbon dioxide to form an energy-rich compound B. https://www.google.com/patents/DE102010009543A1?cl=en

24. Miccio F, Murri AN, Landi E (2016) High-temperature capture of CO_2 by strontium oxide sorbents. Ind Eng Chem Res 55:6696–6707. https://doi.org/10.1021/acs.iecr.6b00184
25. Bagherisereshki E, Tran J, Lei FQ, Yeung NA (2018) Investigation into SrO/$SrCO_3$ for high temperature thermochemical energy storage. J Sol Energy 160:85–93. https://doi.org/10.1016/j.solener.2017.11.073
26. Mendoza EYM, Santos AS, Lopes EV, Drozd V, Durygin A, Chen JH, Saxena SK (2019) Iron oxides as efficient sorbents for CO_2 capture. J Mater Res Technol 8:2944–2956. https://doi.org/10.1016/j.jmrt.2019.05.002
27. Hakim A, Marliza TS, Abu Tahari MN, Wan Isahak WNR, Yusop MR, Hisham MWM, Yarmo MA (2016) Studies on CO_2 adsorption and desorption properties from various types of iron oxides (FeO, Fe_2O_3, and Fe_3O_4). Ind Eng Chem Res 55:7888–7897. https://doi.org/10.1021/acs.iecr.5b04091
28. Hakim A, Yarmo MA, Marliza TS, Abu Tahari MN, Samad WZ, Yusop MR, Hisham MWM, Dzakaria N (2016) The influence of calcination temperature on iron oxide (α-Fe_2O_3) towards CO_2 adsorption prepared by simple mixing method. Malaysian J Anal Sci 20:1286–1298. https://doi.org/10.17576/mjas-2016-2006-07
29. Hakim A, Marliza TS, Abu Tahari MN, Yusop MR, Hisham MWM, Yarmo MA (2016) Development of α- Fe_2O_3 as adsorbent and its effect on CO_2 capture. Mater Sci Forum 840:421–426. https://doi.org/10.4028/www.scientific.net/MSF.840.421
30. Saisuwansiri P, Worathanakul P (2019) A study of CO_2 thermodynamic adsorption and desorption with bi-metal loading on zeolite Y. Mater Today 17:1458–1465. https://doi.org/10.1016/j.matpr.2019.06.168
31. Austin N, Butina B, Mpourmpakis G (2016) CO_2 activation on bimetallic CuNi nanoparticles. Pro Nat Sci-Mater 26:487–492. https://doi.org/10.1016/j.pnsc.2016.08.007
32. Lahuri AH, Yarmo MA, Marliza TS, Abu Tahari MN, Samad WZ, Dzakaria N, Yusop MR (2017) Carbon dioxide adsorption and desorption study using bimetallic calcium oxide impregnated on iron(III) oxide. Mater Sci Forum 888:479–484. https://doi.org/10.4028/www.scientific.net/MSF.888.479
33. Zeldowitsh J (1935) On the theory of the Freundlich adsorption isotherm. Acta Physicochimica URSS 1:961–974
34. Freundlich H (1926) Colloid and capillary chemistry, 3rd edn. Methuen, London, pp 114–122
35. Brunauer S, Emmett PH, Teller E (1938) Adsorption of gases in multimolecular layers. J Am Chem Soc 60:309–319. https://doi.org/10.1021/ja01269a023
36. Hameed BH, Din ATM, Ahmad AL (2007) Adsorption of methylene blue onto bamboo-based activated carbon: kinetics and equilibrium studies. J Hazard Mater 141:819–825. https://doi.org/10.1016/j.jhazmat.2006.07.049
37. Gunay A, Arslankaya E, Tosun I (2007) Lead removal from aqueous solution by natural and pretreated clinoptilolite: adsorption equilibrium and kinetics. J Hazard Mater 146:362–371. https://doi.org/10.1016/j.jhazmat.2006.12.034
38. Webber TN, Chakravarti RK (1974) Pore and solid diffusion models for fixed bed adsorber. J Am Inst Chem Eng 20:228–238. https://doi.org/10.1002/aic.690200204
39. Hutson ND, Yang RT (1997) Theoretical basis for the Dubinin-Radushkevitch (D-R) adsorption isotherm equation. Adsorption 3:189–195. https://doi.org/10.1007/BF01650130
40. Voudrias E, Fytianos F, Bozani E (2002) Sorption description isotherms of dyes from aqueous solutions and wastewaters with different sorbent materials. Global Nest Int J 4:75–83
41. Temkin MI, Pyzhev V (1940) Kinetics of ammonia synthesis on promoted iron catalyst. Acta Physicochimica URSS 12:327–356
42. Shahbeig H, Bagheri N, Ghorbanian SA, Hallajisani A, Poorkarimi S (2013) A new adsorption isotherm model of aqueous solutions on granular activated carbon. World J Model Simulat 9:243–254
43. Elebi OC, Uzum C, Shahwan T, Erten HN (2007) A radiotracer study of the adsorption behavior of aqueous Ba^{2+} ions on nanoparticles of zero-valent iron. J Hazard Mater 148:761–767. https://doi.org/10.1016/j.jhazmat.2007.06.122

44. Cotoruelo LM, Marques MD, Diaz FJ, Rodriguez-Mirasol J, Rodriguez JJ, Cordero T (2012) Adsorbent ability of lignin-based activated carbons for the removal of p-nitrophenol from aqueous solutions. Chem Eng J 184:176–183. https://doi.org/10.1016/j.cej.2012.01.026
45. Azeez SO (2016) Sorption of 4-nitroaniline on activated kaolinitic clay and jatropha curcas activated carbon in aqueous solution. Jordan J Chem 11:128–145
46. El-Sadaawy M, Abdelwahab O (2014) Adsorptive removal of nickel from aqueous solutions by activated carbons from doum seed (*Hyphaenethebaica*) coat. Alex Eng J 53:399–408. https://doi.org/10.1016/j.aej.2014.03.014
47. Mobasherpour I, Salahi E, Pazouki M (2011) Removal of nickel (II) from aqueous solutions by using nano-crystalline calcium hydroxyapatite. J Saudi Chem Soc 15:105–112. https://doi.org/10.1016/j.jscs.2010.06.003
48. Mdlalose WD, Mokhosi SR, Dlamini S, Moyo T, Singh M (2018) Effect of chitosan coating on the structural and magnetic properties of $MnFe_2O_4$ and $Mn0.5Co_{0.5}Fe_2O_4$ nanoparticles. AIP Adv 8:056726. https://doi.org/10.1063/1.5007760
49. Sing KSW, Everett DH, Haul RAW, Moscou L, Pierotti RA, Rouquerol J, Siemieniewska T (1985) Reporting physisorption data for gas/ solid systems with special reference to the determination of surface area and porosity. Pure Appl Chem 54:603–619. https://doi.org/10.1351/pac198254112201
50. Condon JB (2006) Surface area and porosity determinations by physisorption measurements and theory, 1st edn. Elsevier, UK, pp 1–28
51. Hakim A, Abu Tahari MN, Marliza TS, Wan Isahak WNR, Yusop MR, Wahab MWM, Yarmo MA (2015) Study of CO_2 adsorption and desorption on activated carbon supported iron oxide by temperature programmed desorption. Jurnal Teknologi 77:75–84. https://doi.org/10.11113/jt.v77.7010
52. Burghaus U (2014) Surface chemistry of CO_2—adsorption of carbon dioxide on clean surfaces at ultrahigh vacuum. Pro Surf Sci 89:161–217. https://doi.org/10.1016/j.progsurf.2014.03.002
53. Illarionov AG, Shchetnikov NV, Illarionova SM, Popov AA (2017) Effect of heating temperature on the formation of structure and phase composition of a biocompatible alloy Ti–6Al–4V–ELI subjected to equal-channel angular pressing. Phys Met Metallogr 118:272–278. https://doi.org/10.1134/S0031918X1703005X
54. Burghaus U (2009) Surface science perspective of carbon dioxide chemistry-adsorption kinetics and dynamics of CO_2 on selected model surfaces. Catal Today 148:212–220. https://doi.org/10.1016/j.cattod.2009.07.082
55. Ouchabi M, Baalala M, Elaissi A, Bensitel M (2016) Effect of calcination temperature on the structure of copper orthophosphates and their catalytic activity in the decomposition of 2-propanol. J Mater Environ Sci 7:1417–1424
56. Yoshikawa K, Sato H, Kaneeda M, Kondo JN (2014) Synthesis and analysis of CO_2 adsorbents based on cerium oxide. J CO_2 Util 8:34–38. https://doi.org/10.1016/j.jcou.2014.10.001
57. Binet C, Daturi M, Lavalley JC (1999) IR study of polycrystalline ceria properties in oxidised and reduced states. Catal Today 50:207–225. https://doi.org/10.1016/S0920-5861(98)00504-5
58. Chen X (2015) Modeling of experimental adsorption isotherm data. Information 6:14–22. https://doi.org/10.3390/info6010014
59. Elmorsi TM (2011) Equilibrium isotherms and kinetic studies of removal of methylene blue dye by adsorption onto miswak leaves as a natural adsorbent. J Environ Prot 2:817–827. https://doi.org/10.4236/jep.2011.26093
60. Ammendola P, Raganati F, Chirone R (2017) CO_2 adsorption on a fine activated carbon in a sound assisted fluidized bed: thermodynamics and kinetics. Chem Eng J 322:302–313. https://doi.org/10.1016/j.cej.2017.04.037
61. Rashidi NA, Yusup S (2017) Potential of palm kernel shell as activated carbon precursors through single stage activation technique for carbon dioxide adsorption. J Clean Prod 168:474–486. https://doi.org/10.1016/j.jclepro.2017.09.045
62. Dada AO, Olalekan AP, Olatunya AM, Dada O (2012) Langmuir, Freundlich, Temkin and Dubinin-Radushkevich isotherms studies of equilibrium sorption of Zn^{2+} onto phosphoric acid modified rise husk. IOSR—JAC 3:38–45

63. Ringot D, Lerzy B, Chaplain K, Bonhoure JP, Auclair E, Larondelle Y (2007) In vitro biosorption of ochratoxin a on the yeast industry by-products: comparison of isotherm models. Bioresour Technol 98:1812–1821. https://doi.org/10.1016/j.biortech.2006.06.015
64. Djomgoue P, Siewe M, Djoufac E, Kenfack P, Njopwouo D (2012) Surface modification of Cameroonian magnetite rich clay with Eriochrome Black T. Application for adsorption of nickel in aqueous solution. Appl Surf Sci 258:7470–7479. https://doi.org/10.1016/j.apsusc.2012.04.065

Resin Capacity of Gold Adsorption in Hydro Metallurgical Process

Imam Santosa, Anwaruddin Hisyam, Ainaa Izyan Nafsun, and Shukeri Ismail

Abstract The changing capacity of gold adsorption with new and reuse resins in hydro metallurgical process was studied with Purolite and Dowex Minix resin. Fresh resin or regenerated resin is put into 1000 ml of gold mother solution and mixed, and the samples are taken for every 10 min. As a results adsorption capacity for fresh Purolite resin logarithmically increased in 70 min to reach 95.3% capacity and 34.3% regenerated Purolite resin adsorption capacity. Adsorption capacity for fresh Minix resin was 95.7% capacity and barren Minix resin adsorption capacity was 71.9%. Resin Metals loaded after elution process showed Dowex Minix resin had higher Au concentration and Au selectivity in adsorption process than Purolite resin. The total metal adsorption capacity of Purolite was higher than the Dowex Minix resin. Purolite resin seemingly has faster loading rates compared to Dowex Minix resin. Purolite resin seems stronger base resin, higher load, and more difficult for elute than Dowex Minix.

Keywords Resin capacity · Gold · Adsorption

1 Introduction

Adsorption is an equilibrium-diffusion–reaction process. By contacting fluids with adsorbent, the desired objective of separation may be achieved [1]. Adsorption is simple-efficient operation and low cost that suitable for gold recovery [2]. Gold adsorption can perform a lot of separations. Low gold concentration in the leach solution is not suitable with membrane, evaporation, or distillation separation [3]. Some commonly known gold recovery and extraction methods are precipitation, ion

I. Santosa · A. Hisyam (✉) · A. I. Nafsun · S. Ismail
Faculty of Chemical and Natural Resources Engineering, Universiti Malaysia Pahang, 26300 Gambang, Pahang, Malaysia
e-mail: ahisyam@ump.edu.my

I. Santosa · A. Hisyam
Chemical Engineering Department, Ahmad Dahlan University, Ring Road Selatan, Yogyakarta 55166, Indonesia

M. A. A. Zaini et al. (eds.), *Proceedings of the 3rd International Conference on Separation Technology*, Lecture Notes in Mechanical Engineering,
https://doi.org/10.1007/978-981-16-0742-4_13

exchange, solvent extraction, and flotation. These recovery methods are correlated with several benefits and disadvantages. Selecting for a proper method as the optimal technique generally depends on the cuts of the stone, the presence of foreign metal ions, and the type of ores [4–8]. It can be learned from the literatures that, in comparison to the mentioned techniques, resin adsorption is the most effective technique for gold recovery due to its rapid adsorption, high loading capacity, simultaneous elution and regeneration, and ambient temperature activity [9, 10].

Some absorbents were researched to recover gold from the leach solution but showed poor adsorption capacity and selectivity [11]. Minix, a strong-base-resins was produced by Mintex for gold recovery from cyanide solutions [12]. Dowex Minix resins have better selectivity for gold than activated carbon [13]. Purolite A860S has a polyacrylate matrix [14]. The budget estimate for adsorption operation needs quick experimental feasibility test. Some research key for selecting adsorbents are selectivity, capacity, adsorption rate, and regeneration ability [3].

This study investigated Purolite resin activity and Dowex Minix resin activity for their availability to be used in the adsorption process.

2 Experiments

Fresh Purolite resin (3.6 ml) as a standard resin are used for the test. Regenerated Purolite (33,5 ml) and washed regenerated Purolite resin (3.5 ml) are used for the test. Every resin sample was put into 1000 ml of gold mother solution (39.486 ppm Au with 500 ppm CN) in the container.

Fresh Minix resin (3.6 ml) as a standard are used for the test. Barren Minix resin, 3.5 ml are used for the test. Every resin sample was put into 1000 ml of gold mother solution (26.660 ppm Au with 500 ppm CN) in the container. The solution was mixed at 200 rpm, 10 ml of every solution sample are taken for every 10 min. Au ion concentration determined with AAS.

The adsorption capacity of gold on the resin (%) was calculated on the basis of the expression:

$$R = \frac{(C_o - C_t)}{C_0} \times 100\% \qquad (1)$$

where R is the adsorption capacity of gold on resin, while C_o and C_t are the gold concentration in the solution at the time 0 (initial) and t, respectively.

The adsorption capacity about the adsorbent was calculated on the basis of the expression:

$$q_t = \frac{(C_o - C_t)V_o}{m_o} \qquad (2)$$

where q_t is the adsorption capacity, C_0 and C_t are the gold concentration in the solution at the time 0 (initial) and t, respectively, V_o is the volume of cyanide solution (mL), and m_o is the mass of adsorbent (grams).

3 Results and Discussion

It was identified in Fig. 1 that the adsorption capacity (recovery rate) of fresh Purolite resin logarithmically increased in 70 min to reach 95.3% capacity. At the initial process, Au ion adsorption was rapid because a lot of adsorption sites were empty. When residence time was increased, Au-ion adsorption was decrease until approaches equilibrium. Regenerated Purolite resin from stripping products cannot be reused directly because the adsorption capacity is only logarithmically increased to 34.3% in 70 min. Regenerated Purolite resin must be washed with $H2SO_4$ 1 M after the stripping process before it is reused for the adsorption process. It logarithmically increased adsorption capacity to 87% in 70 min.

From Fig. 2, adsorption capacity versus times for recovery rate raised with raised time. It happened when the sorbent-to-metal ratio is high [15, 16]. Purolite fresh resin and Dowex Minix fresh resin showed that the gold adsorption capacity of Purolite resin as the same as Dowex Minix fresh resin, although the gold concentration in solution of Purolite resin is 50% higher than the gold concentration in solution of Dowex Minix resin. Purolite resin has faster loading rates compared to Dowex Minix resin Dowex Minix fresh resin adsorption capacity logarithmically increased adsorption capacity to 95.7% in 70 min. It showed that gold adsorption capacity Dowex Minix resin higher than Purolite resin.

From Fig. 3, adsorption capacity barren Dowex Minix resin was higher than regenerated Purolite resin. Dowex Minix barren resin adsorption capacity logarithmically increased adsorption capacity to 71.9% in 70 min. It showed that after elution, Dowex

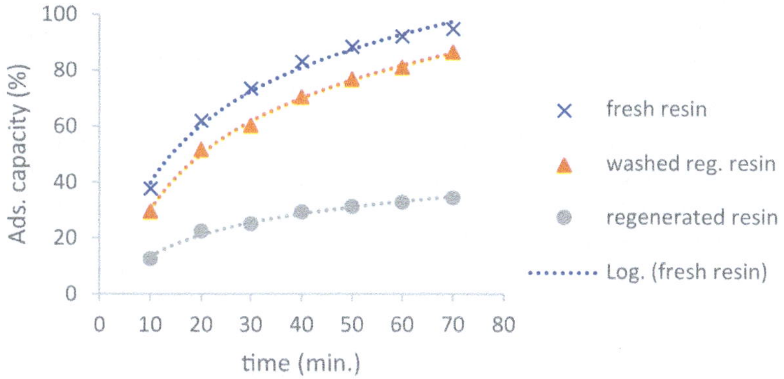

Fig. 1 Plot adsorption capacity versus times for recovery rate Purolite resin

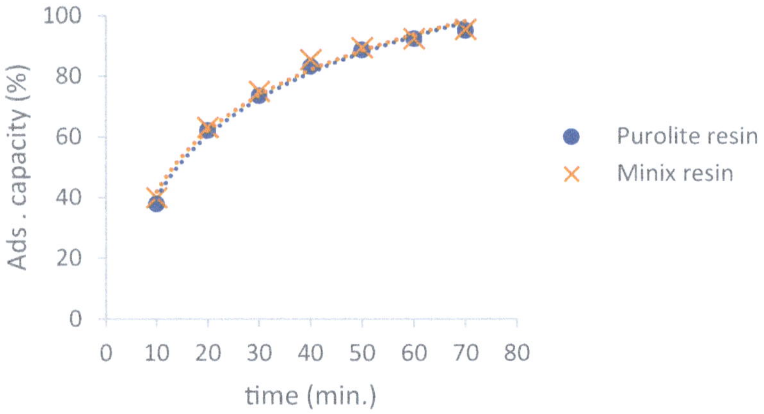

Fig. 2 Adsorption capacity versus times for recovery rate fresh Purolite resin and fresh Dowex Minix resin

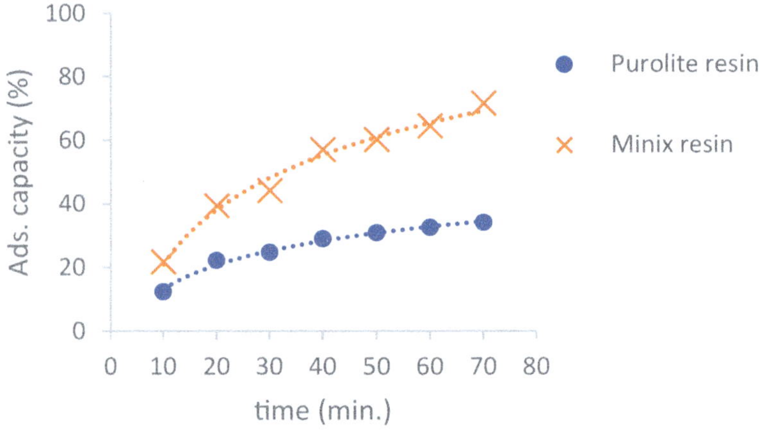

Fig. 3 Plot adsorption capacity versus times for recovery rate regenerated Purolite resin and barren Dowex Minix resin

Minix resin was a cleaner from metal residual than Purolite resin. Metal residual in generates Purolite resin sharply, decreasing gold loading in the next adsorption process. The generated Purolite resin needed to wash with acid before reuse in the adsorption process to improve loading capacity.

Gold loading capacity versus times for recovery rate Dowex Minix fresh resin and Dowex Minix barren resin logarithmically increased in 70 min. Dowex Minix barren resin had lower adsorption capacity than Dowex Minix fresh resin showed that after elution, Dowex Minix resin had residual metal impurities (Fig. 4).

Gold loading capacity rate versus times for Dowex Minix resin in Fig. 5 explained that adsorption rate Dowex Minix fresh resin and Dowex Minix barren resin fresh

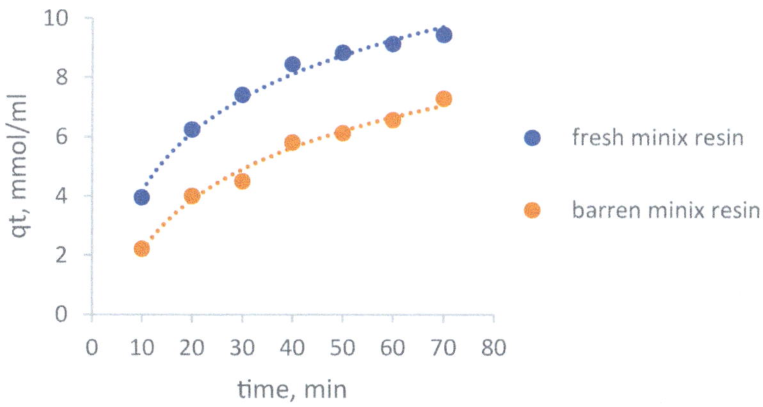

Fig. 4 Gold loading capacity versus times for recovery rate Dowex Minix resin

Fig. 5 Gold loading capacity rate versus times for Dowex Minix resin

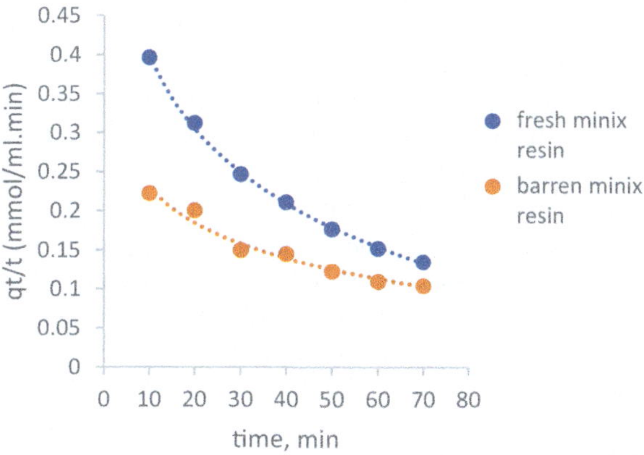

drop logarithmically. Because Dowex Minix resin had metal ion residual impurities, in the early process, its adsorption rate lower than fresh resin Dowex Minix. The longer adsorption was processed, the closer adsorption rate value. Ultimately the adsorption rate value is the same, so the adsorption capacity of Dowex Minix resin is fixed. Dowex Minix resin might be reuse without washing.

Data Table 1 showed Dowex Minix resin had higher Au concentration and Au selectivity in the adsorption process than Purolite resin. Fe loaded was very high because purolite has an affinity for ferrocyanide compounds [14]. Total metal adsorption capacity Purolite apparently higher than Dowex Minix resin. In general, Purolite resin seemingly has faster loading rates compared to Dowex Minix resin. Purolite resin seems stronger base resin, higher load, and more difficult for elute than Dowex Minix [17, 18].

Table 1 Resin Metals loaded after the adsorption process

Type of resin	Sample description	Concentration (ppm)			
		Au	Cu	Fe	Zn
Reg. Purolite	Head Assay	477.75	4807.50	8604.00	66,850.50
Reg. Purolite	After Acid Washing	541.25	6143.94	7929.88	10,801.38
Barren Dowex Minix	Head Assay	626.50	1047.88	194.13	1203.38
Barren Dowex Minix	After Acid Washing	710.38	1310.25	359.00	496.00

4 Conclusions

The adsorption capacity of fresh Purolite resin logarithmically increased in 70 min to reach 95.3% capacity. The regenerated resin from stripping product cannot be reused directly because the adsorption capacity is low. The regenerated resin must be washed before it is reused for the adsorption process. Dowex Minix fresh resin adsorption capacity logarithmically increased the adsorption capacity to 95.7% in 70 min. It showed that gold adsorption capacity Dowex Minix resin is higher than Purolite resin. After elution, Dowex Minix resin was cleaner from metal residual than Purolite resin. Resin Metals loaded after the elution process showed Dowex Minix resin had higher Au concentration and Au selectivity in the adsorption process than Purolite resin. The total metal adsorption capacity Purolite is apparently higher than the Dowex Minix resin. In general, Purolite resin seemingly has faster loading rates compared to Dowex Minix resin. Purolite resin is seems stronger base resin, higher load and more difficult for elute than Dowex Minix.

Acknowledgements This research was supported by UMP and UAD. We thank ICoST 2020 committee for their wisdom and we thank 3 "anonymous" reviewers for their so-called insights.

References

1. Tien C (2019) Introduction to adsorption basics, analysis, and application, Elsevier. Raderwerg, Amsterdam, pp 1–5
2. Li M, Zheng J, Zeng Q, Shang X (2016) Thermodynamic and column studies of Au(I) adsorption on weak alkaline anion exchange fiber from cyanide leaching solution. Sep Sci Tech 51(7):1166–1175. https://doi.org/10.1080/01496395.2016.1148727
3. Albright IF, Knaebel KS (2009) Albright's chemical engineering handbook: adsorption. Taylor & Francis Group, Boca Raton, pp 1119–1137
4. Awual MR, Ismael M (2014) Efficient gold(III) detection, separation and recovery from urban mining waste using a facial conjugate adsorbent. Sens Actuators B Chem 196:457–466. https://doi.org/10.1016/j.snb.2014.02.055

5. Behbahani M, Najafi F, Amini MM, Sadeghi O, Bagheri A, Hassanlou PG (2014) Solid phase extraction using nanoporous MCM-41 modified with 3,4-dihydroxybenzaldehyde for simultaneous preconcentration and removal of gold(III), palladium(II), copper(II) and silver(I). J Ind Eng Chem 20:2248–2255. https://doi.org/10.1016/j.jiec.2013.09.057
6. Jha MK, Gupta D, Lee J, Kumar V, Jeong J (2014) Solvent extraction of platinum using amine based extractants in different solutions: a review. Hydrometallurgy 142:60–69. https://doi.org/10.1016/j.hydromet.2013.11.009
7. Zhang T, Gregory K, Hammack RW, Vidic RD (2014) Co-precipitation of radium with barium and strontium sulfate and its impact on the fate of radium during treatment of produced water from unconventional gas extraction. Environ Sci Technol 48:4596–4603. https://doi.org/10.1021/es405168b
8. Wilson AM, Bailey PJ, Tasker PA, Turkington JR, Grant RA, Love JB (2014) Solvent extraction: the coordination chemistry behind extractive metallurgy. Chem Soc Rev 43:123–134. https://doi.org/10.1039/c3cs60275c
9. Calmon C (2018) Ion exchange pollution control. CRC Press, Boca Raton. https://doi.org/10.1201/9781351073868
10. Khaliq A, Rhamdhani M, Brooks G, Masood S (2014) Metal extraction processes for electronic waste and existing industrial routes: a review and Australian perspective. Resources 3:152–179. https://doi.org/10.3390/resources3010152
11. Wang X, Xu J, Li L, Li H, Yang C (2017) Thiourea grafted PVDF affinity membrane with narrow pore size distribution for Au (III) adsorption: Preparation, characterization, performance investigation and modeling. Chem Eng J 314:700–713. https://doi.org/10.1016/j.cej.2016.12.035
12. Pilsniak M, Trochimczuk AW (2014) Selective recovery of gold on functionalized resins. Hydrometallurgy 146:111–118. https://doi.org/10.1016/j.hydromet.2014.03.016
13. Deventer VJ, Wyethe JP, Kotze MH, Shannon J (2000) Comparison of resin-in-solution and carbon-in-solution for the recovery of gold from clarified solutions. The J South African Inst Mining Metall 221–227. Corpus ID: 104227598
14. Versiane AL, Virgínia ST, Ciminelli (2000) Application of ion exchange resins in gold hydrometallurgy. a tool for cyanide recycling. Solvent Extr, Ion Exch 18(3):567–582. https://doi.org/10.1080/07366290008934698
15. Morcali MH, Zeytuncu B, Yucel O (2013) Platinum uptake from chloride solutions using biosorbents. Mater Res 16:528–538. https://doi.org/10.1590/S1516-14392013005000006
16. Tasdelen C, Aktas S, Acma E, Guvenilir Y (2009) Gold recovery from dilute gold solutions using DEAE-cellulose. Hydrometallurgy 96:253–257. https://doi.org/10.1016/j.hydromet.2008.10.006
17. Marsden J, House I (2006) The chemistry of gold extraction, SME. Colorado, USA, pp 337–338
18. Zhang H, Dreisinger DB (2004) The recovery of gold from ammoniacal thiosulfate solutions containing copper using ion exchange resin columns. Hydrometallurgy 72:225–234. https://doi.org/10.1016/S0304-386X(03)00183-X

Heavy Metal Adsorbent of Carbon from Sago Liquid Biowaste for Sustainable Technology

Zainab Ngaini, Rafeah Wahi, Hasnain Hussain, Nurul Qhalila Bahrin, and Nur Hanani Hasana

Abstract Sarawak is one of the world's largest exporter of sago flour, from which the processing leads to a generation of biowaste in a significant amount. Thus, utilization of the biowaste is crucial to create a zero-waste sago processing industry. In this work, the heavy metal adsorbent was prepared from sago activated sludge via microwave technology. Sago effluent was treated via an activated sludge process to produce biomass, followed by microwave pyrolysis and chemical activation using NaOH. The efficiency of the adsorbent for adsorption of Cr, Pb and Zn in aqueous solution was studied at pH 2, contact time (24 h), adsorbent dosage (0.2–1 g/50 mL), and initial concentration (5–25 mg/L). Physicochemical analyses showed that the adsorbent has an average pore size of 36.29 μm and BET surface area of 471.1 m^2/g. The maximum removal of heavy metals was: Pb (89.8%), Cr (47.0%) and Zn (18.4%) at adsorbent dosage (1 g/50 mL), initial concentration (5 mg/L), mixing speed (150 rpm) and contact time (24 h). The Langmuir and Freundlich isotherm studies showed that Qe for Pb removal by sludge activated carbon was 3.202×10^{-3} mg/g. The results indicated the potential application of sago activated carbon for the removal of heavy metals, especially Pb from wastewater. Further isotherm study for the occurrence of chemisorptions process could be beneficial, which at the same creating a zero-waste sago processing industry for sustainable technology.

Keywords Adsorption · Biowaste · Heavy metal · Microwave pyrolysis · Sago effluent · Sago waste

1 Introduction

Sago (*Metroxylon sagu*) is an indigenous plant of Sarawak, Malaysia and easily grown in freshwater swamps and tropical lowland forests. Malaysia is one of the largest producers of sago flour in the world with annual export est. 41,000–48,000

Z. Ngaini (✉) · R. Wahi · H. Hussain · N. Q. Bahrin · N. H. Hasana
Faculty of Resource Science and Technology, Universiti Malaysia Sarawak, 94300 Kota Samarahan, Sarawak, Malaysia
e-mail: nzainab@unimas.my

M. A. A. Zaini et al. (eds.), *Proceedings of the 3rd International Conference on Separation Technology*, Lecture Notes in Mechanical Engineering,
https://doi.org/10.1007/978-981-16-0742-4_14

tonnes. A significant amount of solid and liquid biowaste is generated during sago flour processing—most are left unutilized or improperly disposed of. Thus, sago biowaste is a potential source for utilization, with an estimated annual generation of more than 300 tonnes of solid sago biowaste in Sarawak [1]. Biowastes from the sago industry, for example, sago *hampas* (final residue generated at the end of sago flour processing) and sago bark are commonly left to degrade in the vicinity of the mills while the sago effluent is commonly discharged into the river. Sago industries consumed an estimated 30,000 L of water for every 1 tonne of sago processed, thus, generating a vast amount of sago effluent. Sago effluent is categorized as wastewater and contained high biochemical oxygen demand (BOD), chemical oxygen demand (COD) [8] and other nutrients [3]. The effluent also contains a high total suspended solids (TSS) [9] and acidic in nature with high organic matter, unpleasant odor and irritating color [10]. The traditional practice is to discharge the effluent into a nearby river, which is meant to allow natural decomposition of the effluent by microorganisms such as fungi and bacteria. However, the disposal of sago effluent into the nearby river could create serious environmental impacts such as unpleasant smells, and high chemical oxygen demand (COD) and biochemical oxygen demand (BOD) [2]. Several studies reported on the use of sago solid bio-waste namely *hampas* and bark as bio-sorbents [3–5] and other value-added products [1, 6, 7]. However, none reported on the development of bio-sorbent from sago effluent.

Sago activated sludge is a potentially feasible and sustainable feedstock for large scale production of biochar for adsorbent purposes. The activated sludge process is mainly regulated by the action of bacteria particularly aerobic bacteria that oxidizing the organic matters and forming the flocs. The size of the floc particles can be determined by the agglutination strength of the floc-forming bacteria, the filamentous organisms that presence in the system and the level of turbulence during the activated sludge process. The common floc-forming bacteria include *Escherichia sp.*, *Bacillus sp.*, *Achromobacter sp.*, *Arthrobacter sp.*, *Pseudomonas sp.*, *Aerobacter sp.* and *Citromonas sp.*, to name a few [11]. A mixture of effluent and activated sludge agitated and aerated to form biomass via sedimentation.

Biochar is a carbon-rich material derived from biomass after thermal decomposition processes via pyrolysis in the presence or absence of oxygen at the elevated temperature [12]. Pyrolysis is a typical thermal decomposition method of converting organic materials into carbon-rich solids namely char and volatile matter [13]. Microwave pyrolysis offers many advantages over conventional pyrolyses, such as the ability to produce a uniform and efficient heating environment [14]. A higher amount of volatile matter released during the pyrolysis of biomass produces biochar with lower densities and higher porosities [1], which in turn, can be used as a precursor for the production of activated carbon and adsorbent at a lower cost. The production of low-cost adsorbents to substitute conventional and costly wastewater treatment methods i.e. reverse osmosis, chemical precipitation, ion-exchange, membrane separation, solvent extraction and electrodialysis has received great interest among researchers [1].

Herein, we report our work on the preparation of sago activated carbon (SAC) via sago activated sludge process followed by microwave pyrolysis and chemical activation method. The physicochemical characterization, morphology and surface area were characterized, and the adsorptive performance in Zn, Cr, and Pb adsorption in aqueous solution were investigated in batch adsorption study. The production of low-cost adsorbents derived from sago effluent in removing heavy metal ions from wastewater is significant towards sustainable technology in producing zero biowastes from the sago industry.

2 Materials and Methods

2.1 Materials and Characterization

Sago effluent was obtained from Herdsen Sago Industries in Pusa, Sarawak. Sodium hydroxide, hydrochloric acid 98%, sulphuric acid 99%, methanol and hexane were obtained from Merck, which dichloromethane from Fisher and acetylsalicylic acid 99% from Acros. All the reagents and solvents were used as received without any further purification. The pH of sago effluent was measured using Thermo Scientific/Orion 2-Star B39285 while chemical oxygen demand (COD), total suspended solids (TSS) and ammoniacal nitrogen (AN) were determined using HACH DR/4000U Spectrophotometer.

2.2 Instrumental Analysis

Fourier transform infrared spectroscopy (FTIR) spectra of sago biomass (SBM), sago biochar (SBC), sago activated carbon (SAC) and solid acid catalyst were characterized using Perkin Elmer 1605 under ambient condition and KBr pellets. The IR spectra were plotted over a frequency range from 400 to 4000 cm^{-1}. The surface morphology of SBM, SBC and SAC were analyzed using a Scanning Electron Microscope (SEM, JEOL 7500F-1) at 0.5 to 30 kV. Pyrolysis of SBM was carried out using a modified convection microwave oven (SHARP R-958A) of 1000 W output power and frequency of 2450 MHz. Heavy metal concentrations were analyzed using Flame Atomic Absorption Spectrophotometer (FAAS, Thermo Scientific iCE 3500). The specific surface area was determined using Brunauer–Emmett–Teller (BET) (Quantachrome ASIQC0000-3) with liquid nitrogen adsorption at 77 K.

2.3 Preparation of Sago Biochar (SBC)

Fresh sago effluent was placed in a 50 L container equipped with a twin gas inlet/outlet valve. After 5 days aeration at room temperature, the sediment was filtered, dried and weighed to afford SBM (20.0 g). SBM (5.0 g) was subjected to microwave pyrolysis (10 min, ~300 °C) with nitrogen flow: 100 mL/min for 30 min prior analysis to afford SBC (2.07 g, 41%) [15, 16].

2.4 Preparation of SAC from Sago Biomass (SBC)

SBC (30.0 g) was drenched into a solution NaOH (20%, 50 mL) for 2 h and oven-dried (105 °C, 24 h). HCl (5 mL, 5 M) was added, stirred, and subsequently washed with deionized water to pH 4.5, and dried to produce approximately 25 g SAC [15, 16].

2.5 Adsorption Study

Effect of adsorbent dosage on Pb, Cr and Zn removal by SAC

Zn, Cr and Pb solution was prepared by dissolving homogenously 5 mL heavy metal aliquot with 1000 mL distilled water and 4.5 mL of 65% of nitric acid at pH 2. Five different adsorbent dosages (0.2, 0.4, 0.6, 0.8 and 1.0 g) were mixed in 100 mL at fixed initial concentration of metal ion solutions of 5 mg/L. The experiment was performed in triplicates where the average values were taken. The mixture was agitated on a rotary shaker (150 rpm, 24 h), gravity filtered, and the filtrate was analyzed using FAAS. A control experiment was prepared for every adsorption batch [16].

Effect of initial concentration on Pb, Cr and Zn removal by SAC

A fixed adsorbent dosage of 0.1 g of SAC1 was mixed with 50 mL of different initial concentration metal ion solutions prepared (5, 10, 15, 20 and 25 mg/L). The experiment was performed in duplicate where the average values were taken. The mixture was agitated on a rotary shaker (150 rpm, 24 h), gravity filtered, and the filtrate was analyzed using FAAS. The heavy metal was diluted 10 times upon analysis. A control experiment was prepared for every adsorption batch [16]. The removal efficiency (E) of adsorbent is calculated following Eq. 1:

$$E\ (\%) = \frac{(C_i - C_f)}{C_i} \times 100\,\% \tag{1}$$

C_i is an initial concentration of metal ion solution and C_f is a final concentration of metal ion solution.

Fig. 1 Activated sludge process of sago effluent

3 Results and Discussion

3.1 Preparation of SBM via Activated Sludge Process

Sago effluent underwent an activated sludge process for 3–5 days. During the sludge process, bacteria and microorganisms degraded the suspended solids and dissolved organic matter via aerobic oxidation. In the presence of dissolved oxygen, the bacteria degraded the carbonaceous and nitrogenous wastes to formed sediment as sludge [11]. The activated sludge process of the sago effluent is shown in Fig. 1. The sludge was collected and filtered to produce SBM. Every 50 L of sago effluent produced approximately 20 g of SBM. The pH of sago effluent after the activated sludge process increased from pH 4 to 7, during which the layers of sludge sediment was formed in the container.

3.2 Assessment of Water Quality After Activated Sludge Process

The COD values of the sago effluent showed an intense decrease from 30.0 ± 0.33 to 16.67 ± 0.17 mg/L which indicated the improvement of water quality after sludge activating process [17]. The TSS of sago effluent was also decreased after the activated sludge process from 176.6 ± 0.65 to 153.3 ± 0.54 mg/L. The low TSS allows light penetration in the water body and photosynthetic activity of the aquatic organism [5]. Besides COD and TSS, AN in the sago effluent was also decreased from 21.0 $\pm$ 0.16 to 0.71 ± 0.55 mg/L after the activated sludge process. The lower level of AN in the wastewater indicated most AN has oxidized in the process and ready to be discharged to the water body [18].

Table 1 Characteristic properties of SAC

Properties	
Moisture content (%)	18.34
Density (g/L)	150 ± 2.0
BET Surface area, S_{BET} (m^2/g)	471.1
Average pore size (μm)	36.29
CHN analysis (%)	
(i) Carbon (ii) Hydrogen (iii) Nitrogen	53.24 3.18 2.51

3.3 Physicochemical Characterisation and Proximate Analysis

The biomass from the sludge activating process underwent microwave pyrolysis and yielded 41.4% of SBC. The remaining weight percentage was lost due to the conversion of the SBM into gases during the pyrolysis process [12]. The dark powdery form of SBC was chemically activated using NaOH to yield 80% of SAC. Sodium hydroxide (NaOH) was used during the activation to produce more porous surfaces and improve sorption properties [18]. The sodium metal intercalated and forces apart lamellae of the crystallites that formed the carbon structure of SAC [1].

The properties of SAC as an adsorbent are shown in Table 1. Large surface area (471.1 m^2/g) and small particle size (36 μm) of SAC indicated its characteristic and accessibility as adsorbent [2, 3].

3.4 Instrumental Analysis

The SBM, SBC, SAC samples were subjected to both FTIR and SEM to identify the functional groups responsible for adsorption and surface morphology.

Functional Group Analysis

The IR absorption spectra of SBM, SBC, SAC and commercial activated carbon (CAC) are shown in Fig. 2. In comparison to CAC, the IR spectra of SBM, SBC and SAC (Fig. 2a–c) showed broad peaks at 3408–3427 cm^{-1} attributed to the $\upsilon_{O\text{-}H}$ stretching vibration of carboxylic acids, phenols and alcohols in lignin [3, 4]. The broad O–H peak in SBM (Fig. 2a) was reduced after the double carbonization and activation process. The pyrolysis caused the decomposition of functional groups and the liberation of volatile matter during carbonisation [1, 8]. A weak absorbance peak at 2950–2850 cm^{-1} (Fig. 2a–c) attributed to $\nu_{(C\text{-}H)}$ bands. The IR spectra of SAC and SBC (Fig. 2b–c) showed the presence of peaks at 1706–1703 cm^{-1} and 1650–1560 cm^{-1} attributed of $\upsilon_{C=C}$ and $\upsilon_{C\text{-}O}$ from lignin and hemicellulose in SBM and

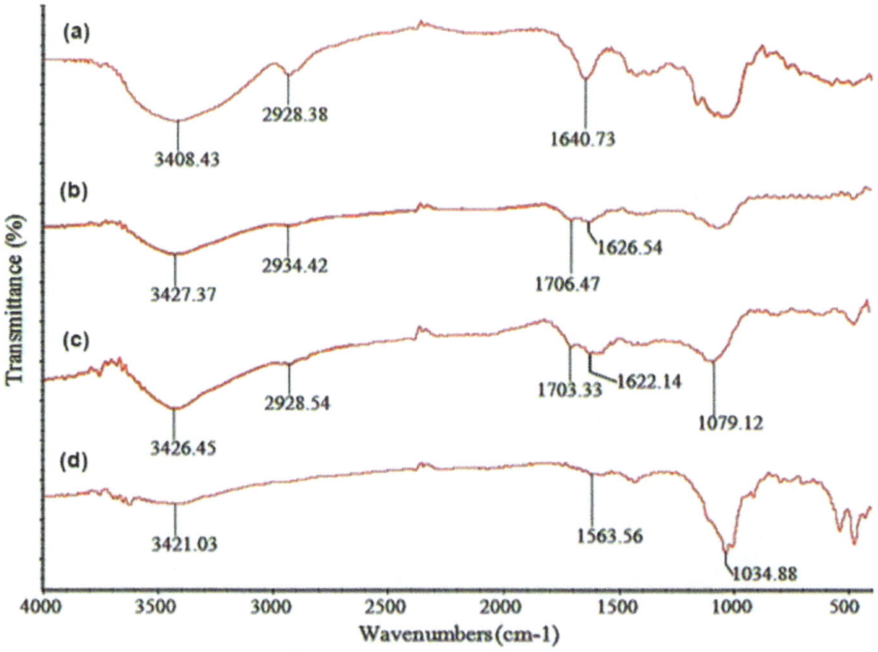

Fig. 2 The IR spectra of **a** SBM **b** SBC **c** SAC **d** CAC

SBC and reduced after activation in SAC. While broad peaks at 1300–1000 cm^{-1} attributed to $\nu_{(C\text{–}O)}$ stretching.

Surface morphology

The surface morphology of SBM, SBC and SAC was characterized and compared to CAC using SEM (Fig. 3). SBM micrograph (Fig. 3a) revealed the existence of starch's granules on its surface and disappeared (Fig. 3b) following microwave pyrolysis due to the removal of the volatile matters [3]. The porosity of the biochar was, however, less accessible as the surfaces of the biochar still covered with mineral matters, which partially blocked the pores during pyrolysis [12]. Chemical activation of SBC using NaOH afforded the microporous structure of SAC (Fig. 3c) due to the decomposition of organic materials and volatile matters and the development of the microporous structure [1, 4]. NaOH minimized the formation of tars and other fluid that clogged up and inhibited the development of pores [19]. The porous carbon structure of SAC resulted from the NaOH removal during washing afforded bigger and more accessible pores size compared to CAC (Fig. 3d). SAC is produced from the activated sludge difference in the raw material.

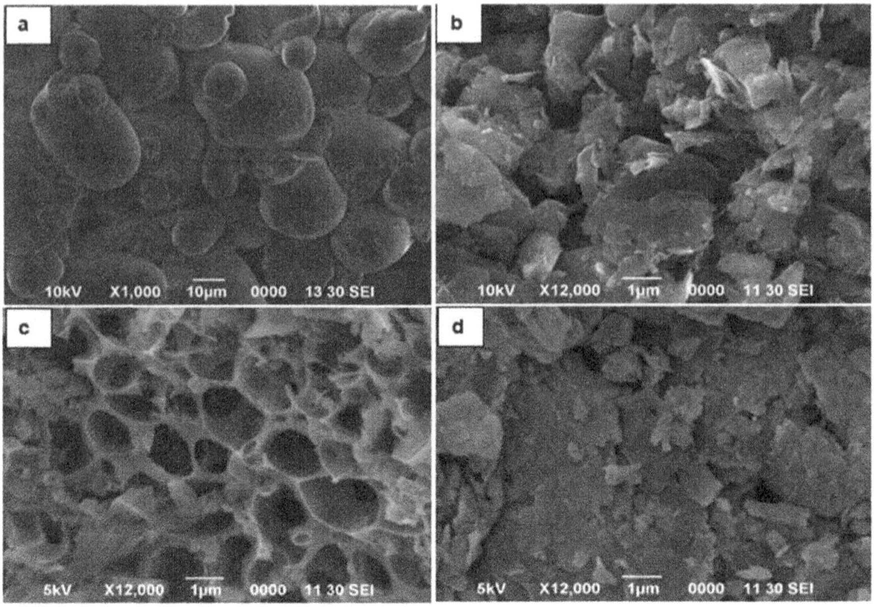

Fig. 3 The SEM micrographs of **a** SBM **b** SBC **c** SAC **d** CAC

3.5 Batch Adsorption Study Heavy Metals

Activated carbon is commonly studied and applied as heavy metal adsorbent because of its high porosity, high surface area, and other important functional groups, which make it optimal for metal adsorption [20]. However, the ability of the activated carbon to adsorb metal depends on the raw material and type of metal to be adsorbed [11, 15]. The prepared SAC was studied for heavy metal adsorption on Zn, Cr and Pb.

Effect of adsorbent dosage

The effect of adsorbent dosages to adsorption of Zn, Cr and Pb by SAC is depicted in Fig. 4. The ability of SAC to adsorb heavy metal was studied by varying the adsorbent dosage at a fixed metal concentration of 5 mg/L. Pb showed higher adsorption compared to Zn and Cr at constant initial metal concentration. Pb, Cr and Zn showed the highest adsorption onto SAC at 1.0 g adsorbent dosage with 82.93%, 38.28% and 14.78%, respectively.

Lower adsorption of Cr and Zn is due to the limited adsorption to monolayer where all the adsorbed molecules were in contact with the surface layer of the adsorbent [15].

The adsorption of metals onto SAC increased with the increasing of the adsorbent dosage. This was due to the presence of higher surface area with more functional groups at higher adsorbent dosage. The surface area of activated carbon as well as the functional groups is important for adsorption capacity. Large surface area contributed

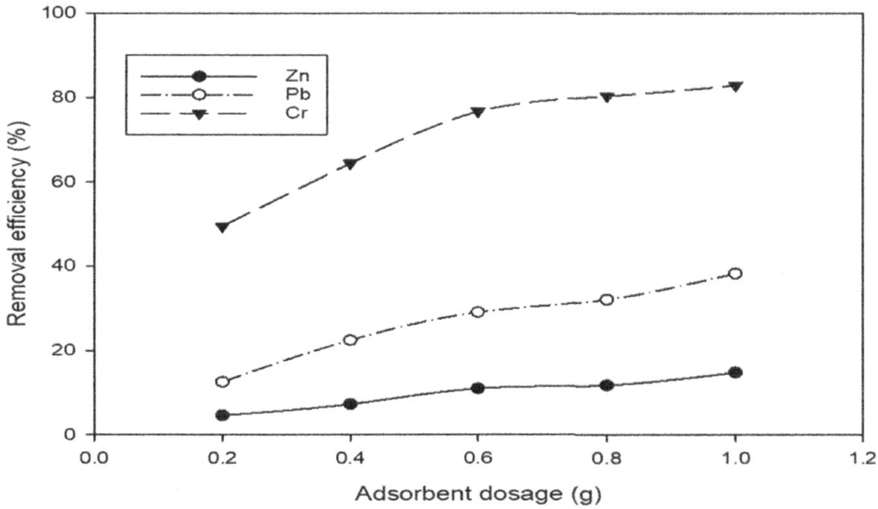

Fig. 4 Effect of adsorbent dosage on adsorption of Zn, Cr and Pb by SAC (Experimental condition: Initial concentration: 5 mg/L, mixing speed: 150 rpm, contact time: 24 h)

to higher adsorption capacity [15, 18]. Besides, the presence of functional groups such as C=O, S=O and –OH was also believed to be embroiled in the formation of chemical bonding which accountable in metal adsorption [20]. Such functional groups were important in the formation of Van der Waals bonding since they bind metals to the adsorbent during adsorption [21]. However, the ability of activated carbon to adsorb metal depends on the raw material used to make the activated carbon and the type of metal that is adsorbed by the activated carbon [15].

Several factors that affect heavy metal adsorption onto activated carbon are the pH of the metal solution, the particle size of activated carbon, agitation time and initial metal concentration. The pH of the solution greatly influences the absorption of heavy metals, as it verifies the surface charges of the adsorbent, the degree of ionization and the speciation of the adsorbent [22].

Effect of initial concentration

The ability of SAC to adsorb heavy metal was also studied with varying initial concentrations of Zn, Cr and Pb, ranging from 5 to 25 mg/L using 1 g of adsorbent dosage. The removal efficiency of SAC on Zn, Cr and Pb is shown in Fig. 5.

The percentage of efficiency of removal decreased with increased concentrations of Pb in solution. The highest adsorption onto SAC is at 5 mg/L initial concentration. At an initial concentration of 10 mg/L, the percentage of removal efficiency was almost similar to the initial concentration of 5 mg/L. From 15 to 25 mg/L initial concentration, there was an intense decrease in percentages of removal efficiency of Pb. The highest adsorption occurs at the lowest initial concentration is the result of more adsorption sites of activated carbon adsorbent accessible at low concentrations

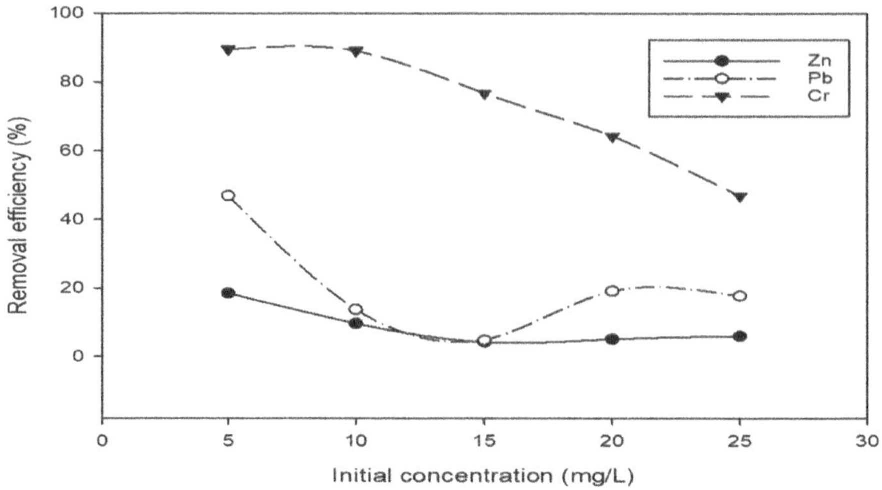

Fig. 5 Effect of initial concentration on adsorption of Zn, Cr and Pb by SAC (Experimental condition: Adsorbent dosage: 1 g/50 mL, mixing speed: 150 rpm, contact time: 24 h)

of heavy metal ions [15, 23]. In this study, Pb showed the highest adsorption compared to Zn and Cr.

The percentage of removal efficiency of Zn and Cr decreased from the initial concentration of 5 to 15 mg/L. However, the percentage of removal of Cr tends to fluctuate from 5 to 15 mg/L initial concentration. This is because the adsorption sites of SAC become more saturated and the occurrence of adsorption and desorption phenomenon [15] at 15 mg/L as the initial metal ions concentration increased. From 15 to 20 mg/L of initial concentration, there was a slight increase in the percentage of removal efficiency of Zn and Cr. The percentage of removal efficiency for Zn was increased at 25 mg/L and slightly decreased for Cr. This finding could be due to the occurrence of bulk metal ions transfer to the surface of SAC [18]. The Langmuir and Freundlich isotherm studies showed the Q_e for Pb removal by SAC was 3.202 $\times 10^{-3}$ mg/g. The details of the findings are also reported [18].

4 Conclusion

The present study concluded that sago activated carbon (SAC) can be successfully prepared from sago effluent via an activated sludge process followed by microwave pyrolysis and chemical activation. SAC afforded high surface area and microporous structure. The SAC can be applied as a low-cost adsorbent to remove Pb from aqueous solution. However, very low removal was observed for Cr and Zn in this work. In general, the activated sludge process can be potentially applied in sago mill before effluent discharge to the nearby stream, and the activated sludge generated can be used as feedstock for the adsorbent. This preliminary work is recommended for further isotherm study for the occurrence of chemisorptions process in creating a zero-waste sago processing industry for sustainable technology.

Acknowledgements This work was funded by Universiti Malaysia Sarawak Tun Openg Sago Research Chair (F07/TOC/1742/2018)

References

1. Wahi R, Zuhaidi NFQ, Yusof Y, Jamel J, Kanakaraju D, Ngaini Z (2017) Chemically treated microwave-derived biochar: an overview. Biomass Bioenergy 107:411–421. https://doi.org/10.1016/j.biombioe.2017.08.007
2. Rajan Y, Ngaini Z, Wahi R (2019) Novel adsorbent from sago-grafted silica for removal of methylene blue. Int J Environ Sci Technol 16(8):4531–4542. https://doi.org/10.1007/s13762-017-1389-9
3. Ngaini Z, Noh F, Wahi R (2014) Esterified sago waste for engine oil removal in aqueous environment. Environ Technol 35(22):2761–2766. https://doi.org/10.1080/09593330.2014.920051
4. Ngaini Z, Wahi R, Abang Kamaluddin DH, Mohd Yusoff NA (2014) Modified sago waste for oil adsorption. Pertanika J Sci Technol 22(1):153–161
5. Wahi R, Chuah AL, Nourouzi MM, Ngaini Z, Choong-Shean-Yaw T (2017) Utilization of esterified sago bark fibre waste for removal of oil from palm oil mill effluent. Int J Chem Eng 5(1):170–177. https://doi.org/10.1016/j.jece.2016.11.038
6. Ngaini Z, Azlin ARKA, Nazlina S, Hasnain H, Norhaizat S, Teng JX, Lawai V (2013) Production of fire-retardant sound-absorbing panels from sago waste. J Trop For Sci 25(4):510–515
7. Ngaini Z, Noh F, Wahi R (2018) Facile sorbent from esterified cellulosic sago waste for engine oil removal in marine environment. Int J Environ Sci Technol 15(2):341–348. https://doi.org/10.1007/s13762-017-1389-9
8. Banu JR, Kaliappan S, Dieter B (2006) Treatment of sago wastewater using hybrid anaerobic reactor. Water Qual Res J Can 41:56–62. https://doi.org/10.1007/BF03326295
9. Abd Rashid W, Musa H, Wong SK, Bujang K (2010) The potential of extended aeration system for sago effluent treatment. Am J Appl Sci 7:616–619
10. Ayyasamy PM, Banuregha R, Vivekanandhan G, Rajakumar S, Yasodha R, Lee S, Lakshmanaperumalsamy P (2008) Bioremediation of sago industry effluent and its impact on seed germination (green gram and maize). World J Microb Biot 24:2677–2684. https://doi.org/10.1007/s11274-008-9796-1
11. Gerardi MH (2006) Wastewater bacteria. Wiley-Interscience, Hoboken, New Jersey
12. Lehmann J, Joseph S (2009) Biochar for environmental management: Science and technology. Earthscan Ltd., London
13. Demirbas A, Arin G (2002) An overview of biomass pyrolysis. Energy Sources 24:411–482. https://doi.org/10.1080/00908310252889979
14. Miura M, Kaga H, Sakurai A, Kakuchi T, Takahashi K (2004) Rapid pyrolysis of wood block by microwave heating. J Anal Appl Pyrol 71:187–199. https://doi.org/10.1016/S0165-2370(03)00087-1
15. Wahi R, Senghie H (2007) The use of microwave derived activated carbon for removal of heavy metal in aqueous solution. J Sci Tech 17:97–108
16. Wahi R, Ngaini Z, Jok VU (2009) Removal of mercury, lead and copper from aqueous solution by activated carbon of palm oil empty fruit bunch. World Appl Sci J 5:84–91

17. Gerardi MH (2002) Nitrification and denitrification in the activated sludge process. Wiley-Interscience, New York
18. Makshut NA, Ngaini Z, Wahi W, Hussain H, Mahmut NI, Bahrin NQ (2020) Nano-sized absorbent from pyrolysed sago activated sludge for removal of Pb (II) from aqueous solution. Pertanika J Sci Technol 28(3):893–916
19. Pechyen C, Atong D, Aht-Ong D, Viboon S (2007) Investigation of pyrolyzed chars from physic nut waste for the preparation of activated carbon. J Solid Mech 1498–1507.https://doi.org/10.1299/jmmp.1.498
20. Rao MM, Rao GPC, Seshaiah K, Choudary NV, Wang MC (2008) Activated carbon from Ceiba pentandra hulls, an agricultural waste, as an adsorbent in the removal of lead and zinc from aqueous solutions. J Waste Manag 28:849–858. https://doi.org/10.1016/j.wasman.2007.01.017
21. Malik R, Ramteke DS, Wate SR (2007) Adsorption of malachite green on groundnut shell waste based powdered activated carbon. J Waste Manag 27:1129–1138. https://doi.org/10.1016/j.wasman.2006.06.009
22. Salam OEA, Reiad NA, Elshafei MM (2011) A study of the removal characteristics of heavy metals from wastewater by low-cost adsorbents. J Adv Res 2:297–303. https://doi.org/10.1016/j.jare.2011.01.008
23. Qian Q, Machida M, Tatsumoto H (2007) Textural and surface chemical characteristics of activated carbons prepared from cattle manure compost. J Waste Manag 98(2):353–360. https://doi.org/10.1016/j.wasman.2007.03.029

Optimization of Porous Carbon Production from Coconut Shell Using Microwave Heating Technique for Adsorption of Lead and Zinc

Aminu Isa Mohammed, Jibril Mohammed, Surajudeen Abdulsalam, and Yusuf Olabode Raji

Abstract Heavy metal ions are responsible for several acute health challenges and environmental degradations. This study was aimed to optimize the preparation of porous carbon from coconut shell using microwave for adsorption of heavy metal ions. Response surface methodology was used to optimize the effects of preparation variables on the properties of activated carbon, and the influences of the process parameters were examined using analysis of variance (ANOVA) to identify the significant parameters. The optimum condition was obtained at microwave power of 100 W, irradiation time of 2.7 min and 1.0 KOH impregnation ratio which resulted the removal of 64.0% lead and 56.0% zinc in batch sorption studies. The experimental data show adsorption capacities of 151 and 112 mg/g for lead and zinc, respectively and fitted well into Freundlich model. The pseudo-second-order kinetic model best described the kinetics for lead and zinc ions.

Keywords Adsorption · Optimization · Coconut shell · Response surface methodology · Microwave

---

A. I. Mohammed · J. Mohammed (✉) · S. Abdulsalam · Y. O. Raji
Department of Chemical Engineering, Faculty of Engineering and Engineering Technology, Abubakar Tafawa Balewa University, Bauchi PMB 0248, Nigeria
e-mail: jibrilmuhammad@gmail.com

A. I. Mohammed
e-mail: acidnana22@yahoo.com

S. Abdulsalam
e-mail: asurajud@gmail.com

Y. O. Raji
e-mail: yoraji411@gmail.com

Y. O. Raji
Advanced Membrane Technology Research Centre (AMTEC), School of Chemical and Energy Engineering, Universiti Teknologi Malaysia, UTM Johor Bahru, 81,310 Johor, Malaysia

M. A. A. Zaini et al. (eds.), *Proceedings of the 3rd International Conference on Separation Technology*, Lecture Notes in Mechanical Engineering,
https://doi.org/10.1007/978-981-16-0742-4_15

1 Introduction

There is significant attention from the scientific community in the recent years on the release of heavy metal ions and inorganic compounds into the environment which causes harmful impacts on human health, contamination of water bodies and degradation of ecological environment. The aim of every researcher is to produce effective and sustainable approaches that are comparable with commercially available ones for effective mitigation to the environment [1]. The rapid industrialization in the urban area has led to increased disposal of heavy metal ions into the environment. The increase in the use of the heavy metal ions over the past few years has inevitably resulted in an increased movement of metallic substances in the aquatic environment. Trace concentrations of zinc are important for the physiological functions of living tissues and regulate many biochemical processes. However, just like other heavy metals, when zinc and lead are discharged into natural water at increased concentrations in industrial wastewater or from mining operations it can have severe toxicological effects on humans and aquatic ecosystems [1].

Hence, it is very important to remove lead and zinc from industrial wastewater before transporting and recycling into the natural water bodies. A number of different technologies have been developed over the past years to remove heavy metal ions from wastewater before discharging it. The most important technology includes coagulation/flocculation and other conventional methods. All the chemical methods have proved to be more cost effective and less efficient than the adsorption process [2]. This can be achieved through the use of various techniques and materials that are available in the environment which are considered as waste or low costs materials. Nowadays, microwave has become an alternative source of energy for material heating and processing, due to its numerous advantages like volumetric internal heating, uniform distribution of temperature, rapid rise in temperature and energy saving [3]. Moreover, activated porous carbon is the most widely used adsorbent in separation and purification application due to its high adsorption abilities for a large number of organic and inorganic compounds. It is used in contaminants removal from water and gases. Nevertheless, the prices of commercially available activated carbons are relatively high and hence limit their usage [4].

Activated carbon is the most widely used adsorbent in separation and purification applications due to its high adsorption abilities for many organic compounds [4] Adsorption of contaminants by activated carbon is efficient, reliable, economical, and environmentally friendly which makes it the most widely used technique. It is also prepared from substances that are high in carbon content and convert to a precursor of high carbon content. Activated carbon filtration is most effective in removing organic contaminants from water, which is responsible for taste, odour and colour problems and can generally be used to improve water quality by removing the remaining traces of chlorine present in the water.

2 Materials and Methods

2.1 *Materials*

Coconut shell was taken from Bauchi state, Nigeria. The shells were washed with water to free the material from soil, dirt and other contaminants and dried at 105 °C for 24 h. After that, the shells undergo milling and separated using shakers and sieves to the size range of 0.5–1.18 mm prior to microwave heating at different microwave power level, impregnation ratio and irradiation time. Nitrogen gas was flowed through the system to create an inert atmosphere. Potassium hydroxide pellets (99%), lead and zinc salts were supplied by Choose bond Nig. Ltd Jos Plateau State. Nitrogen and carbon dioxide were supplied by Nokson Nig. Ltd.

2.2 *Sample Preparation Method*

Coconut shell undergone pre-treatment to be free from soil and any other impurities by washing with water and dried at 105 °C for 24 h. Next, the shells undergone milling to reduce the particle size and separated using sieves to the size range of 0.5–1.18 mm. The resulting char was impregnated with potassium hydroxide (KOH) at an impregnation ratio of 1:1 g/g (mass ratio) of chemical reagent char [5]. The resultant mixture was dried in the laboratory oven for 24 h at 105 °C before it was loaded into a glass reactor which was fixed inside the microwave oven. The material was activated by the heating of the microwave oven under the flow of nitrogen gas 150 cm^3/min at power level of 100 W and irradiation time of 2.7 min [5]. Then, the resultant activated carbon was removed from the microwave and cooled. The resulted activated carbon was washed with distilled water until the pH of the rinse was close to neutral point and then dried at 105 °C for 24 h [6]. The produced activated carbon was stored in desiccators prior to be used. In the design of the experiments, there are factors and responses based on the central composite design (CCD) in the production of adsorbent, a feature of the response surface methodology (RSM). Design Expert software version 7.1.6. (STAT-EASE Inc., Minneapolis, USA) was used to carry out the regression analysis to establish its suitability with the equations generated, and also to statistically determine the significance of the equations.

2.3 *Proximate and Ultimate Analysis*

The proximate analysis determined the fixed carbon, ash content, moisture content and volatile content of the coconut shell. While the ultimate analysis indicates the composition of the raw material in wt% of nitrogen, oxygen, sulphur, carbon and hydrogen.

2.4 Characterization of Samples

Pore textural analysis was characterized with micromeritics ASAP 2020 for full isotherm analysis. SEM analysis with Karl Zeiss (EVO50 XVPSEM, Germany) was carried out to study the surface morphology and development of porosity. The surface organic structures of the raw material and activated carbon were chemically and structurally studied by Fourier Transform Infrared Spectroscopy (FT-IR) using Perkin Elmer Spectrum one model.

2.5 Removal Efficiency

The adsorption capacity and removal efficiency of lead and zinc on the produced porous activated carbon was calculated using Eq. (1):

$$\% \, \text{Removal} = \frac{\text{Co} - \text{Ce}}{Co} \times 100 \tag{1}$$

where Co is the initial concentration of lead and zinc metal ions (mg/L) and Ce is the final equilibrium concentration of lead and zinc (mg/L).

2.6 Adsorption Studies

Batch sorption studies were conducted by shaking 0.1 g of the produced porous carbon with different initial concentrations of the heavy metal ions solution in Teflon covered 30 ml bottles at room temperature (30 °C) in a shaker at 115 rpm. The applied shaking speed allows the surface area to come in contact with heavy metal ions. Simulated wastewater (100 mL) containing Zn (II) and lead independently was placed in a 250 mL flask and 1.5 g/L adsorbent was added. The Zn (II) and lead concentration of the treated simulated wastewater was carried out at time interval between 0 and 180 min using standard methods for examination of water and wastewater. The heavy metal ions concentration was determined using UV–Vis spectrometer (Perkin Elmer lambda 25) at 535–626 nm and its uptake at equilibrium was determined by Eq. (2):

$$\text{Qe} = \frac{(\text{Co} - \text{Ce})\text{V}}{W} \tag{2}$$

where W is the weight of the adsorbent (g) and V is the volume of solution (L).

2.7 *Kinetic and Equilibrium Studies*

Kinetic and equilibrium adsorption were conducted by shaking 0.1 g of the activated carbon with different initial concentrations of the heavy metal ions solution in Teflon covered 30 ml bottles at 30 °C in a shaker at 115 rpm. The applied shaking speed allows the surface area to come in contact with heavy metal ions to attain equilibrium. The heavy metal ions concentration and the uptake at time t, Qt were determined,

$$\mathrm{Qt} = \frac{(\mathrm{Co} - \mathrm{Ct})\mathrm{V}}{W} \tag{3}$$

where Ct is the concentration at a particular time t.

3 Results and Discussion

3.1 *Proximate and Ultimate Analyses*

The ultimate and proximate analyses of coconut shell are summarized in Table 1. High percentage of volatiles of 71.3%, average fixed carbon content of 21.8%, low moisture and ash content of 6.5% and 1.35%, respectively were recorded while the ultimate analysis shows that the coconut shell is composed of 48.3% carbon, 7.12% hydrogen, 1.14% nitrogen, 0.04% sulphur and 43.4% oxygen.

The oxygen content was calculated by difference of the total percentage of other components. Similar results were reported by other researchers.

Table 1 Proximate and ultimate analysis revealing the percentage composition of coconut shell

Proximate (%)	This study	[1]	[7]	[8]
Moisture	6.50	5.60	10.5	3.52
Fixed carbon	21.8	23.3	18.8	20.6
Ash	1.35	1.10	0.66	1.28
Volatiles	70.4	71.3	70.1	74.6
Ultimate (%)	This study	[1]	[3]	
Carbon	48.3	48.7	53.4	
Hydrogen	7.12	6.34	7.50	
Nitrogen	1.14	1.52	1.70	
Sulphur	0.04	0.04	0.00	
Oxygen	43.4	43.4	37.4	

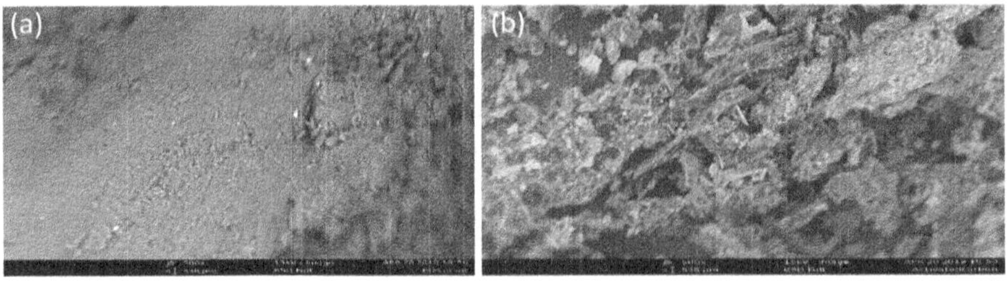

Fig. 1 SEM images of **a** coconut shell and **b** activated carbon

3.2 Characteristics of Materials

SEM micrographs of raw coconut shell and the coconut shell activated carbon are shown in Fig. 1. Figure 1a shows the SEM image of raw material with no obvious pore formation on the surface. It is generally suggested that the surface contains volatiles and functional groups. In Fig. 1b, the produced activated carbon shows cracks and irregular structures because of chemical evaporation and volatiles removal during carbonization.

3.3 Surface Area and Pore Volume

The textural composition of coconut shell porous carbon was analyzed by means of nitrogen adsorption. Table 2 summarizes of the textural properties of activated carbon. The surface area, pore size distribution and pore volume of porous carbon are significant properties attributable to its adsorption capacity which in general is directly proportional to surface area [3]. The coconut shell porous carbon had BET of 332.7 m^2/g and pore volume of 0.12 cm^3/g. It has been showed that the pore volume decreased with carbonization time while surface area increased with carbonization time. This is associated with the effect of potassium hydroxide as activating agent.

Activated carbon was produced by impregnation with potassium hydroxide at different impregnation ratios of 1–3, irradiation times of 0–15 min and microwave

Table 2 Porosity parameters obtained from nitrogen adsorption isotherm

Parameter	
BET surface area (m^2/g)	332.7
Micropore (m^2/g)	414
External surface area (m^2/g)	106
Pore volume (cm^3/g)	0.120
Average Pore width (nm)	5.30
Langmuir surface area (m^2/g)	776.9

Table 3 Experimental design matrix as generated by Design Expert software and the responses

Run	Power level (W), nX_1	Irradiation time (min), X_2	Impregnation ratio, X_3	Lead adsorption (%), Y_1	Zinc adsorption (%), Y_2	Yield (%), Y_3
1	450	6.00	2.00	51.0	77.2	71.2
2	400	15.0	3.00	80.0	75.0	72.0
3	450	6.00	2.00	41.0	69.8	87.4
4	450	6.00	2.00	65.0	86.0	93.0
5	500	15.0	1.00	80.0	80.0	84.6
6	450	8.00	2.00	95.0	94.0	90.0
7	450	8.00	2.00	77.0	93.0	81.2
8	400	8.00	2.00	60.0	91.0	82.0
9	100	1.00	3.00	46.0	92.0	88.0
10	450	6.00	1.50	71.0	93.0	65.0
11	100	15.0	3.00	75.0	82.0	69.1
12	450	6.00	1.50	86.0	88.0	82.0
13	450	6.00	1.50	73.0	93.0	88.0
14	400	6.00	3.00	42.0	47.0	65.0
15	400	6.00	3.00	80.0	86.0	80.0
16	450	8.00	2.00	89.0	90.0	84.0
17	450	8.00	2.00	47.0	70.0	88.6
18	450	8.00	2.00	80.0	92.0	84.0
19	600	8.00	1.00	85.0	89.0	80.0
20	450	6.00	2.00	62.0	89.0	62.0

power levels of 100–800 watts [5]. The resultant mixture was dried in an oven over a period of 24 h and 105 °C. The dried mixture was loaded into a glass reactor which was fixed inside the microwave oven at power level of 100 W and irradiation time of 2.7 min. The resultant activated carbon was removed from the microwave oven and cooled (Table 3).

3.4 Analysis of Variance (ANOVA)

ANOVA shows the model variables used in analyzing the factors that are significant to lead and zinc adsorption. Table 4 shows the analysis of variance for response surface model for lead and zinc adsorption.

Below are the equations generated after considering the significant terms of model, the final empirical equations for yield (Y_3), lead (Y_1) and Zinc (Y_2) removal efficiency are given in equations,

Table 4 Analysis of variance for response surface model for lead and zinc adsorption

Source	Sum of square	df	Mean square	F-Value	P-Value, Prob > F
Lead adsorption					
Model	1551.1	6	258.52	3.49	0.0280
X_1	148.14	1	148.14	2.00	0.1810
X_2	6.6500	1	6.6500	0.09	0.7694
X_3	175.71	1	175.71	2.37	0.1477
X_1X_2	209.10	1	209.10	2.82	0.1169
X_1X_3	451.50	1	451.50	6.09	0.0283
X_2X_3	559.45	1	559.45	7.55	0.0166
Residual	963.83	13	74.140		
Zinc adsorption					
Model	2872.2	9	319.13	1.34	0.3269
X_1	125.77	1	125.77	0.53	0.4844
X_2	410.82	1	410.82	1.72	0.2187
X_3	983.58	1	983.58	4.12	0.0297
X_1X_2	50.000	1	50.000	0.21	0.6782
X_1X_3	19.220	1	19.220	0.08	0.7823
X_2X_3	537.92	1	537.92	2.26	0.0440
X_1	329.11	1	329.11	1.38	0.2673
X_2^2	614.27	1	614.27	2.58	0.1396
X_3^2	528.15	1	528.15	2.21	0.1675
Residual	2384.8	10	238.48		

$$Y_1 = 69.32 - 4.07x_1 + 0.79x_2 - 3.59x_3 + 5.11x_1x_2 - 7.51x_1x_3 + 8.36x_2x_3 \quad (4)$$

$$\begin{aligned} Y_2 = 58.93 - 3.80x_1 + 7.06x_2 - 8.49x_3 + 2.5x_1x_2 + 1.55x_1x_3 \\ + 8.2x_2x_3 + 9.88x_12 - 9.88x_22 - 6.19x_32 \end{aligned} \quad (5)$$

$$Y_3 = 79.91 \quad (6)$$

From Eqs. (4)–(6), the positive sign shows synergetic effect, while the negative sign indicates antagonistic effect. The correlation coefficients showing the relationship between optimization factors and responses are 64% and 56% of the total change in the removal efficiency of lead and zinc, respectively. All the factors play a vital role either independently or by way of reaction to the removal of lead and zinc [9].

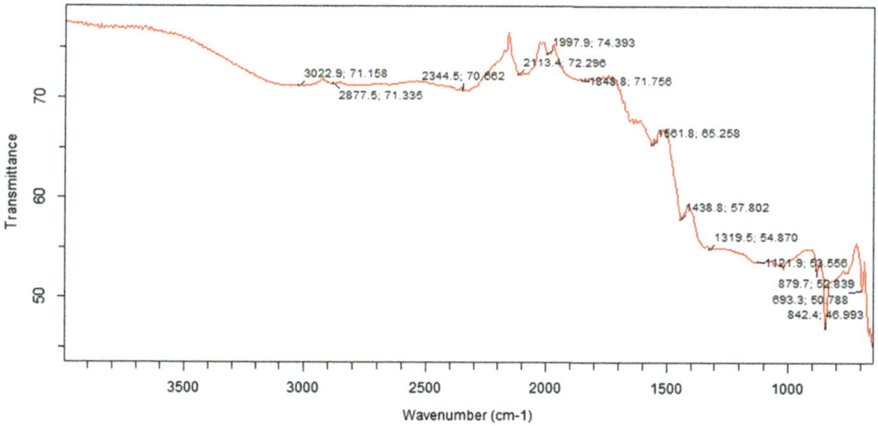

Fig. 2 The FT-IR for coconut shell activated carbon

3.5 *Fourier Transform Infrared Spectroscopy (FTIR)*

The FTIR spectrum analysis of the coconut shell activated carbon was depicted in Fig. 2. The peak observed shows a wide band at 3500–3000 cm^{-1} which correspond to O–H hydroxyl groups stretching vibrations. However, this peak is an evident that shows the presence of phenolic group which is more attributed to activated carbon. The band peak of 2877.6 cm^{-1} derived from the asymmetric aliphatic saturated group causing vibration of CH groups. The peak between 1997.9 and 1501.8 cm^{-1} could be attributed to alkenes conjugate (C=C) which are attributed to cellulose with raw material having the highest strength. The band 1718.8 cm^{-1} which correspond to C=O stretch in aldehydes, saturated aliphatic present in the precursor but absent in the activated carbon and it is as a result of thermal instability of aldehyde and ketone groups at high temperature. The band at 1438.8 to 1121.9 cm^{-1} is associated with C–O stretching vibrations in ethers and alcohol.

3.6 *Effect of Initial Concentration on Adsorption*

The adsorption uptake increased with initial concentration and faster equilibrium was attained at lower concentration. This was assigned to availability of enough vacant active sites at lower concentrations which had limited adsorbate molecules to occupy on the adsorbents. The unoccupied vacant sites present in the adsorption process at low concentrations transformed into lower adsorption uptake of pollutant from solution. More so, at the initial concentrations of 50–100 mg/L for zinc equilibrium condition was achieved after 40 min while 150–250 mg/L heavy metal concentration it took 120–180 min before equilibrium was established. In addition, heavy metal ions concentration at equilibrium is measured at the point where the amount of heavy metal

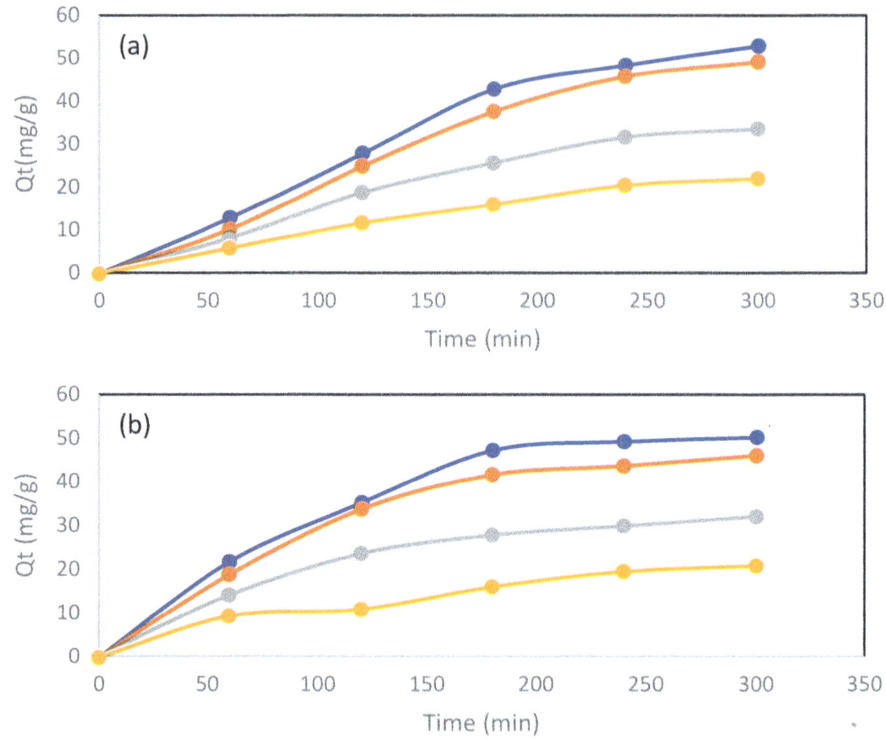

Fig. 3 Effect of initial concentration on **a** lead and **b** zinc adsorption at concentrations 50–250 mg/L

adsorbed on the adsorbents is in a state of dynamic equilibrium with the amount of heavy metal ions desorbing from it and it reveals the maximum adsorption capacity of the adsorbents. The results show that equilibrium position was attained more rapidly at lower initial heavy metal concentrations than at higher initial concentrations of heavy metal ions for the adsorbents. At lower concentrations, there is a smaller number of heavy metal ions competing for the vacant active sites available on the activated carbon (Fig. 3).

3.7 *Effect of Carbon Dosage on Adsorption*

The percentage removal increases significantly with increasing dosage from 0.1 to 1.0 g. The observed behavior can be attributed to the availability of active site on the surface of activated which allows easy penetration of lead and zinc.

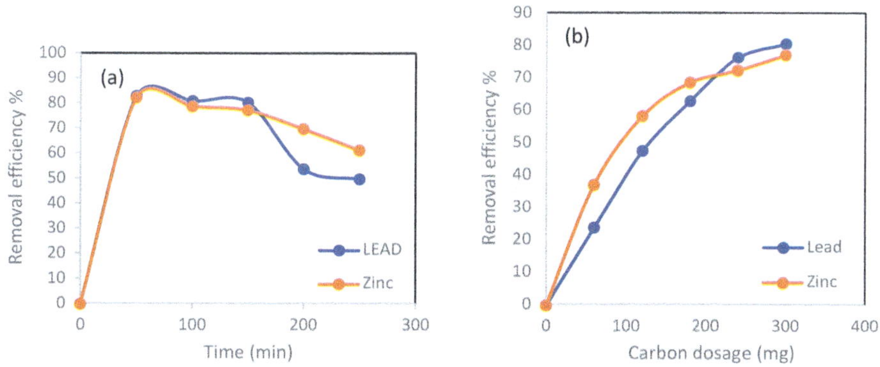

Fig. 4 Removal efficiency for lead and zinc at **a** different contact times and **b** carbon dosage

3.8 *Effect of Contact Time on Adsorption*

The amount of contact surface between an adsorbent and adsorbate solution play an important role in adsorption process. Figure 4 demonstrates that there was an increase in lead and zinc heavy metal ion removal efficiency with increasing dose of the adsorbent because of increase in surface area and availability of more active sites for heavy metals adsorption. Also, the decrease in adsorption capacity is due to the fixed initial concentration, while the dose increased for the fixed solute load resulting the availability of ions per unit mass of adsorbent.

3.9 *Adsorption Isotherm*

Adsorption isotherm is employed in describing the interactions between solutes and adsorbent. The three most common isotherm models, namely Temkin, Freundlich and Langmuir were used to analyze the data as presented in Table 5. The Langmuir

Table 5 Isotherm parameters for lead and zinc adsorption

Langmuir	Q_m(L/mg)	b (L/mg)	R^2
Lead	151.52	0.0021	0.9952
Zinc	112.36	0.0023	0.9917
Freundlich	K_f(mg/g)	1/n	R^2
Lead	1.342	0.5958	0.9954
Zinc	0.750	0.198	0.9977
Temkin	A (L/g)	B	R^2
Lead	0.0285	5.77	0.9509
Zinc	0.0288	7.38	0.8783

isotherm assumes a monolayer adsorption and is given as,

$$\mathrm{Qe} = \frac{QmbCe}{1 + bCe} \tag{7}$$

where Qm is the monolayer adsorption capacity (mg/g) and b is the constant related to the free adsorption energy (L/mg) [10]. The correlation coefficient R^2 of 0.99 and above indicate that the adsorption data of lead and zinc on the activated carbon is fitted to Langmuir isotherm. The values of these parameters were evaluated from the slope and intercept of the graph (Fig. 5).

The Freundlich isotherm is applicable to adsorption process that occur on heterogeneous surface. This isotherm gives an expression that is non-linear form and is defined by Eq. (8),

$$Qe = K_{\mathrm{f}} Ce^{1/n} \tag{8}$$

where K_{f} is the adsorption capacity and 1/n is the adsorption intensity. The value of n correlates to the favourability of adsorption process. For a favourable adsorption process 1/n should be less than 1, both K and 1/n were calculated from the intercept and slop of the graph. The values of 1/n at 0.5958 and 0.198 were less than unity which indicated good adsorption intensity for both lead and zinc, respectively. The regression coefficient of 0.9954 and 0.9977 were very close to unity which indicates that Freundlich isotherm model was favourable and best fitted model.

Temkin isotherm assumes the heat of adsorption of all molecules in the layer would decrease linearly with surface coverage rather than logarithmically due to adsorbent- adsorbate interactions. The Temkin model equation is defined as,

$$Qe = \frac{RT}{b} \mathrm{lnAce} \tag{9}$$

where b is the Temkin constant which stands for heat of sorption, A is the maximum binding energy, R is the ideal gas constant (8.314 J/mol.K) and T is the absolute temperature (K). However, Freundlich isotherm had the best fitting with maximum adsorption capacity of 151.52 and 112.36 mg/g for lead and zinc adsorption. Table 5 shows the adsorption parameters for Langmuir, Freundlich and Temkin isotherm models.

From the Table 5, it is evident that potassium hydroxide activated carbon produced from coconut shell had a very good adsorption capacity. The rate of adsorption b shows that the coconut shell activated carbon had better affinity for lead and zinc.

The favourability of Langmuir isotherm can be expressed in terms of a dimensionless constant separation factor (R_{L}) defined by,

$$R_{\mathrm{L}} = \frac{1}{1 + bC_{\mathrm{o}}} \tag{10}$$

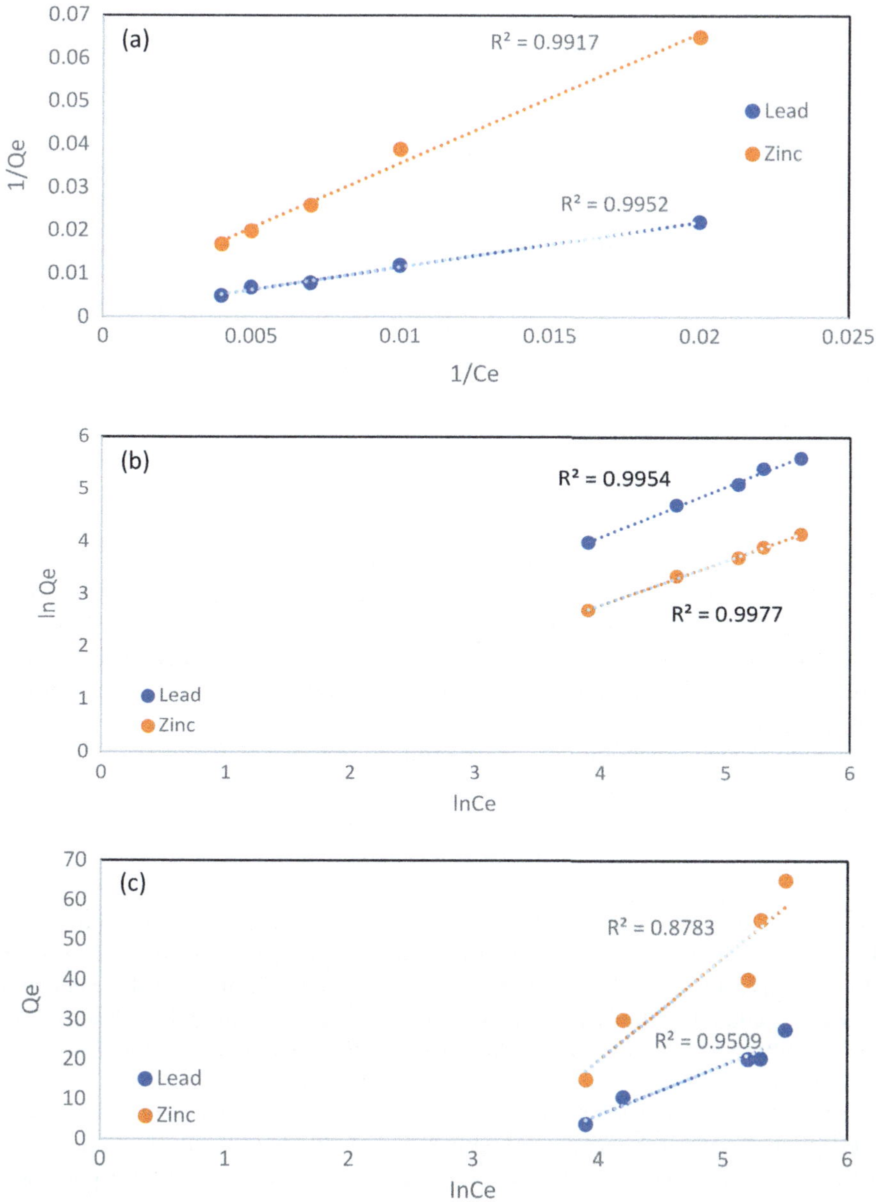

Fig. 5 Linear regression of **a** Langmuir, **b** Freundlich and **c** Temkin isotherms for lead and zinc adsorption

where b is constant, and Co is the highest heavy metal concentration (mg/L). R_L value indicates the adsorption isotherm is unfavourable ($R_L > 1$), linear ($R_L = 1$), favourable ($0 < R_L < 1$) and irreversible ($R_L = 0$) [11]. The values of R_L in this study were found to be 0.142 and 0.083 for adsorption of lead and zinc respectively. This indicates that Langmuir and Freundlich isotherms model was favourable.

3.9.1 Adsorption Kinetics

The kinetic models employed in this study are the pseudo-first-order and pseudo-second-order equations which were used to determine the model that best described adsorption of lead and zinc heavy metal. Pseudo-first-order model is defined by Eq. (11),

$$\log(q_e - q_t) = \log(q_e) - \frac{k_1}{2.303}\mathrm{t} \tag{11}$$

where q_t is the amount of lead and zinc per mass of sorbent at any time, q_e is the amount of lead and zinc per mass of sorbent at equilibrium (mg/g) and k_1 is the rate constant of first order adsorption (l/min). Values of k_1 were evaluated from the slopes of the linear plots of log ($q_e - q_t$) against t. The variations of the calculated q_e values from the experimental q_e values were significantly high, showing that the model was not appropriate.

Pseudo-second-order equilibrium adsorption is defined by Eq. (12),

$$\frac{t}{q_t} = \frac{1}{k_2 q_e 2} + \frac{1}{q_e}\mathrm{t} \tag{12}$$

where k_2(g/mg.min) is the second-order adsorption rate constant, q_e and k_2 can then be determined from the slope and the intercept of plot $\frac{t}{q_t}$ versus t. The linear plot demonstrates a good correlation between the experimental and the calculated q_e values indicating the applicability of the second order model to describe the adsorption process. Table 6 shows the constants of kinetic models for lead and zinc adsorption.

The Pseudo-second-order equation fitted the experimental data with a correlation coefficient close to unity as illustrated in Fig. 6. In Table 6, the deviation obtained between the experimental and calculated values for lead and zinc adsorption was larger, rendering a lower regression coefficient in the pseudo-first-order equation. Conversely, the pseudo-second-order model fitted well with the experimental data, when compared to the pseudo-first-order model.

Table 6 Kinetic models for lead and zinc adsorption

Initial conc. (mg/L)	q_e exp	Pseudo-first-order model			Pseudo-second-order model		
		q_e cal	k_1	R^2	q_e cal	k_2	R^2
Leadadsorption							
250	63.0	72.1	0.011	0.815	63.7	0.0266	0.963
200	51.3	55.7	0.013	0.996	56.8	0.0246	0.965
150	40.8	36.3	0.012	0.987	41.8	0.0438	0.981
100	26.1	27.7	0.011	0.958	27.9	0.0241	0.930
50	14.0	13.9	0.008	0.907	14.1	0.0234	0.896
Zinc adsorption							
250	64.0	67.6	0.0132	0.918	65.4	0.0363	0.971
200	54.5	83.5	0.0134	0.943	69.4	0.0092	0.708
150	41.5	39.9	0.0127	0.990	45.3	0.0280	0.951
100	27.3	33.2	0.0115	0.962	31.0	0.0159	0.882
50	14.9	20.0	0.0076	0.778	17.6	0.0057	0.467

4 Conclusion

Coconut shell is a solid waste material that could be used for activated carbon production using chemical activation. The properties of porous carbon were influenced by activation conditions. Proximate analysis reveals that the precursor is rich in carbon content. CCD of response surface methodology was used to optimize the preparation condition under microwave. The optimum condition of 100 W microwave power, 1.0 impregnation ratio and 2.7 min irradiation time, gives the BET area of 332.79 m^2/g. The activated carbon was used for the removal of lead and zinc, which translated to 64% and 56% removal efficiency, respectively.

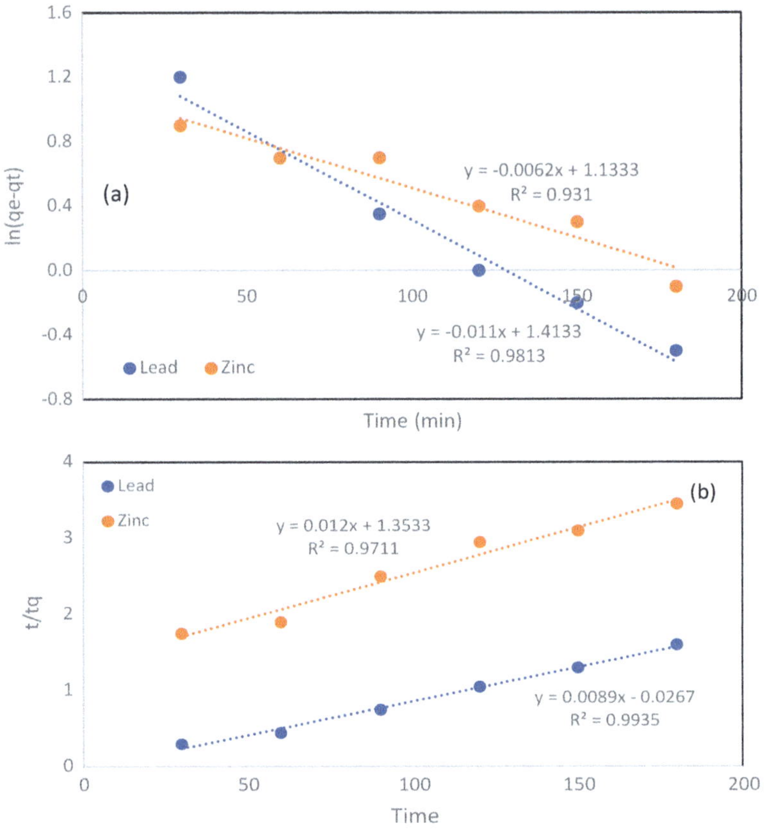

Fig. 6 Curve fittings of **a** Pseudo-first-order and **b** pseudo-second-order models

References

1. Mohammed J, Nasri NS, Zaini MA, Hamza UD, Ahmed MM (2015) Preparation and surface modification of activated carbon from coconut shells for benzene and toluene adsorption from water. Appl Mech Mater 1–207. https://doi.org/10.1016/j.ibiod.2015.02.012
2. Amuda OS, Giwa A, Bello IA (2007) Removal of heavy metal from industrial wastewater using modified activated coconut shell carbon. Biochem Eng J 36:174–181
3. Hesas HR, Arami-Niya A, Daud WWMA, Sahu JN (2013) The effects of a microwave heating method on the production of activated carbon from agricultural waste. A Rev J Anal Appl Pyrol 100:1–11
4. Hamza DU, Nasri NS, Abdul KA, Ahmed MM, Jibril M (2014) Modification of activated carbon for enhancement of gas contaminant removal. *A review:* UMT 11th international annual symposium on sustainability science and management, Terengganu, Malaysia, pp 1336–1342
5. Mohammed J, Nasri NS, Zaini MA, Hamza UD, Ahmed MM (2014) Optimization of preparation microwave irradiated bio-based materials as porous carbons for VOCs removal using response surface methodology. Appl Mech Mater 554:175–179. https://doi.org/10.4028/www.scientific.net/AMM.554.174

6. Gratuito MKB, PanyathanmapornT CRA, Sirinuntawittaya N (2008) Production of activated carbon from coconut shell: optimization using response surface methodology. Bioresour Technol 99(11):4887–4895
7. Yang K, Peng J, Srinivasakannan C, Zhang L, Xia H, Duan X (2010) Preparation of high surface area activated carbon from coconut shells using microwave heating. Bioresour Technol 101:6163–6169
8. Carter K, Nascimento M, Soares PSM, SouzaVP De (2005) Adsorption of heavy metal cations using coal fly ash modified by hydrothermal method. Fuel 88(9):1714–1719
9. Ahmad Hammed MA (2010) Optimization of preparation conditions for mango steen peel-based activated carbon for the removal of remazol brilliant blue using response surface methodology. Chem Eng J 165:883–890
10. Auta M, Hammed BH (2015) Preparation of waste tea activated carbon using potassiumacetate as an activating agent for adsorption of acid blue 25 dye. Chem Eng J 171:502–509
11. Ahmedna M, Marshall ME, Rao RM (2000) Production of granular activated carbon from select agricultural by-products and evaluation of their physical, chemical and adsorption properties. Bioresour Technol 71:113–123

Improvement of the Mechanical Properties by Guar Gum Addition of Epoxidized Natural Rubber/Polyethylene Glycol Composite Membranes

Kritsada Phatcharasit and Wirach Taweepreda

Abstract The aim of this work was improvement of guar gum (GG) as filler on mechanical properties of epoxidized natural rubber/polyethylene glycol (ENR/PEG) composite membranes. In this work, the effects of guar gum in various contents (at 0, 10, 20, 30 and 40 phr) on the membrane structure were investigated to determine the tensile strength, swelling properties, dielectric constant and scanning electron microscope (SEM). ENR/PEG/GG composite membranes were successfully prepared by using two-roll mill, followed by compression molding at 150 °C and directly mixed with crosslinking agents to prepare membranes. The results obtained for tensile properties of membrane increased with the addition of guar gum for 20 phr and decreases with increasing of the guar gum content due to the aggregation and poor dispersion. On the other hand, the addition of guar gum was found that increased of the water uptake properties because guar gum was containing several hydroxyl groups. The dielectric constant showed that increases with increasing the amount of guar gum. The SEM micrographs showed a small pores and fracture surfaces of ENR/PEG/GG composite membranes.

Keywords Composite membranes · ENR · Guar gum · Mechanical properties

1 Introduction

In recent years, polymeric membranes have constituted an increasingly important and cost effective treatment for a widely such as wastewater treatment, separation ability and air pollution but with weakness such as chemical resistance, thermal resistance and low mechanical properties [1]. Polymeric membranes are usually in two main

K. Phatcharasit (✉)
Department of Rubber and Polymer Engineering, Faculty of Engineering, Thaksin University, Phatthalung Campus, Songkhla, Thailand
e-mail: krit_off@hotmail.com

W. Taweepreda
Department of Materials Science and Technology, Faculty of Science, Prince of Songkla University, Songkhla, Thailand

M. A. A. Zaini et al. (eds.), *Proceedings of the 3rd International Conference on Separation Technology*, Lecture Notes in Mechanical Engineering,
https://doi.org/10.1007/978-981-16-0742-4_16

groups bases on the structure of the membrane: porous and dense membranes. The improvement of polymer membranes has been advance the quality can be made by incorporating the organic and inorganic fillers (e.g. zeolite, silica, guar gum) into the polymer matrix [2].

Nowadays, blending of two polymers has a reforming technique to develop new material. The physical properties of polymer blends depend on the dispersion, stiffness and morphology of each constitutive in the properties of blends [3]. Natural rubber (NR) has been known as raw material for many various applications good mechanical properties such as low heat build-up, high mechanical strength, high abrasion resistance and good elasticity [4]. NR is a poor thermal resistance and ozone due to a high unsaturation backbone, while low chemical and oil resistance due to non-polarity. Thus, NR has been chemically modified to epoxidized natural rubber (ENR) for its applications [5]. Epoxidation is prepared with natural rubber (NR) latex processing technology and procedure to increase the polarity of NR by consolidate polar epoxide function groups which are randomly into the molecular chains, increase the polarity of ENR such as oil resistance and reduced air permeability. The mechanical properties increase with increasing ENR composition in the blends but scorch time (T_{s2}) and cure time (T_{c90}) were decrease with increasing ENR composition [3, 6]. Guar gum is a polymeric structure and a natural of hydrocolloidal polysaccharide at high molecular weight, containing several hydroxyl groups.

Guar gum is popularly used as an emulsifying, guar gum for food that is thickening additives, food ingredients, emulsifying additives and stabilizing additives. Guar gum dissolves in polar solvent on dispersion and hydrogen bonds are strong but nonpolar solvents, guar gum forms weak hydrogen bonds [6]. In addition, the crosslinking ability of guar gum was assessed using various crosslinking agents [7], such as borate, glutaraldehyde [8], metal ions, bisacrylamide and metaphosphate. It has been reported that chitosan (1%, w/v) with guar gum (1%, w/v) mixture was 85:15% v/v was found that better mechanical properties [9]. Guar gum films with 10% cloisite 20A were subjected to gamma radiation resulted improved physical properties of guar gum films due to polymer structures [10] and used guar gum into polyester fabric mixed matrix membrane for permeability and water treatment [11].

In this work, the effect of guar gum addition (at 0, 10, 20, 30 and 40 phr) on the tensile properties, water uptake properties, dielectric constant and morphological of epoxidized natural rubber/polyethylene glycol composite membranes were investigated.

2 Experimental

2.1 Materials

Epoxidized natural rubber with 50 mol% epoxidation (ENR-50) was purchased from the Muang Mai Guthrie Public Co., Ltd., Thailand, Polyethylene glycol (PEG) 400

Table 1 The compositions of ENR/PEG/GG composite membranes

Ingredients	phr (part per hundred of rubber)				
ENR/PEG	90/10	90/10	90/10	90/10	90/10
Guar gum (GG)	0	10	20	30	40
ZnO	3	3	3	3	3
Stearic acid	1	1	1	1	1
MBT	2	2	2	2	2
Winstay®L	1	1	1	1	1
Sulfur	2.5	2.5	2.5	2.5	2.5

was purchased from Sigma-Aldrich Co., Ltd., Guar gum powder (purity 99.5% and moisture 9.31%) was purchased from the Chemipan Corporation Co., Ltd. Thailand All chemicals for vulcanization in the including activators (zinc oxide, ZnO) and stearic acid), accelerator (2-mercapto benzothiazole, MBT), antioxidant ((3-3-tert-butyl-4-hydroxy-5-methylphenyl) propionate, Wingstay®L) and sulfur as a vulcanizing agent were supplied from Kij Paiboon Chemical Ltd., Thailand.

2.2 *Preparation of ENR/PEG/GG Composite Membranes*

The ENR/PEG blends loads with guar gum (GG) as filler were prepared by two-roll mill. The compositions of ENR/PEG/GG composite membranes prepared using in the present study are given in Table 1. After mixing, the blends were compression by compression machine at 150 °C to formed plate with thickness of plate approximately 0.4 mm. The samples were constrained at room temperature for 16 h before characterizations.

2.3 *Characterization*

Tensile properties of ENR-GG membranes were determined according to ASTM D 412 using Zwick Roell tensile tester. Five specimens were analyzed to obtain the median value. Water uptake test was carried out by making samples of 2 × 2 × 0.4 cm^3. Three specimens were placed in a deionized water for 24 h at room temperature. The percentage of water uptake was calculated according to the following Eq. (1).

$$\%\text{Water uptake} = [(W_t - W_0)/W_0] \times 100 \qquad (1)$$

Dielectric measurements were carried out at 25 °C using precision LCR meter (Agilent 4285A). The operating conditions were the frequency range 75 kHz up to 30 MHz by. Three specimens were analyzed to obtain the median value and

Studies on the morphology of ENR-GG membranes were carried out using field emission scanning electron microscope (FE-SEM, JEOL JSM-5800 LV) operating at an acceleration voltage of 20 kV. The membranes samples were aluminum stubs with a layer of gold prior to examination.

3 Results and Discussion

3.1 Tensile Properties

The tensile strength and elongation at break of the ENR/PEG/GG composite membranes at room temperature with different blends loading are shown in Fig. 1. the tensile strength of the ENR/PEG with guar gum (GG) blends increased with

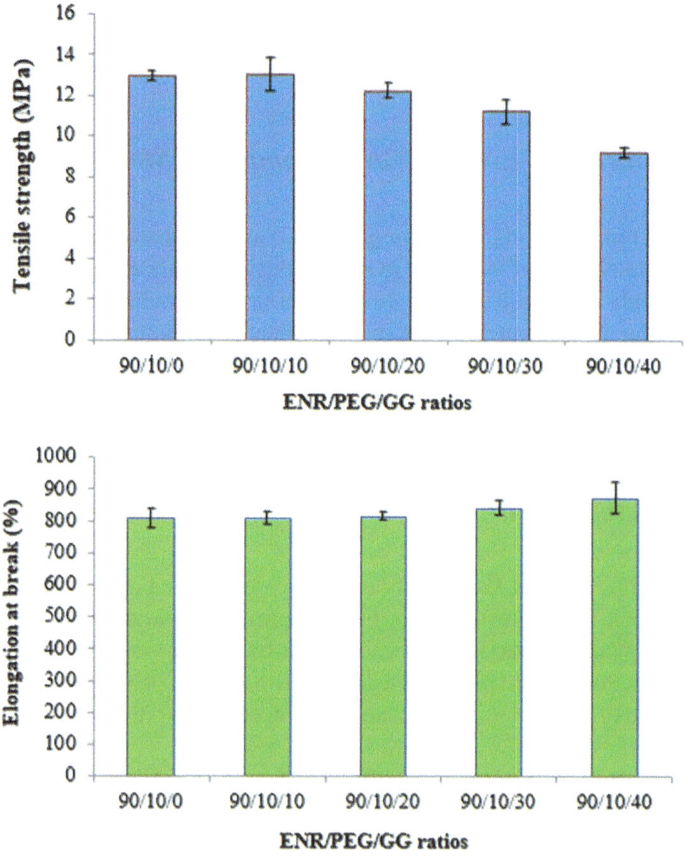

Fig. 1 Tensile and elongation at break properties of the ENR/PEG/GG composite membranes

increasing GG loading to 10 phr at 13 MPa when compared to ENR/PEG alone and tensile strength is likely to decrease. The tensile strength of composite membranes increased with increasing in GG content. This indicated that vulcanized GG might be dispersed phase in ENR matrix and formation of resistant network structure among ENR and guar gum [12] and hydrogen bonds interaction among hydroxyl groups of guar gum and ENR reported that 15% GG increased tensile properties of chitosan film by twofold [9]. On the other hand, decrease of tensile properties value with addition of GG due to aggregates of GG in ENR matrix, it can easily be inserted into crosslink network structure resulting decreased in tensile properties.

The elongation at the break of the ENR/PEG increased with increasing GG content. Possible reason for this may be absorption of more water with increasing GG content due to water which as disrupted intermolecular hydrogen bonds of ENR/PEG and GG. Thus, the chain mobility enlarged that increased the elongation at the break value of the composite membranes [12].

3.2 *Water Uptake*

The water uptake of the composite membranes continuously increased with the addition of GG into ENR/PEG, as shown in Fig. 2. These changes are due to the increased hydrophilicity of the composite membranes when the amount of GG is increased and high water uptake may be due to physical effect occurring in porous structure. By the addition of amounts of GG, the hydrophilicity of ENR/PEG/GG increased which good membranes. Thus, the membrane can absorb water molecules increased, suitable for using in the membrane [11].

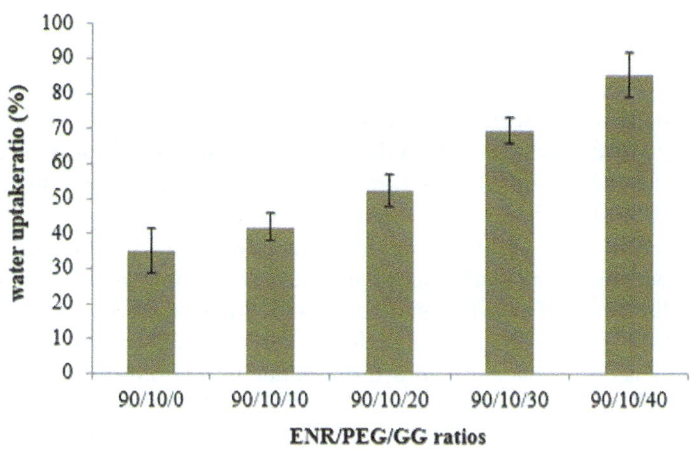

Fig. 2 Water uptake of the ENR/PEG/GG composite membranes

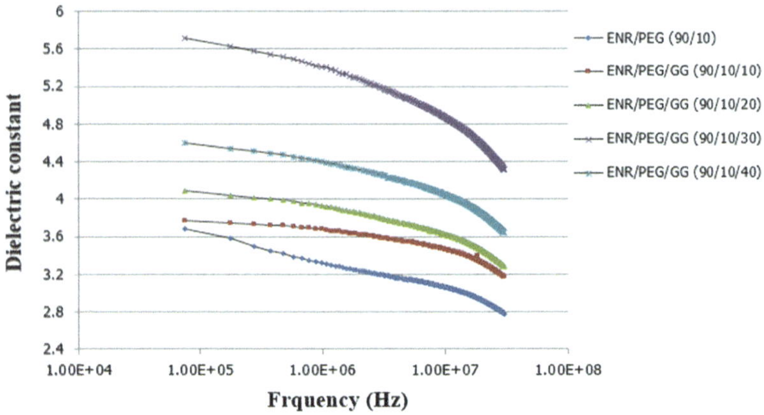

Fig. 3 Dielectric constant of ENR/PEG/GG composite membranes

3.3 Dielectric Constant

The dielectric constant results are shown Fig. 3. It was found that dielectric constant of composite membranes increases with increasing the amount of guar gum due to ENR/PEG blends with increasing GG indicated the hydrophilic improvement and good electrical conductivity. Dielectric constant properties reflect to the polarization from both composition and structure due to the polar functional groups associated and the formation of hydrogen bonds between functional groups of ENR/PEG and GG [13].

3.4 Morphology of Composite Membranes

The SEM images of the surfaces of the ENR/PEG/GG composite membranes were shown in Fig. 4b–e. It was found that the pores increased with the increasing of GG loading as filler and porous surfaces were clearly seen and gave smooth surface, when compared to ENR/PEG. The pores might have generated due to formation of channels in presence of GG [14], it has the ability to permeable membrane well. The GG fillers were well distributed and some aggregates of GG particles were still observed because of the being of conventional composite structure while, fracture surfaces were observed Fig. 4a–e illustrates that the top surface of the composite membranes were heterogeneous because fracture surfaces due to the impact of electron beam during observation [15]. Furthermore, it is seen that GG content increased at 20 phr. This is consistent with the results showing that the tensile properties where reduce in tensile strength was observed with the increase in GG content.

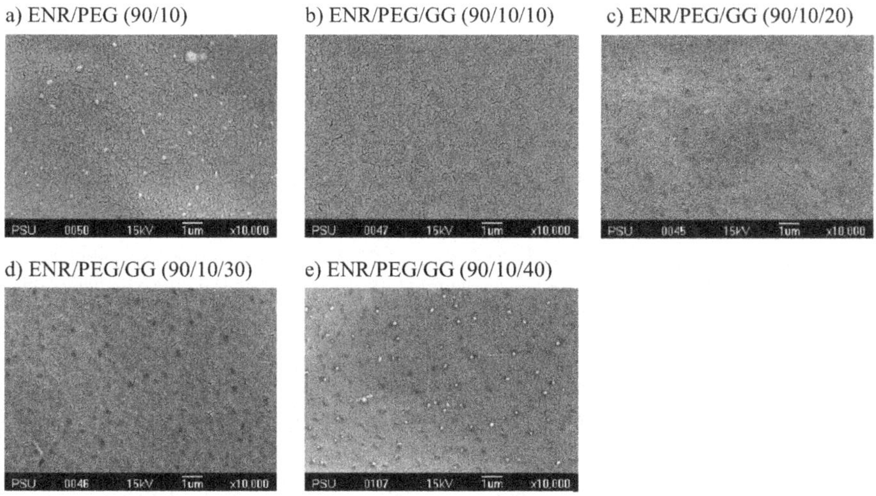

Fig. 4 SEM micrographs of the surface of the ENR/PEG/GG composite membranes

4 Conclusion

In summary, the ENR/PEG/GG composite membranes can be prepared using two-roll mill. The presence of GG as filler to the increases of water uptake properties and dielectric constant due to guar gum was containing several hydroxyl groups with indicated the hydrophilic. The tensile properties reached maximum values increased at 10 phr when more than 20 phr of GG, tensile properties decreased due to the aggregation. The addition of SEM micrographs showed the surfaces of the ENR/PEG/GG composite membranes become smooth and pore, the pores increased with the increasing of GG content. However, this work was improvement to the mechanical properties for composite membranes, the best loading of GG at 30 phr that significantly improved the mechanical properties of the membranes.

Acknowledgements This research was supported by Thailand Research Fund and Thaksin university. Authors would like to thank Department of Rubber and Polymer Engineering, Faculty of Engineering, Thaksin University and Department of Materials Science and Technology, Faculty of Science, Prince of Songkla University.

References

1. Nikolay K, Jorge G, Freek K, Emie JMH (2016) Recent developments in zeolite membranes for gas separation. J Membr Sci 499:65–79. https://doi.org/10.1016/j.memsci.2015.10.049
2. Nazwa J, Norfarhana AS, Nurul AA, Ibrahim A, Rizafizah O (2013) Influence of silica addition on the properties of epoxidised natural rubber/polyvinyl chloride composite membrane. J Appl Polym Sci 129:2789–2795. https://doi.org/10.1002/app.38997

3. Hazwani SJ, Ismail H, Azura AR (2015) ENR-50 compatibilized natural rubber/recycled acrylonitrile-butadiene rubber blends. Sains Malaysiana 44:835–842. https://doi.org/10.17576/jsm-2015-4406-09
4. Manisara P, Sawanya L (2013) Influence of poly(vinyl chloride) on natural rubber/chlorosulfonated polyethylene blends. Open J Org Polym Mater 3:81–86. https://doi.org/10.4236/ojopm.2013.34013
5. Cuong MV, Huong TV, Hyoung JC (2015) Fabrication of natural rubber/epoxidized natural rubber/nanosilica nanocomposites and their physical characteristics. Macromol Res 23(3):284–290. https://doi.org/10.1007/s13233-015-3040-2
6. Saowaroj J, Thongchai N, Onusa S, Anyapong B (2015) Preparation and characterization of epoxidized natural rubber and epoxidized natural rubber/carboxylated styrene butadiene rubber blends. J Metals Mater Miner 25(1):27–36. https://doi.org/10.14456/jmmm.2015.4
7. Mudgil D, Barak S, Khatkar BS (2014) Guar gum: processing, properties and food applications-A Review. J Food Sci Technol 51(3):409–418. https://doi.org/10.1007/s13197-011-0522-x
8. Dai L, Zhang L, Wang B, Yang B, Khan I, Khan A, Ni Y (2017) Multifunctional self-assembling hydrogel from guar gum. Chem Eng J 330:1044–1051. https://doi.org/10.1016/j.cej.2017.08.041
9. Rao MS, Kanatt SR, Chawla SP, Sharma A (2010) Chitosan and guar gum composite films: preparation, physical, mechanical and antimicrobial properties. Carbohyd Polym 82(4):1243–1247. https://doi.org/10.1016/j.carbpol.2010.06.058
10. Saurabh CK, Gupta S, Bahadur J, Mazumder S, Variyar PS, Sharma A (2015) Mechanical and barrier properties of guar gum based nano-composite films. Carbohyd Polym 124:77–84. https://doi.org/10.1016/j.carbpol.2015.02.004
11. Mofradi M, Karimi H, Dashtian K, Ghaedi M (2020) Processing Guar Gum into polyester fabric based promising mixed matrix membrane for water treatment. Carbohydr Polym. In press, J Pre-proof. https://doi.org/10.1016/j.carbpol.2020.116806
12. Nandi S, Guha P (2018) Modelling the effect of guar gum on physical, optical, barrier and mechanical properties of potato starch based composite film. Carbohyd Polym 200:498–507. https://doi.org/10.1016/j.carbpol.2018.08.028
13. Manikoth R, Kanungo I, Fathima NN, Rao JR (2018) Dielectric behaviour and pore size distribution of collagen-guar gum composites: effect of guar gum. Carbohyd Polym 88:628–637. https://doi.org/10.1016/j.carbpol.2012.01.031
14. Al-Hassan AA, Norziah MH (2012) Starch-gelatin ediblefilms: water vaporpermeability and mechanical properties as affected by plasticizers. Food Hydrocolloids 26(1):108–117. https://doi.org/10.1016/j.foodhyd.2011.04.015
15. Jiménez A, Fabra MJ, Talens P, Chiralt A (2012) Edible and biodegradablestarchfilms: a review. Food Bioprocess Technol 5(6):2058–2076. https://doi.org/10.1007/s11947-012-0835-4

Utilizing Natural Fibre as a Sustainable Acoustic Absorber

Mohd Syafiq Syazwan Mustafa, Mohammad Amin Nurasyid, Kamarul Aini Mohd Sari, Fatimah Yusop, Mariah Awang, M. A. A. Rahman, Nuramidah Hamidon, Mohd Kamaruzaman Musa, and Faridahanim Ahmad

Abstract The availability of agricultural waste as sound absorptive materials can be used in indoor and outdoor application. This study aims the potential of natural fibre from agricultural waste as materials for measurement of sound absorption coefficient (SAC). The fibre selected in this study are oil palm, coconut and banana. The natural fibre samples were treated by alkaline treatment during processing stage. The mixture proportion with 95% of natural fibre and 5% of sodium hydroxide (NaOH) as a binder and all the samples follow the fixed ratio of 95:5 for every weight of selected natural fibre. Then, the treated and mixed natural fibre were put into a mould for compaction process to produce diameter of 100 mm and 30 mm with thickness of 20 mm and 30 mm, respectively. All natural fibre samples were tested using Impendence Tube Method (ITM) based on ASTM E1050-09. The tests were conducted according to standard ISO10534-2 with high frequency range and low frequency. As a result, samples of coconut fibre with 20 mm thickness showed a good SAC value at low frequency which is 0.31, meanwhile oil palm fibre with 30 mm thickness attained 0.41. For high frequency testing, coconut fibre sample of 20 mm thickness resulted the highest SAC value at 0.41 and oil palm fibre sample of 30 mm thickness had the highest SAC value which is 0.81. Therefore, natural fibre showed respectable potential as an absorption material for sound reduction.

Keywords Impendence Tube Method (ITM) · Natural fibre · Sound Absorption Coefficient (SAC)

M. S. S. Mustafa (✉) · M. A. Nurasyid · K. A. M. Sari · F. Yusop · M. Awang · M. A. A. Rahman · N. Hamidon · M. K. Musa
Faculty of Engineering Technology, Universiti Tun Hussein Onn Malaysia, Pagoh Kampus, 84600 Muar, Johor, Malaysia
e-mail: mohdsyafiq@uthm.edu.my

F. Ahmad
Faculty of Engineering, Universiti Teknologi Malaysia, Skudai, Johor, Malaysia

M. A. A. Zaini et al. (eds.), *Proceedings of the 3rd International Conference on Separation Technology*, Lecture Notes in Mechanical Engineering,
https://doi.org/10.1007/978-981-16-0742-4_17

1 Introduction

Noise pollution is unwanted or excessive sound that can have adverse effect on human health and environment quality [1]. Noise pollution can lead to a great influence on the environment, human health, economy and noise pollution has become the third pollution resource [2]. As the issue of undesirable and possibly hazardous noise has turned out to become serious, the interest of better environment and residential security is expanded and turns into a major necessity. Different studies focusing acoustic properties have been performed. Acoustical material plays significant roles in acoustic building, for example, the control of room acoustics, industrial noise control, studio acoustics and automotive acoustics [3]. Noise from building can be controlled by implementing a good design for the building. Overall building interior can reduce noise by the extensive use of materials that can absorb sound. Moreover, the advancement study of the acoustic attenuation technique of porous materials offers a wide opportunity [4]. The available usage of agricultural waste as sound absorptive materials can be used in indoor and outdoor application. Agricultural wastes that contain numerous amounts of fibre selected in this study are oil palm, coconut and banana. Overall, 998 million tons of agricultural waste are produced annually, and 1.2 million tons of agricultural waste are disposed of annually in Malaysia in landfills [5].

As indicated by researcher, Malaysia is at present the world's biggest producer and exporter of palm oil [6]. Palm tree is Malaysia biggest agricultural plantation with the aggregate of 4.69 million hectares with existence cycle of 25–30 years for beneficial oil palm. It is estimated that in dry load there would be in excess of seven million tons of oil palm trunk wastage available every year [5]. Malaysia was ranked ninth, before Tanzania, in the top ten coconut producers in the world in 2016 by producing 555,120 tons per year, according to the United Nations Food and Agriculture Organization. Besides coconut wastes generation through processing operations in coconut-based industry, abundant coconut waste is also generated in Malaysia from stalls where coconut juice is sold and the inner part is ground for coconut milk. This leaves the whole part of the coconut head as waste. Malaysia has different type of natural materials that drive the development of small art businesses with eco-friendly idea. Being a potential material, banana bark as the waste delivered by banana plantation is yet underestimated and less famous contrasted with other materials [7]. Despite its plentiful availability in Malaysia, banana bark usage is yet lacking in further improvement.

Most of the building structure used synthetic materials for sound absorption.

panel and still a regular practise for them. The higher amount usage of synthetic materials can cause pollution to the environment and human health in a very short time. The objectives of this study are to investigate alternative materials for sound absorption using natural fibre and selecting the best natural fibre to be used as a replacement for synthetic materials. Due to the problem, natural based materials are selected to replace the synthetic materials because the materials are naturally biodegradable, less hazardous to human health and non-harmful to the environment.

Therefore, the urgency of this study is to produce a practical and high quality of sound absorber for selected natural fibre.

2 Literature Review

Literature review discuss on benefits of natural fibre composites for sound absorber, sustainable panel, alkaline treatment, SAC and fibre size.

2.1 Benefits of Natural Fibre Composites for Sound Absorber

The use of natural fibre, for example, coconut, palm oil and banana reinforced composites have drawn a lot of attention for sound absorption material. Due to their biodegradable, lightweight, less expensive, nontoxic and nonabrasive characteristics, natural fibre are accepting much consideration in composites as a substitute for synthetic fibre for acoustic absorption purposes [8]. The natural fibre with desirable physical and mechanical properties are shown as high performance composites with environmental and economic advantages [4].

The needs of natural fibre to substitute synthetic fibre such as glass wool and rock wool had been discussed among the acoustician today. There are a lot of benefits of natural fibre and can be utilized for sound absorption materials. Natural fibre has been known as a green material because of its biodegradability. Natural fibre additionally known to be lower density contrasted with synthetic fibre, less expensive and renewable [3].

Aside from that, natural fibre processing is more economical and greener towards environment contrasted with synthetic material because of the advancement of technology [5]. Moreover, natural fibre is likewise a lot more secure for human health contrasted with synthetic fibre as it doesn't require any dealing with safety measures. Production of natural fibre includes fundamentally lower carbon emission contrasted with synthetic fibre [8]. In this way, natural fibre has been sorted as environmentally green material, and may hold the key to produce greener sound absorption later on.

2.2 Sustainable Sound Absorption Panel

In recent time, green technology proposed used manufacture materials from agricultural as sustainable material for noise absorption purposes. Natural fibre like oil palm fibre, coconut fibre and banana fibre have numerous advantages, for instance low weight, low density, low cost, acceptable specific properties and recyclable or biodegradable [9]. These materials have shown great highlights from the both of part of sound qualities and mechanical. Previous study showed, at low recurrence, 0 to

500 Hz, sound absorption of rice husk was higher than virgin polyurethane with the estimation of 0.899 at 250 Hz [8].Though, the virgin polyurethane recorded higher absorption at a higher frequency, 2000 Hz with an estimation of 0.679. This demonstrates natural fibre have the potential as filler material of sound absorbent material [8].

Nowadays, in agro-industrial and estate of crop industry, high value of crop waste, which is right now treated as solid waste were created. In practice this residue is burned which contributes to environmental pollution issues in almost regions and offers restricted value to the industry [10]. By thinking about this situation, an alternative practice should be considered by the crop industry to commercialize the residue from crop waste to recycle without causing environmental pollution and produce another product like sound absorption panel. This practice will require less energy, and lessens pollution in modern effluents, as well as being financially advantageous because of its reduced expenses [7].

2.3 Alkaline Treatment

Improvement of hydroxyl groups by hydrophilization may increase the mechanical properties of fibre and produce a stronger bond by enhancing the wetting effect of the resin on the fibre [7]. The research also found an increment in fibre surface roughness and surface area after alkaline treatment. Removal of some impurities through alkaline treatment has made every single fibre clearly observed and exhibits excellent mechanical properties compared to others. Therefore, a treatment is applied. Alkaline treatment increases the surface roughness of fibre by evacuating some important substances like lignin, gelatine, and hemicelluloses of the fibre. In spite of the fact that the removal of these substances brings down the acoustic absorption coefficient of the material, it permits better fibre-binder interface bond, fibre fitness, life span, and anti-fungus quality and in particular reduce the diameter of the fibre [11].

2.4 Sound Absorption Coefficient (SAC)

Sound Absorption Coefficient (SAC) is defined as ratio of absorbed energy to incident energy noting the amount of sound being absorbed by a material [12]. The absorption coefficient ranges between zero and one, one meaning no sound energy is reflected and the sound is either absorbed or transmitted. The higher the percentages, the better the absorption is, indicating most of the sound being absorbed and less is being reflected and is portrayed in the work [7]. As illustrated in Fig. 2.1, it is observed that there are three types of waves encountered by a certain material which are incident sound, reflected sound and transmitted sound. Incident waves are being projected onto a material and it can be reflected or transmitted [1].

Sound absorption coefficient can be calculated, the absorbing coefficient can be mathematically presented as follows in Eq. (1) and (2) [13]:

$$\alpha = \mathrm{EA}/\mathrm{EI} \quad (1)$$

$$\alpha = 1 - |\,\mathrm{r}\,|^2 \quad (2)$$

where;

α = sound absorption coefficient, EI = incident energy, EA = absorbed energy, r = incident reflection factor, IAbs = Sound intensity absorbed.

2.5 *Fibre Size*

Fibre diameter is the most significant physical parameter for enhancement of the sound absorption performance of any fibrous material. The decreasing amount in fibre distance leads to an increased value of the sound absorption coefficient. This is due to more fibre are required to achieve a similar volume density at a similar thickness of the sample material. This results in a higher airflow resistance inside the sample material. Therefore, the acoustical performance of the sample material increases because of the viscous friction through air vibration [11]. The diameter of coconut fibre is 397 μm, oil palm fibre 300 μm and banana fibre 25 μm respectively [14]. All the natural fibres were cut with same length which is 20 mm for sample preparation.

3 Methodology

The method of study involved from selection of natural fibre, then preparation of samples and finally was testing the sample. Details of each process were discussed as below.

3.1 *Selection of Natural Fibre*

Many studies focused on a similar group of natural fibre, which have been discovered and suggested to be used as acoustical panel. Natural based materials are selected because the materials are naturally bio-degradable, less hazardous to human health and non-harmful to the environment [10]. Malaysia has produced a large amount of agricultural waste [5]. This study aims to study about sound absorption of three different natural fibre materials that are abundance in Malaysia. The chosen materials

Table 1 Chemical properties of natural fibre

Fibre	Cellulose (%)	Hemicellulose (%)	Lignin (%)	Tensile strength (MPa)	Diameter (μm)
Coconut	68	22	20.6	88.63	397
Oil palm	48	22	25	350 ±7	300
Banana	63	19	5	550 ± 6.8	25

are natural fibre which are coconut coir fibre, oil palm fibre and banana trunk fibre. All of the natural fibre were obtained from Koperasi Pekebun Kecil Daerah Kluang (KOPEDAK) in Simpang Renggam. Table 1 shows the properties of selected natural fibre [14].

3.2 Raw Materials Preparation

Natural fibre waste was collected at KOPEDAK and extracted by using hand scrapping method. Then, the processed fibre were sun-dried for two weeks with an average temperature, 31.5 °C and average relative humidity of 96%. Sun-dried method was used to ensure the natural fibre went completely dried and ready to be used before panel formation. Figure 2 shows the flow of raw natural fibre preparation.

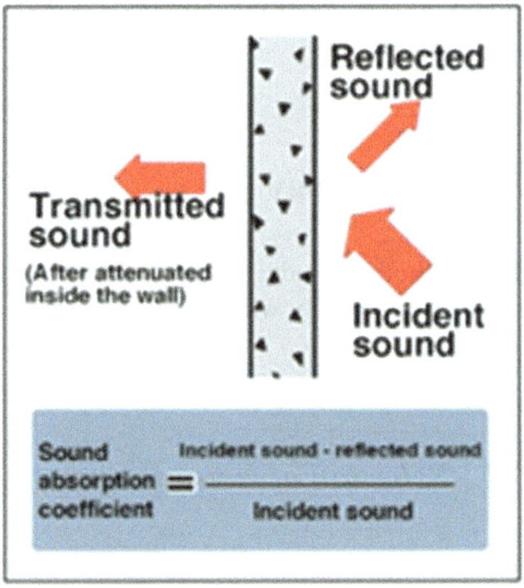

Fig. 1 Schematic representation of sound absorption coefficient

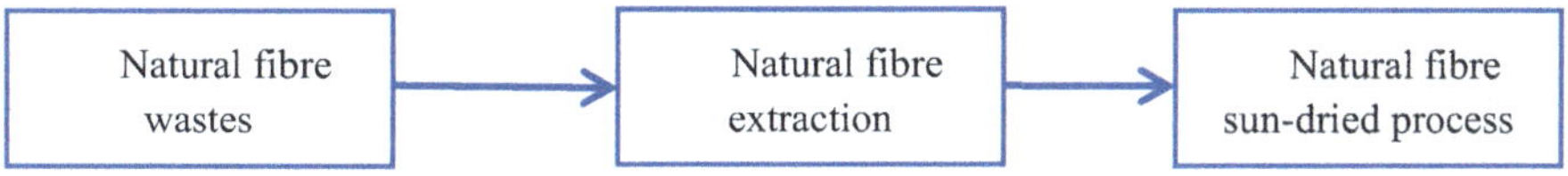

Fig. 2 Raw natural fibre preparation process

3.3 Natural Fibre Sample Preparation

Figure 3 shows the flow of natural fibre samples preparation. There were 6 steps for the sample preparation.

3.3.1 Weighting of Natural Fibre

The materials were being weighted separately and mixed with 50% sodium hydroxide (NaOH) solution for alkaline treatment. The mixing proportion ratio for each natural fibre sample was fixed at 95:5 with 30 and 20 mm thickness as shown in Table 2 for sample 100 mm diameter and Table 3 for 30 mm diameter. All the natural fibre went through the same procedure for sample preparation.

3.3.2 Alkaline Treatment

After raw natural fibre went through drying process, the natural fibre were cut with the same length of 20 mm. Mixing proportion of each natural fibre and NaOH solution was fixed at ratio 95:5. Alkaline treatment act as a binder which increases the surface roughness of fibre by evacuating some important substances like lignin, gelatine, and hemicelluloses of the fibre.

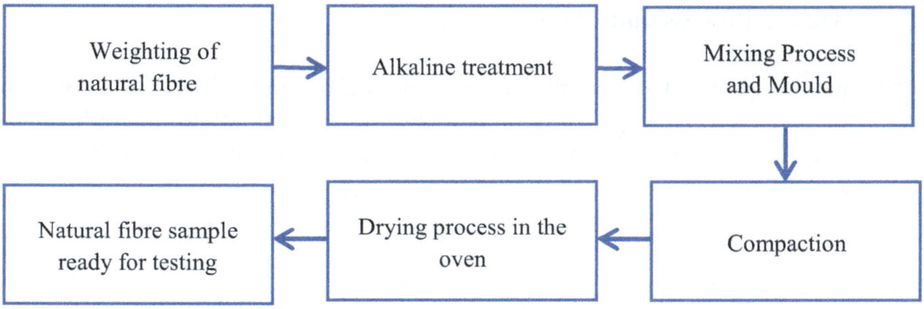

Fig. 3 Flow of natural fibre sample preparation

Table 2 Mixing proportion of 100 mm diameter of natural fibre sample

100 mm diameter samples	Ratio of NaOH to natural fibre
Oil palm fibre sample (30 mm thickness)	150 g: 8.33 g
Coconut fibre sample (30 mm thickness)	150 g: 8.33 g
Banana fibre sample (30 mm thickness)	150 g: 8.33 g
Oil palm fibre sample (20 mm thickness)	100 g: 5.55 g
Coconut fibre sample (20 mm thickness)	100 g: 5.55 g
Banana fibre sample (20 mm thickness)	100 g: 5.55 g

Table 3 Mixing proportion of 30 mm diameter of natural fibre sample

30 mm diameter samples	Ratio of NaOH to natural fibre
Oil palm fibre sample (30 mm thickness)	50 g: 2.63 g
Coconut fibre sample (30 mm thickness)	50 g: 2.63 g
Banana fibre sample (30 mm thickness)	50 g: 2.63 g
Oil palm fibre sample (20 mm thickness)	30 g: 1.57 g
Coconut fibre sample (20 mm thickness)	30 g: 1.57 g
Banana fibre sample (20 mm thickness)	30 g: 1.57 g

3.3.3 Mixing Process and Mould

For Sound Absorption Coefficient (SAC) testing inside Impendence Tube, 30 mm sample diameter was required for high frequency testing and 100 mm sample diameter was required for low frequency testing. Each sample had two different thickness, 20 mm and 30 mm (Fig. 4).

3.3.4 Compaction

Compaction process was conducted by using soil compactor. The average blow pressure of the rammer into the mould is 3.7 kPa to ensure the natural fibre bind and compact together to produce panel absorption board.

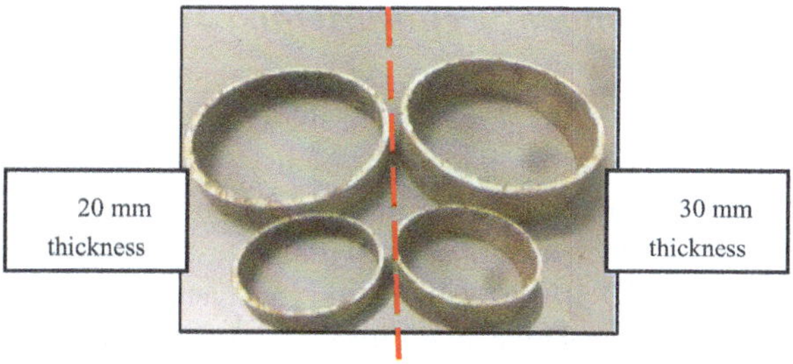

Fig. 4 Mould with diameter of 100 mm and 30 mm for sample preparation

3.3.5 Drying Process in the Oven

After all the natural fibre samples were fully compacted, the samples were demoulded and placed into dry oven for 2 h at 100 °C. To ensure all the natural fibre samples completely dried, all the samples were placed again into the oven for 3 days at temperature of 35 °C. Drying process of natural fibre samples was conducted at Highway Laboratory, UTHM Pagoh Campus.

3.3.6 Natural Fibre Samples

Figure 5 shows natural fibre samples after completely dried in the oven. After that, the samples were exposed at room temperature for 24 h to ensure it totally dried and ready for Impendence Tube Method (ITM) testing.

Fig. 5 Natural fibre samples with 100 mm diameter and 20 mm thickness

3.4 Impendence Tube Method Testing

Sound absorption properties of the samples were tested at Building Services Laboratory, UTHM Pagoh Campus by using impendence tube method (ITM). This ITM used two microphones to transfer the sound wave and it required two sample size diameter which are 30 and 100 mm to test the frequency from 75 to 5000 Hz as indicated by ISO 10,534-2 standard [15]. Factors such as the surrounding setting of air, temperature, relative humidity, and atmospheric pressure inside the room at the start of the procedure and the sample's insertion are important and must be considered [16]. The test sample was placed at one end of a straight, rigid, and smooth impedance tube. Two microphones position will transfer the frequency and the SAC value will be determined. The absorption coefficient can be assessed from the direct measurement of the reflection coefficient. Plane waves are produced in the tube by a sound source, and the sound pressures are assessed at two locations from a short distance to the sample [17]. The complex acoustic transfer function of the two microphone signals is identified and used to compute normal-incidence absorption coefficient of test material. "AED 1001" is a computer software that is used to interpret sound absorption coefficient of samples.

4 Result and Discussions

After all the testing being done for three types of fibre, the results were recorded and analyse for two different frequency (low and high) with two thickness (20 mm and 30 mm).

4.1 Low Frequency Testing

Samples of 100 mm diameter were used to test low frequency of SAC ranging from 75 to 500 Hz. The sound pressure will be measured in two microphones positions and the transfer function between them will be determined. The absorption coefficient can be assessed from the direct measurement of the reflection coefficient.

4.1.1 Natural Fibre 20 mm Thickness

Figure 6 shows the comparison of the sound absorption coefficient (SAC) of three different fibre which are oil palm fibre, coconut fibre and banana fibre for thickness of 20 mm. Coconut fibre sample shows the highest SAC value among all other natural fibre. The highest SAC value for the three natural fibre samples at low frequency

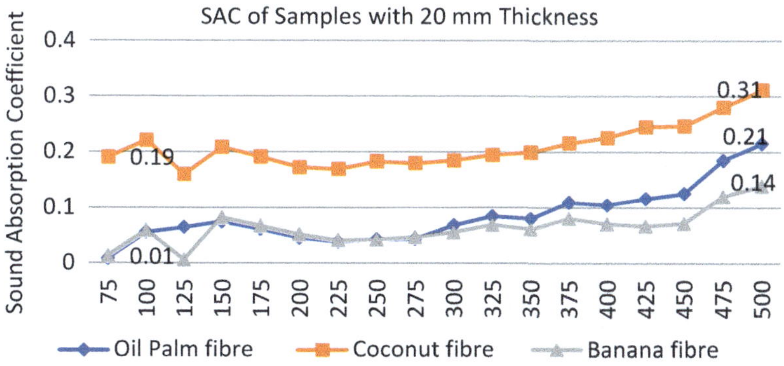

Fig. 6 Low frequency of SAC for three samples with 20 mm thickness

range were at 500 Hz. Coconut fibre sample had the highest SAC value which is 0.31 followed by oil palm fibre sample; 0.21 and banana fibre sample at 0.14.

4.1.2 Natural Fibre 30 mm Thickness

Figure 7 shows the graph SAC value against frequency of three different fibre which are oil palm fibre, coconut fibre and banana fibre for thickness of 30 mm. SAC of banana fibre was the highest at the beginning which was 0.19 compared to oil palm and coconut fibre which was 0.008 and 0.006 respectively. Banana fibre had higher potential of SAC at frequency 75–325 Hz, but oil palm fibre had higher potential of SAC at frequency 350–500 Hz. Maximum value of SAC reached by the three natural fibre sample was at frequency 500 Hz. Oil palm fibre had the highest SAC value at low frequency range which is 0.53 followed by coconut fibre at 0.35 and banana fibre at 0.31.

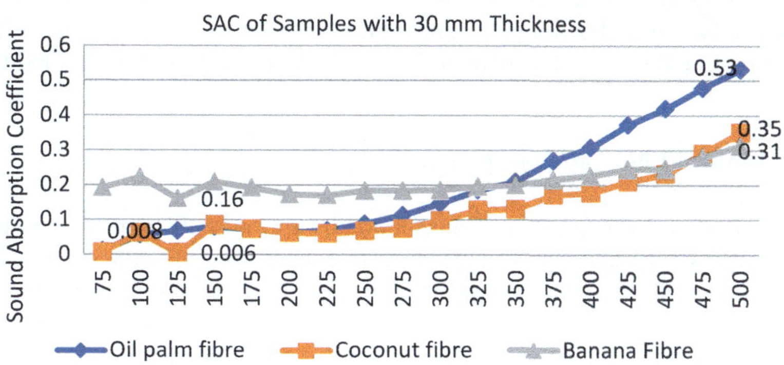

Fig. 7 Low frequency of SAC for three samples with 30 mm thickness

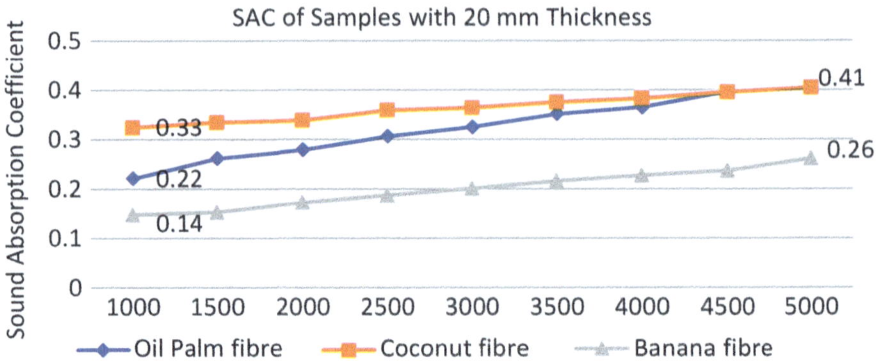

Fig. 8 High frequency of SAC for three samples with 20 mm thickness

4.2 *High Frequency Testing*

Samples with 30 mm diameter were used to test high frequency of SAC ranging from 1000 to 5000 Hz. For absorption coefficient measurements using impedance tube method, it is important to consider the surrounding setting of air temperature, relative humidity, and atmospheric pressure inside the room from beginning of the procedure and the sample's insertion.

4.2.1 Natural Fibre 20 mm Thickness

Figure 8 shows the comparison of the sound absorption coefficient (SAC) of three different fibre for thickness of 20 mm at high frequency range. Coconut fibre sample shows the highest SAC value for high range frequency compared to oil palm fibre and banana fibre. SAC of banana fibre is the lowest at high range frequency as its maximum value does not reach minimum value of coconut fibre sample. At frequency of 4500 Hz and 5000 Hz, oil palm fibre and coconut fibre meet the same value of SAC. Both oil palm fibre and coconut fibre reach SAC value of 0.41 at 5000 Hz.

4.2.2 Natural Fibre 30 mm Thickness

Figure 9 shows the comparison of the sound absorption coefficient (SAC) of three different fibre for thickness of 30 mm at high frequency range. For the graph obtained from Impendence Tube Method (ITM), oil palm fibre shows great performance at high frequency range compared to other natural fibre. The SAC increases with frequency and reach maximum value at 5000 Hz. The highest SAC value obtained by banana fibre which is 0.52 did not reach the minimum SAC value of oil palm fibre at high frequency range. The highest SAC value that obtained by oil palm fibre

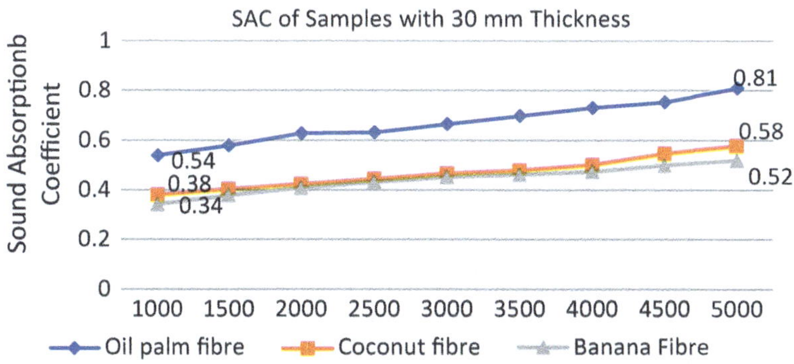

Fig. 9 High frequency of SAC for three samples with 30 mm thickness

and coconut fibre sample is 0.81 at 5000 Hz. SAC of oil palm fibre is the highest at 5000 Hz compared to coconut and banana which are 0.58 and 0.52 respectively.

5 Conclusion

This study showed that by using the Impedance Tube Method (ITM) test to determine sound absorption coefficients of selected natural fibre have been successfully carried out. Samples of sound absorber have been made from natural fibre with 5% alkaline treatment. The sample testing showed that these natural fibre can be good alternatives for sound absorber panel to be installed in the building. Coconut fibre sample with thickness of 20 mm showed great SAC value at low frequency but oil palm fibre sample showed great SAC value with thickness of 30 mm. The highest value obtained by coconut fibre sample with thickness of 20 mm in low frequency range was 0.31. For high frequency range for the sample thickness of 20 mm, SAC value of coconut fibre sample was 0.41 which is the highest. For the sample thickness of 30 mm at high frequency range, the highest SAC value was obtained by oil palm fibre sample which is 0.81. For overall data obtained from ITM testing, the SAC value for the samples might be different according to the type of material used and the thickness of sample. Compared to the study carried out by Yang et al. [2], The result showed that the highest SAC that can be achieved by synthetic fibre for fibre carbon and fibre glass were 0.43 and 0.41 respectively. This showed that natural fibre composite had greater SAC value compared to fibre carbon and fibre glass due to their unique hollow lumen. Thus, natural fibre has high potential to be used as a material for sound absorption panel and alternative replacement for the current synthetic fibre materials.

Acknowledgements Researchers would like to express gratitude and appreciation to Tun Hussein Onn University of Malaysia (UTHM) and all the research partners that contributed their knowledge and expertise in assisting to complete this study.

References

1. Ghorbani K, Hasani H, Zarrebini M Saghafi R (2016) An investigation into sound transmission loss by polypropylene needle-punched nonwovens. Alexandria Eng J 55(2):907–914
2. Yang WD, Li Y (2012) Sound absorption performance of natural fibers and their composites. Sci China Technol Sci 55
3. Zulkifli R, Zulkarnain, Nor MJM (2010) Noise control using coconut coir fiber sound absorber with porous layer backing and perforated panel. Am J Appl Sci 7(2), 260–264, ISSN 1546-9239
4. Shahani P, Soltani ZM (2014) The analysis of acoustic characteristics and sound absorption coefficient of needle punched nonwoven fabrics. J Eng Fibers Fabr 9(2):84–92
5. Agamuthu P, A review of Malaysia Agricultural wastes. Institute of Biological Sciences, Faculty of Science University of Malaya, vol 7(8), pp 1656–1661
6. Wahid MB, Abdullah SNA, Henson IE (2004) Oil palm: aachievements and potential in new directions for a diverse planet. In: Proceeding of the 4th international crops sciences congress, Brisbane, Australia
7. Kuczmarski MA, Johnston JC (2011) Acoustic absorption in porous materials. NASA, Ohio
8. Julia E, Segura J, Nadal A, Gadea JM, Crespo JE (2013) Study of sound absorption of multilayer panels made from ground tyre rubbers. Annals of the Oradea University, pp 147–150
9. Nor MJM, Amin N, Zulkifli R, Fouladi MH, Ismail AR (2009) Analysis on sound absorption of natural coir fiber using delany-Bazley model. Int Conf Mech Eng, 26–28
10. Rahman LA, Raja RI, Roslan AR, Ibrahim Z (2014) Comparison of acoustic characteristics of date palm fibre and oil palm fibre. Res J Appl Sci, Eng Technol 7(8):1656–1661
11. Koizumi T, Tsujiuchi N, Adachi A (2002) The development of sound absorbing materials using natural bamboo fibers. High Perform Struct Compos: The Built Environ, 157–166
12. Bai MR, Lo YY, Chen YS (2015) Impedance measurement techniques for one-port and two-port networks. J Acoust Soc Am, 2279
13. Breysse PN, Lees PS (2006) Noise. The Johns Hopkins University. Bureau of Indian Standards (2006). Fiber hardboards—specification (third revision). Wood and Other Lignocellulosic Products Sectional Committee, CED 20. IS 1658:2006F
14. Mat Hassan NN, Rus AZM (2013) Influences of thickness and fabric for sound absorption of biopolymer composite. Appl Mech Mater, 102–107
15. ISO 10534-2 (1998) Acoustics—determination of sound absorption coefficient and impedance in impedance tubes—Part 2: Transfer-function method
16. Knapen E, Lanoye R, Vermeir G, Van Gemert D (2003) Sound absorption by polymer-modified porous cement mortars. In: 6th international conference on materials science and restoration, MSR-VI edification publishers, pp 347–358
17. Adnan SH, Kamarulzaman NA, Mohd Sari KA, Osman MH, Jamellodin Z, Abdul Hamid NA, Alisibramulisi A, Abdul Roni N, Mohd Yassin NI, Wahee Anuar MNA (2020) Sound absorption of lightweight brick containing expanded polystyrene beads and palm oil fuel ash. IOP Conf Ser Earth Environ Sci 476, 1–10

Effect of Percent Shrinkage and Adsorption of Zinc for Coconut Coir Based Aerogels and Xerogels

Nik Muhammad Faisal Mat Yasin, Noor Aliya Binti Muhammad Kamal, Ana Najwa Mustapa, and Azil Bahari Alias

Abstract Recently, heavy metal contamination is a serious issue due to degradation of health. Various methods for heavy metal treatment in contaminated water have been developed. Adsorption method using low-cost adsorbents is one of the promising techniques. Internal gelation is one of methods to produce hydrogel by cross linked chemically to produce material that consider promising adsorbents in adsorbing heavy metal. This study focus on combining coconut coir with sodium alginate. In this study, alginate hydrogels of 0, 2, 5, 10 and 15 wt% of coconut coir mass ratio were produced via sol–gel processing method and dried by supercritical CO_2 drying to obtain high porous structure of alginate aerogels. Meanwhile, structure obtained from oven drying is termed as xerogels. Both aerogels and xerogels are then characterize by performing several analyses such as Thermogravimetric Analyzer (TGA), Brunaeur-Emmett Teller (BET) and Fourier Transform Infrared Spectroscopy (FTIR). Their percent of shrinkage after drying and adsorption efficiency towards zinc ions are observed. Aerogels yield less amount of shrinkage compared to xerogels. From the analyses, composition of 5 wt% of coconut coir mass ratio gave the best characteristics as excellent adsorbent for both aerogels and xerogels and this is also correlated with their porous characteristics. Although high percent of shrinkage experienced by xerogels, it did not affect its adsorption efficiency towards zinc ions. Hence, coconut coir-based aerogels and xerogels have a great potential to be applied as adsorbent to remove zinc in wastewater.

Keywords Supercritical drying · Heavy metal · Aerogels · Coconut coir · Xerogels · Oven drying · Adsorption · Zinc

N. M. F. M. Yasin · N. A. B. M. Kamal · A. N. Mustapa · A. B. Alias (✉)
Faculty of Chemical Engineering, Universiti Teknologi MARA, Selangor, Malaysia
e-mail: azilbahari@uitm.edu.my

A. B. Alias
Centre of Industrial Process Reliability & Sustainability (INPRES), UiTM Shah Alam, Selangor, Malaysia

M. A. A. Zaini et al. (eds.), *Proceedings of the 3rd International Conference on Separation Technology*, Lecture Notes in Mechanical Engineering,
https://doi.org/10.1007/978-981-16-0742-4_18

1 Introduction

Good water quality is important to human health, social and economic development. Nowadays, to ensure there are sufficient and safe water supplies for everyone is becoming a challenge as population growth and natural environments become degrade. One of the issues that contributed to environmental issues today is heavy metal pollution which is due to the rapid development of industries. Due to the cost and effectiveness of processes such as precipitation and coagulation, these processes are considered inefficient to treat wastewaters hat contain heavy metals [1]. Hence, studies of heavy metals removal especially in conventional method such as carbon adsorption has gained many researchers' attention. Some of the common heavy metals that are generated during manufacturing process of industrial sectors are mercury (Hg), lead (Pb) and zinc (Zn). Zinc is essential for human health but consume large quantities of zinc can cause stomach cramps, vomiting, anemia, and skin irritation. While lead and mercury is harmful to human health toward to organ and nervous system [2, 3]. Hence it is crucial to remove zinc before discharging water for daily uses.

There are several methods in removing heavy metal, these methods can be divided into two types which are physical and chemical. The common practice in industry for removing heavy metals are chemical precipitation, ion exchange, electrolytic recovery and electrodialysis, membrane separation, coagulation and adsorption [4]. Among these methods, adsorption is the most efficient as the other techniques have inherent limitations such as the generation of a large amount of sludge, low efficiency, sensitive operating conditions and costly disposal [3].

In adsorption method, adsorbent plays an important role in determine effectiveness of the method. There are two types of adsorbent which are commercial adsorbent and natural adsorbent. Commercial adsorbent such as activated carbon and silica based materials have complex production, expensive cost and usually from non-renewable sources [5]. Natural adsorbents are those obtained from biological material such as coconut coir, rice husk and empty fruit bunch. According to Behera [6], biomass has captured global attention due to their widespread availability, low in cost, renewable and environmental friendly [6]. Coconut is widely known to be able to adsorb heavy metal such as Cr, Cu, Zn, Ni and Pb, but all the adsorbent used was modified either in physical or chemical treatment to improve adsorption [7]. This research focus on turning coconut coir into aerogels and xerogels by using internal gelation and different drying.

In theory, aerogels and xerogels will improve structure of coconut coir. Aerogels can only produce by drying using supercritical CO_2 while xerogels can only produce by drying using thermal drying method such as oven. Alginate aerogels and xerogels are a class of biopolymer polysaccharides characterized by high surface area, open porosity and good compatibility.

Thus, the goal of this research is to synthesize xerogels and aerogels using internal gelation and drying method. The second goal is to compare xerogels and aerogels in terms of percent shrinkage and adsorption of zinc by varying mass ratio of coconut coir. The main focus of characterization is toward to surface area, pore size and pore volume.

2 Methodology

2.1 Material

In this experiment, the coconut coir was obtained from Padang Jawa, Shah Alam. Chemicals such as Sodium Carbonate ($CaCO_3$), solvent which is Ethyl Alcohol (C_2H_5OH) and carrier material such as Sodium Alginate was obtained from Reichle & De-Massari (R&M). Glucono-δ-lactone (GDL) was obtained from Alfa/US.

2.2 Mechanical Pre-Treatment of Coconut Coir

The coconut coir collected was dried and cut into smaller pieces before grinded using laboratory blender model 8010ES/8010EG at constant speed of 5000 revolutions per minute (rpm) for 20 min. The grinded coconut coir will be sieved so that the size less than 200 microns. This method of preparation was adapted with minor modification from Asim et al. [8].

2.3 Preparation of Coconut Coir-Based Sodium Alginate Composite

0.5 g Sodium Alginate was added to 50 mL distilled water and stirred using magnetic stirrer until homogenized mixture was achieved. 0.1 g of $CaCO_3$ was added to homogeneous mixture and stirred until it was homogenized followed by 0.5 g of grind coconut coir. This step was repeated for 0.75, 1.0, 1.25 g of grind coconut coir. Next, 0.15 g of GDL was added to the mixture and stirred until it was homogenized before transferred to the square mold and placed in the refrigerator until they were completely gel. After that, hydrogel was removed from the square mold and transferred into petri dish. Then, dimension of the samples was recorded. The samples were immersed in ethanol for solvent exchange purpose. The hydrogel will be immersed in 50%, and 75% v/v of ethanol for 2 h, then immersed in 100% v/v for 24 h. The sample will be termed as alcogel after solvent exchange occurred. This method was adopted from Co et al. with some minor modifications [9].

2.4 Supercritical CO_2 ($sCCO_2$) Drying

Supercritical CO_2 drying was used to remove ethanol in the alcogel thus changing alcogel into aerogels. The method was adapted from Mustapa et al. (2017) with minor modification. The alcogel will be dried at pressure 120 bar and temperature of 40 °C for 5 h. After drying, the system needs to be depressurized at 2 bar per minute.

2.5 Oven Drying

All the steps were repeated except for solvent exchange and the hydrogel of 0, 2, 5, 10 and 15 wt% were dried using oven drying at 60°C for 24 h. After dried, xerogels were placed inside incubator to slowly reduce temperature into room temperature.

2.6 Materials Characterization

The characterization analysis such as Brunauer Emmett-Teller (BET), Fourier Transform Infrared Spectroscopy (FTIR) and Thermal Gravimetric Analyzer (TGA) will be used to determine whether coconut coir-based aerogels and xerogels are an adsorptive material. The textural properties of the materials were characterized by nitrogen adsorption–desorption analysis using Brunaeur-Emmett Teller (BET) 3Flex Micromeritics at degassed temperature of 70°C to determine the specific surface area, pore diameter and pore distribution. The surface area was calculated based on the method of Brunauer, Emmett and Teller (BET) meanwhile the pore diameter and its distribution were calculated from the desorption isotherms. The chemical properties of raw coconut coir, 5 wt% of coconut coir-based aerogels before adsorption of zinc, and 5 wt% of coconut coir-based aerogels after adsorption of zinc was determined with the FTIR analysis using Perkin Elmer Spectrum One FTIR Spectrometer. Thermogravimetric Analyzer TGA/SDTA51 manufactured by Mettler Toledo TGA SC is used to determine the proximate analysis of raw coconut coir, blank xerogels and 5 wt% of coconut coir based xerogels at 600°C, 2°C min^{-1} under nitrogen atmosphere at 25 mL/min flow velocity.

2.7 Adsorption of Zinc Ions

In order to analyze the concentration of heavy metals, we use Atomic Adsorption Spectroscopy (AAS) Hitachi z2000. Zinc chloride is used for the preparation of salt solution that consist Zn (II) ions. 50 mL of zinc solution is prepared in a beaker with initial concentration of 4 mg/L. 2 g of sample was added into the beaker that containing 50 mL of zinc solution. The beaker was shaken at 100 rpm for 10 min. After that, 10 mL sample of zinc solution were taken and analysed using AAS. In order to get accurate results, the steps were repeated three times to get average. This method was adapted from Nurul Farhana et al. [10] with some minor modifications. The amount of zinc adsorbed at time were analysed by using Eq. (1)

$$\text{Zinc ion removal}\,(\%) = \frac{C_0 - C_e}{C_0} \times 100 \quad (1)$$

where: *Co* is initial concentration of zinc chloride solution, *Ce* is final equilibrium of zinc chloride solution.

3 Results and Discussion

3.1 Materials Characterization

Proximate analysis is performed to measure the moisture content, volatile matter, ash contained and fixed carbon in the raw coconut coir, coconut coir-based aerogels and xerogels. The results of proximate analysis of three samples which are raw coconut coir, blank xerogels and 5 wt% of coconut coir based xerogels are presented in Table 1.

When aerogels is subjected to high temperatures in a thermogravimetric analyzer, it is found out that its fixed carbon value is 13.42%. Moisture content value obtained for coconut coir in this study 11.01% which is almost similar to the value reported by Wang S et al [11] which is 12.94%. According to Bryan et al. (2019) if biomass has a moisture content less than 30%, then it is suitable for thermochemical conversion [12]. Volatile matter value for coconut coir obtained from this study is 61.55% and the ash content value is 13.74%. According to Zhengang Liu et al. (2016), higher amount of volatile matter lead to lesser amount of ash content [11]. Lower ash content value is could be due to low mineral content in the coconut coir which majorly contribute to ash composition [13]. From Table 1, it can be seen that blank xerogels sample

Table 1 Proximate analysis (wt% dry basis)

Properties	Raw coconut coir	Blank xerogels	5 wt% coconut coir
Moisture content	11.01	8.50	12.07
Volatile matter	61.55	65.72	63.02
Ash	13.74	3.5	13.81
Fixed carbon	13.42	21.54	10.83

which contained mixture of GDL and $CaCO_3$ has the highest amount of fixed carbon which is 21.54% compared to raw coconut coir and 5 wt% coconut coir samples which are 13.42% and 10.83% respectively. Blank xerogels sample also has the lowest ash content and moisture content which is 3.50% and 8.50% respectively. However, blank xerogels sample has the highest amount of volatile matter which is 65.72%. When comparing proximate analysis value of blank xerogels sample and raw biomass sample, it is noticed that there is decreasing trend in moisture content and ash content value, meanwhile, increasing trend in volatile matter content and fixed carbon content. Decomposition and devolatilization reaction occurred during decomposition of lignocellulose cell contained in biomass affected trend in amount of fixed carbon [14].

The proximate analysis was repeated for a sample of 5 wt% coconut coir after undergo oven drying. It was noticed that the fixed carbon value is decreased from 21.54 to 10.83% when comparing with the blank xerogels sample value. The volatile matter contains also showed the same trend which decreasing from 65.72 to 63.02% when comparing with the blank xerogels sample which might be due to the influence of presence of the coconut coir composition. According to Soumya et al. (2016), decreasing trend in volatile matter content for biomass when subjected to activation indicates that there is increasing in total number of pores on the adsorbent surface. It is also believed that there is a relationship between mineral contained in the cross linker (GDL and $CaCO_3$) with the ash content and fixed carbon content value. Due to the mineral contain in hydrogel component, it gives more ash content value from 3.50% for blank sample to 13.81% for 5 wt% coconut coir sample and less fixed carbon. The presence of coconut coir increasing the moisture content value for blank sample to 5 wt% coconut coir sample which is from 8.50% to 12.07% as what has been find out from this analysis.

Table 2 provides the quantitative results derived from the nitrogen adsorption–desorption isotherms. The sample used for analysis were raw coconut coir, alginate xerogels, different ratio of xerogels and aerogels which were 2 and 5 wt%. For the xerogels, only BET surface area increases from 1.40 to 1.85 m^2/g while pore volume adsorption and pore size adsorption decrease from 0.00375 to 0.00355 cm^3/g and 92.43 nm to 80.61 nm respectively. According to The International Union of Pure and Applied Chemistry (IUPAC), pores size greater than 50 nm falls into the category of macropores materials. Blank alginate xerogels has the highest BET surface area which is 2.56 m^2/g. Maria et al. (2017), reported that alginate content caused the specific surface area of the aerogels and the crosslinking to increase. It is important to have high surface area of aerogels as it will increase the loading capacity for the purposes of impregnation with the coconut coir [15]. Generally, several past studies had shown that supercritical $sCCO_2$ drying is more effective that freeze and ambient drying as it gave high surface area and porosity of the aerogels. BET surface area value was decreased to 0.74 m^2/g for 2 wt% coconut coir-based aerogels. However, the BET surface area for 5 wt% coconut coir-based aerogels showed sudden increase to 2.19 m^2/g. This might be due to the increasing of coconut coir weight ratio. This scenario is supported by proximate analysis of volatile matter content. High BET surface area relates with high volatile matter content as it produces more pores due

Table 2 Porous characteristic of raw coconut coir, blank alginate xerogels and coconut coir-based aerogels of 2 and 5 wt%

Sample	BET surface area (m^2/g)	Pore volume adsorption (cm^3/g)	Pore Size adsorption (Å)	Method of drying
Raw coconut coir	2.44	0.00481	78.75	N/A
Blank alginate xerogels	2.56	0.00372	57.99	Oven drying
2 wt% coconut coir-based aerogels	0.74	0.00076	34.15	$sCCO_2$ drying
5 wt% coconut coir-based xerogels	2.19	0.00376	60.93	$sCCO_2$ drying
2 wt% coconut coir-based xerogels	1.40	0.00375	92.43	Oven Drying
5 wt% coconut coir-based xerogels	1.85	0.00355	80.61	Oven Drying

to volatile release from the biomass will create numerous micro-pores on the surface of the adsorbent [16]. Besides, it can be observed that pore size adsorption for 5 wt% coconut coir-based aerogels has the highest value which is 60.93 nm. According to Kumar et al. (2019), in order to increase the rate of adsorption, it is recommended that the pores of the adsorbent need to be bigger than 10 nm [17]. It is to be expected that xerogels and aerogels can be used as adsorbent to adsorb zinc.

Other than physical structure, the adsorption capacity of adsorbents also influenced by the chemical structure of their surfaces. FTIR spectra of raw coconut coir, 5 wt% coconut coir aerogels before and after adsorption of zinc ions are shown clearly in Fig. 1.

Based on Fig. 1, spectra of raw coconut coir and 5 wt% coconut coir-based aerogels obtained were almost similar in terms of placement of peaks. Meanwhile, for 5 wt% aerogels after adsorption of zinc ion, there are only two peaks that can be observed which the stretching vibration of O–H band of material surface was obtained at 3279 cm^{-1}. Another peak observed at 1635 cm^{-1} which indicate C=C bonds. There is no CH and C–O bonds present in the 5 wt% coconut coir sample after adsorption. Furthermore, study conducted by Bhatnagar A et al. [7], also claimed that after adsorption, the CH and C–O bonds becomes weak proving there is attraction between these bonds with heavy metal ions.

Each range of wavelength indicates different chemical bonds which have different functions. Table 3 explains on the range wavelength of the peaks and chemical bond exist on that range.

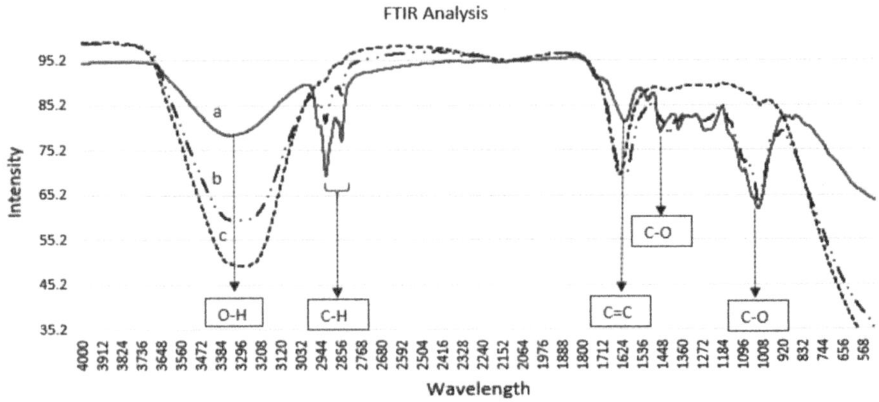

Fig. 1 FTIR spectra of (**a**) raw coconut coir (**b**) 5 wt% coconut coir-based aerogels before adsorption (**c**) 5 wt% of coconut coir-based aerogels after adsorption

Table 3 FTIR spectra

Wavelength	Chemical bond	Explanation
3277–3350 cm^{-1}	O–H	This wavelength associated with stretching vibration of O–H band of material surface for raw coconut coir sample. The change in intensity and peak at 3277 cm^{-1} indicates involvement of O–H group for aerogels before adsorption
2922–2852 cm^{-1}	C–H	This range of wavelength attributed to the stretching vibration of aliphatic C–H bands
1615 and 1616 cm^{-1}	C=C	Both wavelengths indicate there is strong stretching C=C bonds in aromatic rings
1415 and 1030 cm^{-1}	C–O	Both wavelengths correspond to the C–O carboxyl bands

3.2 *Comparison of Coconut Coir-Based Aerogels and Xerogels in Terms of Percent Shrinkage*

Drying method of hydrogel using supercritical CO_2 is known for its advantages as in giving low degree of gel shrinkage, good at preserving a gel's structure, high porosity and low density compared to xerogels. Apart from all these advantages, this drying technique is somehow complicated. Therefore, this study also examined the effects of different drying technique which is oven drying in terms of percent shrinkage and adsorption of zinc. The influence of the drying method on the gel may be seen in Fig. 2. From Fig. 2, it can be seen that xerogels has been proved to shrink notably and also deformed. Meanwhile, aerogels seems to maintain their majority shape and shrunk least after undergo drying process, besides the shrinkage due to the solvent exchange steps (from ethanol to CO_2).

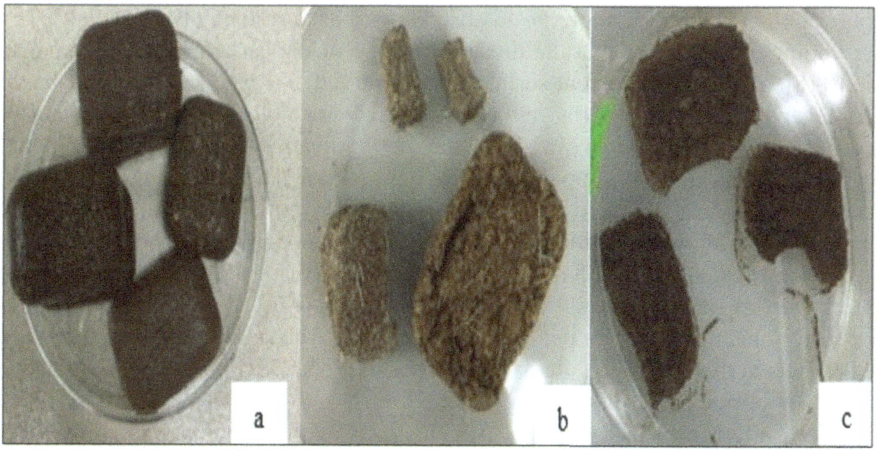

Fig. 2 Comparison of the effect of drying technique (**a**) hydrogel before drying (**b**) aerogels after sCO_2 drying (**c**) xerogels after oven drying

The linear shrinkage of the samples was calculated from the lengths of the samples before and after sCO_2 drying as well as oven drying. The data of percent shrinkage obtained from sCO_2 drying as well as oven drying was plotted in Fig. 3. From Fig. 3, it can be seen that the percent shrinkage of aerogels after sCO_2 were in the

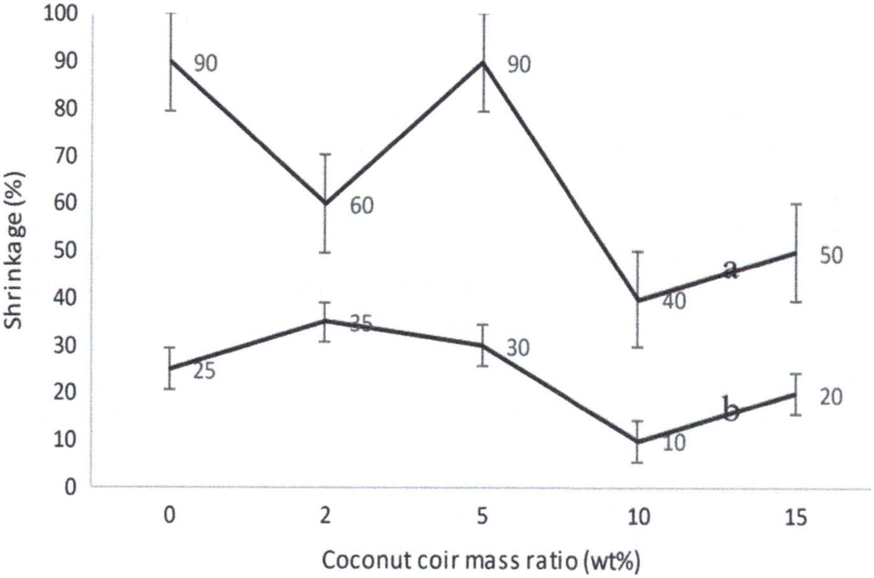

Fig. 3 Graph of percent shrinkage against coconut coir mass ratio (wt%) for (**a**) oven drying (**b**) sCO_2 drying

range of 10–35%. Furthermore, in an observation-based study by Feng et al. (2011), the findings have shown that the linear shrinkage of the aerogels varied from 5 to 30% [18]. The highest percent shrinkage is observed at 2 wt% coconut coir-based aerogels which is 35%. Similar research done by Feng et al. [18] stated that smaller pore size will be generated between the smaller particles, hence, resulting in higher surface tension during the supercritical drying process. Then, there would be more severe shrinkage of the aerogels. This reason is proved by the BET surface area analysis which showed that the pore size for 2 wt% of coconut coir-based aerogels is the smallest which was 34.15 nm. This relationship can be proved further as percent shrinkage was reduced to 30% for 5 wt% coconut coir-based aerogels in which its pore size value obtained was increasing to 60.93 nm. The smallest percent shrinkage of aerogels in this study was observed at 10 wt% of coconut coir weight ratio which yield 10% of percent shrinkage.

Meanwhile for oven drying, the percent shrinkage of the xerogels were in the range of 40–90%. The blank xerogels has the highest percent of shrinkage which was 90%. The xerogels consist of 5 wt% of coconut coir also has undergone the most noticeable shrinkage too. Despite of the highest percentage of shrinkage for blank xerogels, large pore volume and large pore size are still obtained by this oven drying technique. Job et al. (2005) claimed that the pore texture is not completely destroyed despite the capillary forces exerted upon the pores during drying process [19]. In addition, similar to sCO_2 drying, the smallest percent of shrinkage is observed at the same weight ratio of coconut coir which is 10 wt%. This might be due to the increasing of the mass ratio of coconut coir resulting an increase of its stiffness, hence, smaller degree of shrinkage.

3.3 Comparison of Coconut Coir-Based Aerogels and Xerogels in Terms of Adsorption of Zinc

After undergo the drying process of different drying techniques, both the xerogels and aerogels were tested for its adsorption efficiency towards zinc metal. The experiment was repeated three times in order to find the average value. Figure 4 shows the adsorption efficiency for coconut coir-based aerogels and xerogels. From Fig. 4 above, the highest adsorption efficiency towards zinc metal for aerogels can be observed at 5 wt% of coconut coir mass ratio which was 81% percentage removal. The adsorption efficiency increases as coconut coir mass ratio increases. However, the adsorption started to decrease at 10 and 15 wt% coconut coir mass ratio which percentage removal of zinc obtained was 75% and 72% respectively. According to Co et al. (2016) explained that as the coconut coir being too concentrated inside the aerogels, thereby being "crowded" and reducing the surface area and pore volume of the hybrid composite aerogels [20].

This analysis found evidence for too small pore diameters caused difficulty for adsorbates to enter the pore and be adsorbed. Meanwhile, too large diameters caused

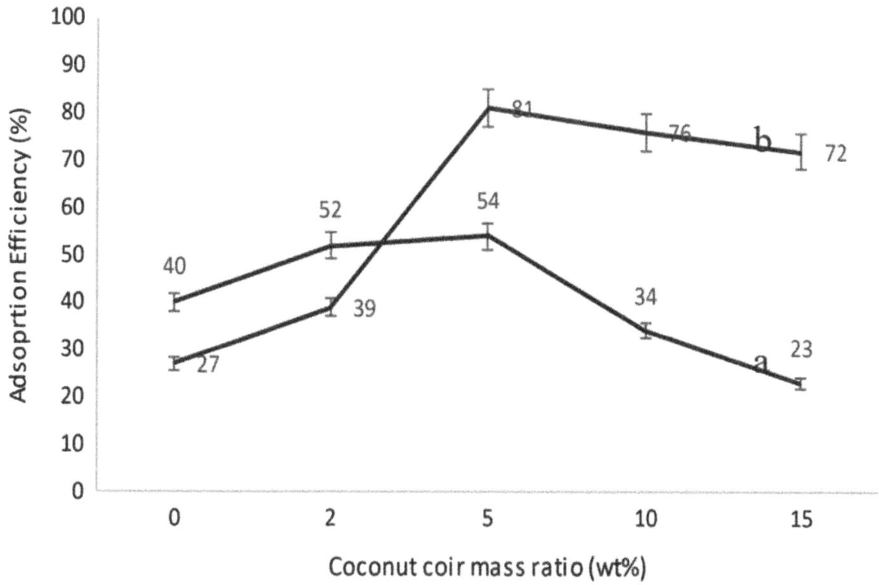

Fig. 4 Graph of adsorption efficiency against coconut coir mass ratio (wt%) for (**a**) oven drying (**b**) sCO_2 drying

the lowering of selectivity and adsorption activity of the material. Hence, 5 wt% coconut coir mass ratios in which its pore size is 60.93 nm and surface area of 2.19 m^2/g is ideal for this application.

Meanwhile, the similar trend of adsorption efficiency also can be observed when the coconut coir based xerogels is dried using oven drying technique as in Fig. 4. The highest adsorption efficiency towards zinc metal can be observed at 5 wt% of coconut coir mass ratio which was 54% percentage removal. However, the adsorption started to decrease at 10 and 15 wt% coconut coir mass ratio which percentage removal of zinc obtained was 34% and 23% respectively.

Besides, another promising finding was that percent of shrinkage did not influence the adsorption efficiency of the coconut coir based adsorbent. This can be explained by adsorption efficiency observed at 5 wt% of coconut coir which gave the highest value despite of its highest percent shrinkage which was 90% as reported in experimental results before.

Hence, from the results, it is clear that the highest adsorption efficiency can be observed at 5 wt% of coconut coir mass ratios for both aerogels and xerogels. When compared, xerogels have lower adsorption efficiency than aerogels. Another observation gained from this study is that, although oven drying technique gave higher percent of shrinkage, it did not give major influence towards adsorption efficiency of zinc metal.

4 Conclusion

The main objective of this research is to synthesize and characterize xerogels and aerogels coconut coir and investigate the capability of those two materials as adsorbent for zinc, were carried out successfully. Xerogels and aerogels were produced by using same process but different drying method which are oven and supercritical drying respectively. Both were characterized using FTIR, TGA and BET. By using oven drying, only surface area increases by comparing xerogels and raw coconut coir, while pore volume adsorption and pore size adsorption decreases. For supercritical drying, BET surface area, pore volume adsorption and pore size show decreasing value from raw coconut coir and aerogels. After drying process, wavelength and peaks at 1615 cm^{-1} that indicates hemicellulose and cellulose. The type of drying used, influenced the adsorption of zinc and percentage shrinkage adsorbent.

Besides, the adsorption efficiency and percentage shrinkage are determined by varying parameter of ratio of coconut coir used in synthesize xerogels and aerogels. For adsorption efficiency, xerogels shows majority decreasing trend of efficiency while aerogels shows majority increasing trend of adsorption efficiency. The highest adsorption efficiency for both adsorbents are at the mass ratio 5 wt% of coconut coir that yield 81% for aerogels and 54% for xerogels. For percentage shrinkage, both adsorbents show decreasing trend. The lowest percentage shrinkage is at 10% for aerogels while for xerogels is 40%. Both adsorbents able to get lowest percentage shrinkage at the same ratio which is 10%. Overall, aerogels and xerogels are capable of adsorbing zinc thus both are valid adsorbent.

Acknowledgements The author would like to express their deepest gratitude to all those who have provided the means and possibility for completing this work. The research was conducted at the Faculty of Chemical Engineering, Universiti Teknologi MARA (UiTM) under the support of FRGS grant (600-IRMI/FRGS 5/3 (189/2019)). Special thanks to Felda Palm Industries Sdn Bhd (FPI) for providing EFB samples to be used in this research.

References

1. Sharma YC, Upadhyay SN (2009) Removal of a cationic dye from wastewaters by adsorption on activated carbon developed from coconut coir. Energy and Fuels 23(6):2983–2988. https://doi.org/10.1021/ef9001132
2. Naseem R, Tahir SS (2001) Removal of Pb(II) from aqueous/acidic solutions by using bentonite as an adsorbent. Water Res. 35(16):3982–3986. https://doi.org/10.1007/BF03326015
3. Agarwal M, Singh K (2017) Heavy metal removal from wastewater using various adsorbents: a review. J Water Reuse Desalin 7(4):387–419. https://doi.org/10.2166/wrd.2016.104
4. Jeon C (2012) Review for physical and chemical process for heavy metal treatment. Clean Technol. 18(4):341–346. https://doi.org/10.7464/ksct.2012.18.4.341
5. Anirudhan TS, Sreekumari SS (2011) Adsorptive removal of heavy metal ions from industrial effluents using activated carbon derived from waste coconut buttons. J Environ Sci 23(12):1989–1998. https://doi.org/10.1016/S1001-0742(10)60515-3

6. Behera S, Arora R, Nandhagopal N, Kumar S (2014) Importance of chemical pretreatment for bioconversion of lignocellulosic biomass. Renew Sustai Energy Rev 36:91–106. https://doi.org/10.1016/j.rser.2014.04.047
7. Bhatnagar A, Vilar VJP, Botelho CMS, Boaventura RAR (2010) Coconut-based biosorbents for water treatment-A review of the recent literature. Adv Colloid Interface Sci 160(1–2):1–15. https://doi.org/10.1016/j.cis.2010.06.011
8. Asim N et al (2019) Development of effective and sustainable adsorbent biomaterial from an agricultural waste material: Cu(II) removal. Mater Chem Phys 249(September 2019):123128. https://doi.org/10.1016/j.matchemphys.2020.123128
9. Co CJU, Quitain AT, Borja JQ, Dugos NP, Takafuji M, Kida T (2017) Synthesis and characterization of hybrid composite aerogels from alginic acid and graphene oxide. IOP Conf. Ser. Mater. Sci. Eng. 206(1). https://doi.org/10.1088/1757-899X/206/1/012053
10. Nurul FA, Alias AB, Norhayati T, Zulkifli AR, Wan A, Ghani AK (2018) Characteristics of rice husk biochar blended with coal fly ash for potential sorption material. Malaysian J Anal Sci 22(2):326–332. https://doi.org/10.17576/mjas-2018-2202-19
11. Wang S et al (2020) Facile synthesis of microporous carbon xerogels for highly selective CO_2 adsorption. J Clean Prod 253:120023. https://doi.org/10.1016/j.jclepro.2020.120023
12. Carre B, Hebrant M, Brosse N, Latif NHA, Hussin MH (2019) Effect of different Prehydrolysis processes on Lignin extractability of coconut husk fibres. J Phys Sci 30:207–219. https://doi.org/10.21315/jps2019.30.s2.18
13. Yerizam M, Faizal M, Marsi M, Novia N (2013) Characteristics of composite rice straw and coconut shell as biomass energy resources (Briquette) (Case study: Muara Telang Village, Banyuasin of South Sumatra). Int J Adv Sci Eng Inf Technol 3(3):232. https://doi.org/10.18517/ijaseit.3.3.326
14. Thaim T, Abdul Rasid R (2016) Improvement empty fruit bunch properties through torrefaction. Australian J Basic 10(17), 114–121. ISSN:1991-8178
15. Maria J et al (2017) Preparation of cellulose aerogels from ionic liquid solutions for supercritical impregnation of phytol. J Supercrit Fluids 130(May):17–22. https://doi.org/10.1016/j.supflu.2017.07.018
16. Masto RE, Ansari MA, George J, Selvi VA, Ram LC (2013) Co-application of biochar and lignite fly ash on soil nutrients and biological parameters at different crop growth stages of Zea mays. Ecol Eng 58:314–322. https://doi.org/10.1016/j.ecoleng.2013.07.011
17. Suresh Kumar P, Korving L, Keesman KJ, van Loosdrecht MCM, Witkamp GJ (2019) Effect of pore size distribution and particle size of porous metal oxides on phosphate adsorption capacity and kinetics. Chem Eng J 358:160–169. https://doi.org/10.1016/j.cej.2018.09.202
18. Feng J, Feng J, Zhang C (2011) Shrinkage and pore structure in preparation of carbon aerogels. J Sol-Gel Sci Technol 59(2):371–380. https://doi.org/10.1007/s10971-011-2514-8
19. Job N et al (2005) Carbon aerogels, cryogels and xerogels: Influence of the drying method on the textural properties of porous carbon materials. Carbon N. Y. 43(12):2481–2494. https://doi.org/10.1016/j.carbon.2005.04.031
20. Banerjee S, Mukherjee S, LaminKa-ot A, Joshi SR, Mandal T, Halder G (2016) Biosorptive uptake of Fe2+, Cu2+ and As5+ by activated biochar derived from Colocasia esculenta: Isotherm, kinetics, thermodynamics, and cost estimation. J Adv Res 7(5):597–610. https://doi.org/10.1016/j.jare.2016.06.002

The Role of Nanoparticle-Gemini Surfactant to Improve the Flowability of the Malaysian Crude Oil

Shamala Devi VijayaKumar, Junaidi Zakaria, and Norida Ridzuan

Abstract Wax deposition on the inner walls of transportation pipelines and production equipment of Malaysian crude oil had been identified as one of the main encounters in the crude oil industry which leads to serious problem in crude oil flow assurance. In this research, the performance of a Gemini surfactant, nanoparticle and their blends are assessed to study their impacts on the viscosity of crude oil using rheometer. Silicon dioxide (SiO_2) (200–600 ppm) along with Gemini surfactant (ethoxylated-2,5,8,11-tetramethyl-6-dodecyn-5,8-dio) (200–1000 ppm) at different temperature (10–30 °C) were studied to discover the effect on viscosity. From the results, the most efficient viscosity reducing concentration of Gemini surfactant, nanofluid and nanoparticle-Gemini surfactant was found at the concentration of 400 ppm, 300 ppm and 200 ppm, respectively. At 30 °C and 80 rpm, the viscosity of the crude oil is reduced from 3.75 to 3.5 cP compared to the blank crude oil. Addition of SiO_2 nanoparticle to Gemini surfactant, resulted in reduction of viscosity of the crude oil from 3.75 to 1.5 cP. At temperature 10 °C, Gemini surfactant reduced the viscosity of crude oil from 51.95 to 4.5 cP, while the addition of SiO_2 nanoparticle only reduced the viscosity from 51.95 to 3.75 cP. As a conclusion, the addition of nanoparticle-Gemini blend shows the significant result at low temperature where it shows the best reduction of crude oil viscosity.

Keywords Nanoparticle · Shear rate · Temperature · Viscosity · Wax inhibitor

1 Introduction

Crude oil industry is one of the major industries in Malaysia where the continuity of production and supply is crucial in order to fulfill the crude oil demand. However, the most common challenge faced by the industry during the production and transportation of Malaysian crude oil is wax deposition and this poses a large problem

S. D. VijayaKumar · J. Zakaria · N. Ridzuan (✉)
Faculty of Chemical and Process Engineering Technology, Universiti Malaysia Pahang, Lebuhraya Tun Razak, 26300 Kuantan, Pahang, Malaysia
e-mail: norida@ump.edu.my

M. A. A. Zaini et al. (eds.), *Proceedings of the 3rd International Conference on Separation Technology*, Lecture Notes in Mechanical Engineering,
https://doi.org/10.1007/978-981-16-0742-4_19

when it deposits inside the pipelines. Crude oil is transported and processed at low temperatures which causes precipitation of wax particles that leads to pressure drop in the production pipeline and processing equipment which results in plugging [1–3]. The buildup of wax in crude oil pipeline and process equipment is a main concern in the petroleum industry. The process of cleaning wax buildup requires shutdown and therefore, the maintenance cost is extremely expensive.

Wax deposition has been a critical operational challenge and imposes major flow assurance problem in the oil industries including the offshore and onshore oil field around the world for many years. Wax precipitation is caused by a few factors such as crude oil composition, viscosity, temperature gradient, production time and shear stress. However, viscosity is recognized as an extremely significant cause of wax deposition rate and prompts wax precipitation and thus need to be reduced in order to make the transportation of crude oil easier [4]. As the bulk oil temperature decreases below its wax appearance temperature [5], wax particles start to precipitate and forms high yield wax gel, that eventually covers the pipeline. The continuation of this process results in gelation where the wax particles aggregate and precipitate in oil cluster as the temperature and pressure changes [3, 6]. This condition leads to poor fluidity of crude oil due to the gelation of wax layer.

Current solutions to overcome wax deposition issue are mainly heating techniques and mechanical removal. However, these methods are not economical as heating requires the use of steam and frequent replacement of electrical heaters while mechanical removal causes temporary shutdown of production [7]. Past researches showed chemical wax inhibitors are commonly used to overcome wax deposition in crude oil as they are more efficient and economical. Heretofore, several chemicals have been used as to improve flowability of crude oil such as, N,N-Dimethyl-N-[2-hydroxy-3-sulfo-propyl]-N′-phenyloctadecanoyl-1,3-diaminopropane [8], (3-(2-methoxyethoxy)propyl-methyl-bis (trimethylsilyloxy) silane) with silicon dioxide [9], Polycarboxilate/poly acrylate/poly vinyl acetate [10], Poly acrylate ester copolymer [11], Ethylene vinyl acetate-co-diethanolamine [12], N,N-bis(dimethyl alkyl)-a, ω-alkanediammoniumdibromide with zinc oxide [15] and Poly (hexyl oleate-co-hexadecyl maleamide-co-n-alkyl oleate) [13]. The advantage of wax inhibitor addition to the crude oil sample is that the deposition can be mitigated without stopping production. The addition of ethylene-co-vinyl acetate and Poly maleic anhydride-alt-1-octadecene with sodium cloisite on crude oil showed that nanoparticle alone reduced the crude oil viscosity approximately 92.5%, EVA and MA reduced up to 88% and 86.4%, while EVA and MA with nanoparticle reduced to 94% and 89.2% respectively [5].

Investigation of interfacial tensions of fatty alcohol polyoxyethanol carboxylate (C_nEO_mC) which is an anionic-nonionic surfactant on crude oil showed that mixed solution of $C_{14}EO_3C$and $C_{14}EO_7C$ with mass fraction ratio of 1:1 can reduce the interfacial tension of crude oil to ultralow values [16]. This can be achieved by adjusting the hydrophilic-lipophilic balance and proportions compatibility. In a study of enhanced oil recovery, N,N′-bis((dimethyltetradecyl)-1,6-hex-anediammonium bromide) Gemini surfactant, partially hydrolyzed poly-acrylamide polymer and/or silica nanoparticle were characterized in terms of rheology and surface tension [17].

The results found that nanoemulsions generally showed shear-thinning behavior that enables injectivity at high shear rates and for improving recovery efficacy at low shear rates. Nanoparticles efficiently adsorbs onto the nanoemulsion surface, decreasing the surface energy. This proves that SiO_2 addition upsurges the adsorption of emulsifier molecules at the interface [17].

Research on plant seed oil also had been done to study their effect on wax deposition where Jatropha seed oils and castor seed oils (CSO) are used and the result showed that the blend of CSO with 40% JSO reduced the wax deposition and gave best impacts on the pour point with paraffin inhibition efficiency of 79.1% [18]. There are also researches being done on the use of nanotechnology in overcoming wax deposition problem. As such, polyacrylamide (PAM) with varying concentrations of nanoclay for an optimum viscous solution was investigated and the findings revealed that a specific range of nanoclay content had encouraging effect on PAM viscosity, aluminium and silicon containing nanofluids caused the highest and lowest viscosity at all temperature, respectively, and the increase in viscosity was noticeable when silicon oxide nanoparticle concentration was increased [19]. However, the effects of nanoparticles, wax inhibitor and their blends on the viscosity of Malaysian crude oil is uncertain, and there are limited researches on this field. This research is aimed to evaluate the effect of Gemini surfactant (ethoxylated-2,5,8,11-tetramethyl-6-dodecyn-5,8-dio), nanoparticle silicon dioxide (SiO_2) and their blend on the viscosity of crude oil at various temperature and shear rate.

2 Methodology

2.1 Materials

Raw crude oil was supplied by PETRONAS Penapisan Terengganu Sdn. Bhd. The crude oil wax appearance temperature (WAT) is 28 °C [5]. Gemini surfactant 2,5,8,11-tetramethyl-6-dodecyn-5,8-diol ethoxylate was used as wax inhibitor which was supplied by Evonik Corporation. The nanoparticle used was silicon dioxide (SiO_2) nanoparticle with a particle size of 20 nm purchased from Nanou Nanotechnology.

2.2 Preparation of Gemini Surfactant

To prepare 250 mL Gemini surfactant solution with 1000 ppm concentration, 0.25 mL of Gemini with 1,000,000 ppm concentration was diluted into cyclohexane in 250 mL volumetric flask. The solution was stirred for 1 h at 25 °C and 700 rpm to get a homogenous solution. 1000 ppm Gemini solution was then diluted using cyclohexane into 800, 600, 400 and 200 ppm concentrations.

2.3 Preparation of Nanoparticles Fluid (Nanofluid)

To prepare 250 mL of SiO_2 solution with 5000 ppm concentration, 1.25 g of SiO_2 powder was mixed with 250 mL of cyclohexane and stirred for 2 h at 60 °C and 700 rpm followed by ultrasonication treatment for a time period of 30 min at 25 °C. SiO_2 solutions of 5000 ppm was then diluted using cyclohexane into 200, 300, 400, 500 and 600 ppm concentrations.

2.4 Preparation of Nanoparticle and Gemini Surfactant Blends

To prepare nanoparticle Gemini surfactants blend, the ratio of Gemini surfactant solution to nanofluid used was 3:1. 3 wt% of Gemini surfactants was mixed with 1 wt% of nanofluid. The total volume for the blends was 10 mL. The mixture was continuously stirred for 1 h at 50 °C and 700 rpm to get a homogenous solution.

2.5 Viscosity Analysis

The Brookfield Viscometer model DV-III was used to measure the crude oil viscosity at various shear rates (80–200 rpm). The viscosity measurement was conducted using nanofluid and Gemini surfactant solution with and without the addition of nanofluid. The samples were heated above wax appearance temperature before starting experiment to completely dissolve wax crystals. The temperature varied within the range of 10–30 °C.

3 Results and Discussion

3.1 Effect of Temperature on the Viscosity of Blank Crude Oil

Figure 1 demonstrates the blank crude oil viscosity against temperature at a fixed shear rate of 80 rpm. The viscosity of crude oil is lower at the higher temperature which was 3.75 cP. From the figure, when the temperature approaches WAT, the viscosity starts to increase and below WAT, the viscosity becomes as high as 51.95 cP at 10°C, where wax begins to form, causing poor fluidity.

Below the WAT, wax particle aggregates initiate the nucleation process and forms crystals. The crystals grow in size with the increasing accumulation of atoms [20] and leads to wax precipitation. Wax precipitation occurs when dissolved wax molecules in crude oil solidify with respect to the function of temperature and enthalpy of

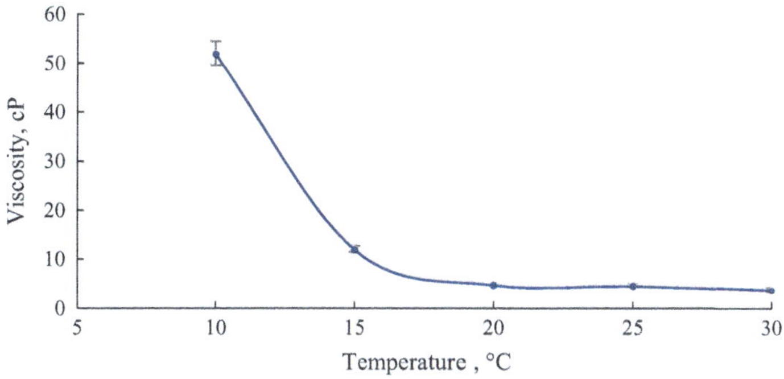

Fig. 1 Viscosity of blank crude oil at different temperatures

crystallization [21]. Thus, the wax particles begin to crystallize and causes rise in blank crude oil viscosity.

3.2 Effect of Addition of Nanoparticles on the Viscosity of Crude Oil

The addition of SiO_2 nanoparticle had significant effect on the crude oil viscosity reduction, specifically at lower temperature. Figure 2 shows the viscosity of crude oil at different concentrations of nanofluid against shear rate at different temperature. At 30 °C and 80 rpm, the lowest crude oil viscosity was obtained at 3.0 cP while the highest was 7.5 cP at 20 °C and 100 rpm. Below WAT, the viscosity of crude oil decreased with incorporation of silicon dioxide but there was not much effect as the temperature approached WAT. As shown in Fig. 2, at a shear rate of 80 rpm, the lowest viscosity for all nanofluid concentration was obtained. Apart from temperature, shear rate also influenced the viscosity of crude oil. Above 100 rpm, the viscosity decreased as the shear rate increased. However, at the shear rate of 100 rpm, the viscosity of crude oil increased slightly. This happens due to the shear thickening fluid behavior resulting from a very viscous liquid. At higher shear rate, the non-Newtonian behavior promotes wax formation in crude oil through crystallization process [22]. This leads to wax crystal aggregation resulting in higher viscosity reading. The concentration of nanofluid influences the effectiveness of viscosity reduction. From Fig. 3, the most efficient concentration of nanofluid to generate the lowest viscous crude oil is found at 300 ppm and 30 °C. While at 10 °C, the viscosity of the crude was drastically reduced from 51.95 to 5.25 cP.

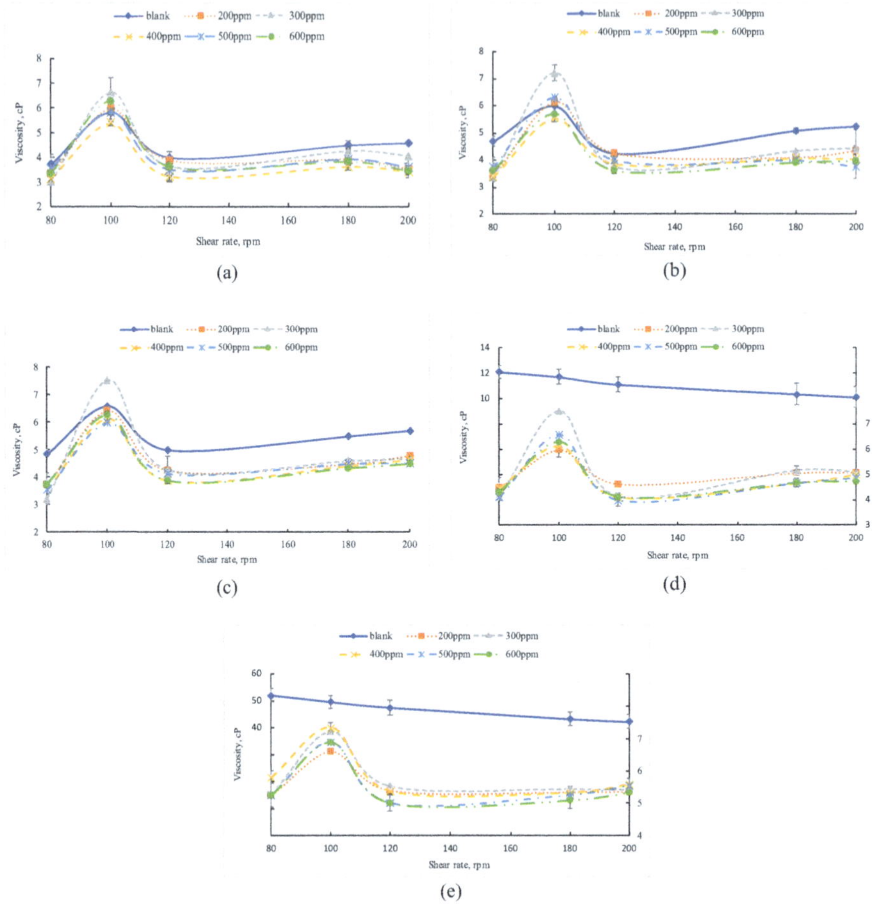

Fig. 2 The effect of viscosity at different range of shear rate and nanofluids concentrations at various temperature; **a** 30 °C, **b** 25 °C, **c** 20 °C, **d** 15 °C, **e** 10 °C

3.3 Effect of Gemini Surfactant with and Without the Incorporation of Nanoparticle on the Viscosity of Crude Oil

The viscosity of crude oil was reduced upon the addition of Gemini surfactant and additional reduction was observed with the addition of nanofluid. Addition of SiO_2 showed significant viscosity reduction and it works well with Gemini surfactant. Figures 4 and 6 demonstrates the impact on crude oil viscosity at different Gemini surfactant concentrations and at different concentration of nanofluid with the most efficient viscosity reducing concentration of wax inhibitor which is 400 ppm respectively. Gemini surfactant as wax inhibitor in this research gave significant effect as the

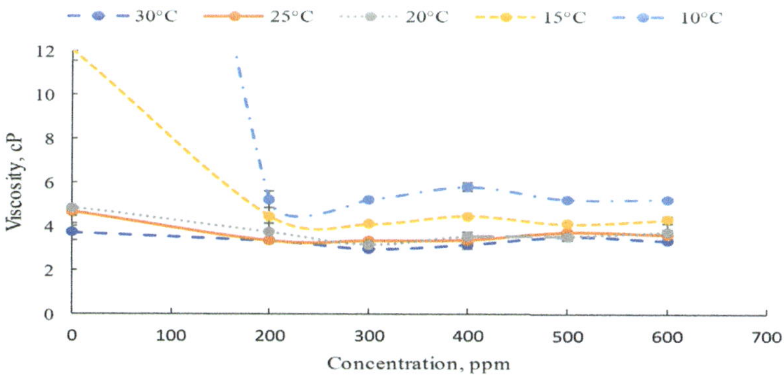

Fig. 3 The effect of different nanofluid concentrations on the viscosity of crude oil at various temperature and at shear rate of 80 rpm

addition of wax inhibitor into crude oil reduced the viscosity at temperature as low as 10 °C, from 51.95 to 4.5 cP as shown in Fig. 4. The crude oil viscosity decreases when temperature rises from 10 to 30 °C. When the wax molecules start to drop below WAT, the crude oil viscosity started to increase, as wax started to form. Wax inhibitor adsorbs onto wax crystals, forming wax crystal lattice assembly in the crude oil that reduces the growing crystals' morphology and interrupts the development of a three-dimensional crystal [6]. A wax dispersant absorbs itself onto the surface of wax to prevent crystal aggregation by preventing the crystal nucleation between crystal surfaces [21].

From Fig. 5, the most efficient viscosity reducing concentration of Gemini surfactant is 400 ppm. The graphs validate that the crude oil viscosity is not impacted by Gemini surfactant concentration as higher concentration showed higher viscosity reading as compared to wax inhibitor of 400 ppm. This can be due to precipitation of higher concentration wax inhibitor [5].

Gemini surfactants consists of two hydrophilic head groups with two hydrophobic tails linked by a spacer at or near the head groups [23]. The spacer can pose rigidity or flexibility that determines their hydrophobicity or hydrophilicity. It is a vital part of Gemini surfactant that regulates the surface adsorption, and the self-assembly aggregation. Some of the ultimate qualities of surfactants is their ability to adsorb at interfaces, high surface activity, low critical micelle concentration (CMC) values and good synergy [14]. A wax dispersants adsorbs onto wax crystals and forms a wax crystal lattice assembly in the crude oil which delays the formation of wax crystals [6]. Figure 6 shows that there is a further reduction in viscosity after the addition of nanofluid together with wax inhibitor. At temperature of 30 °C, above WAT, the viscosity reduced from 3.75 to 1.5 cP after the addition of wax inhibitor and SiO_2 with a load ratio of 3:1, noting the lowest viscosity at nanoparticle concentration of 200 ppm as shown in Fig. 7. Wax inhibitor with most efficient viscosity reducing nanoparticle concentration reduced the viscosity of crude oil from 51.95 to 3.75 cP at 10 °C.

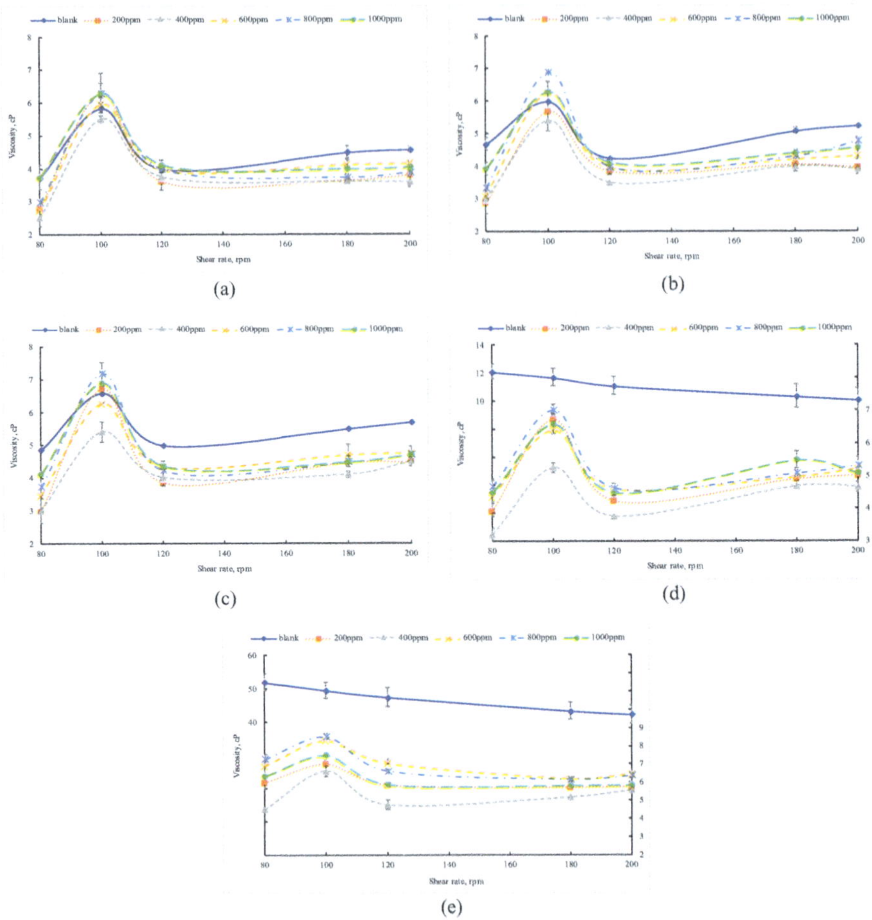

Fig. 4 The effect of viscosity at different range of shear rate and wax inhibitor concentrations at various temperature; **a** 30 °C, **b** 25 °C, **c** 20 °C, **d** 15 °C, **e** 10 °C

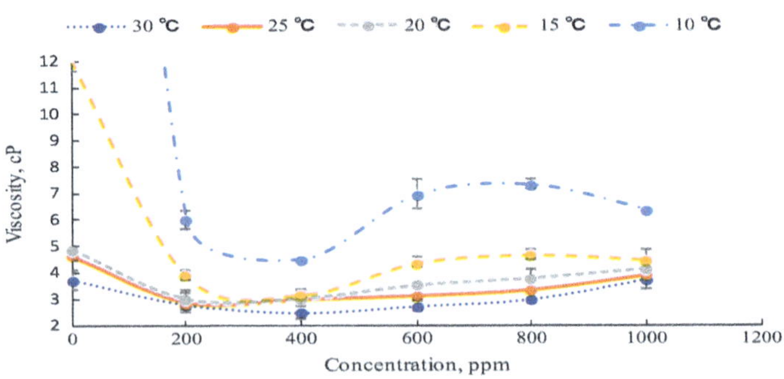

Fig. 5 The effect of different Gemini surfactant concentrations on the viscosity of crude oil at varying temperature and at shear rate of 80 rpm

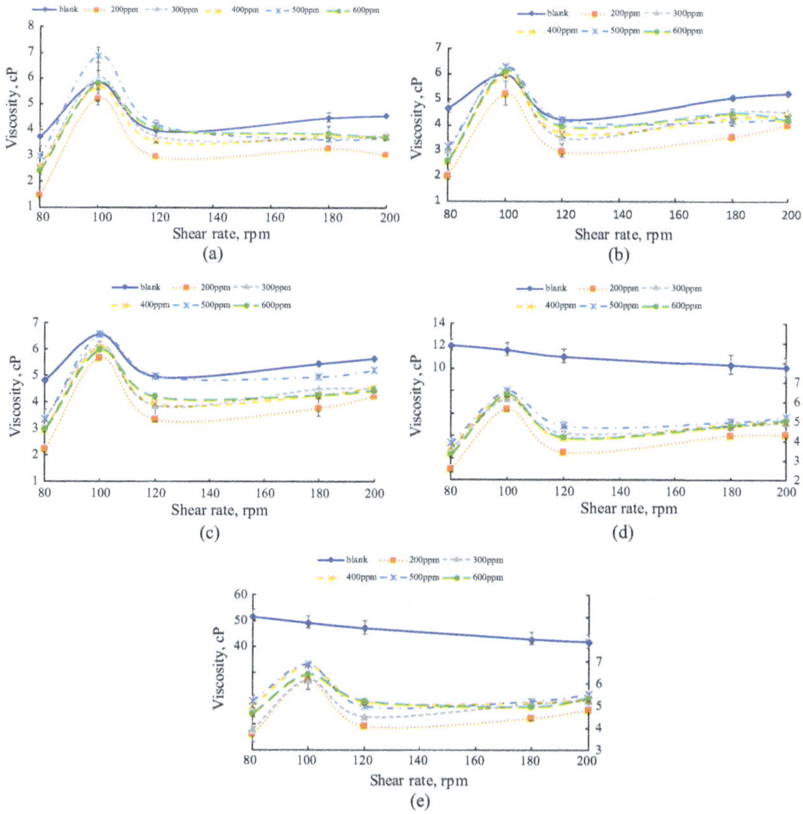

Fig. 6 The effect of viscosity at different range of shear rate and different concentrations of nanofluid with wax inhibitor 400 ppm at various temperature; **a** 30 °C, **b** 25 °C, **c** 20 °C, **d** 15 °C, **e** 10 °C

4 Conclusion

The addition of Gemini surfactant and silicon dioxide nanoparticle enhances the continuity flow of crude oil through viscosity reduction. The performance of the Gemini surfactant, however, depends on the concentration and shear rate, where the most efficient viscosity reducing concentration and shear rate were found to be 400 and 80 rpm. The addition of silicon dioxide nanoparticle and Gemini surfactant together gave the best effect of viscosity reduction which is 1.5 cP at the ratio of 3:1 to Gemini-nanoparticle. Nanotechnology can contribute to more effective, cheaper, and more environmentally friendly solutions than current solutions. Nanoparticle alone gave a significant reduction in viscosity.

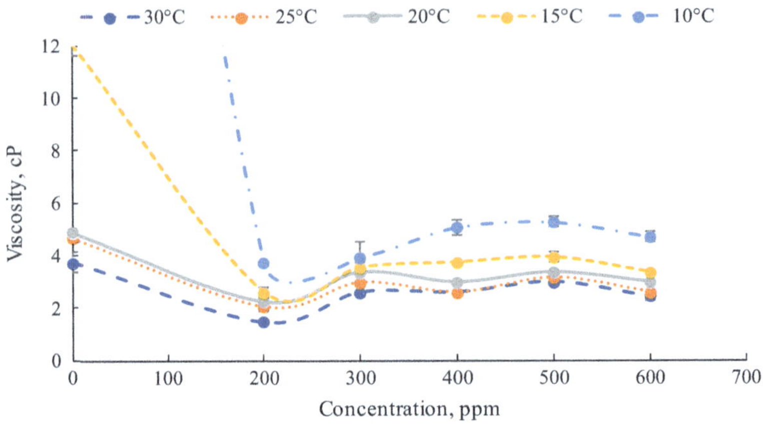

Fig. 7 The effect of different nanofluid concentrations with wax inhibitor 400 ppm on the viscosity of crude oil at varying temperature and at shear rate of 80 rpm

Acknowledgements The authors acknowledge the financial support from the Ministry of Education (MOE), Malaysia, through the Fundamental Research Grant Scheme (RDU 1901060 with reference number FRGS/1/2018/TK02/UMP/03/1) and Universiti Malaysia Pahang (RDU200302 and PGRS1903113) and for having this research project to be successfully conducted. We also thank to Petronas Penapisan Terengganu, Malaysia for providing the crude oil samples for the research project.

References

1. Theyab MA, Diaz P (2016) Experimental study of wax deposition in pipeline—effect of inhibitor and spiral flow. Int J Smart Grid Clean Energy, 174–181. https://doi.org/10.12720/sgce.5.3.174-181
2. Cabanillas JP, Leiroz AT, Azevedo LFA (2016) Wax deposition in the presence of suspended crystals. Energy Fuels 30(1):1–11. https://doi.org/10.1021/acs.energyfuels.5b02344
3. Harun A, Ab Lah NKIN, Husin H, Hassan Z (2016) An overview of wax crystallization, deposition mechanism and effect of temperature & shear. In: ICIMSA 2016—2016 3rd Int. Conf. Ind. Eng. Manag. Sci. Appl. https://doi.org/10.1109/ICIMSA.2016.7503992
4. Ali Theyab M, Diaz P (2016) Experimental study on the effect of inhibitors on wax deposition. J Pet Environ Biotechnol 7(6). https://doi.org/10.4172/2157-7463.1000310
5. Subramanie PAP, Padhi A, Ridzuan N, Adam F (2019) Experimental study on the effect of wax inhibitor and nanoparticles on rheology of Malaysian crude oil. J King Saud Univ - Eng Sci, pp 0–4. https://doi.org/10.1016/j.jksues.2019.05.003
6. Ragunathan T, Husin H, Wood CD (2020) Wax formation mechanisms, wax chemical inhibitors and factors affecting chemical inhibition. Appl Sci 10(2):1–18. https://doi.org/10.3390/app10020479
7. Sousa AL, Matos HA, Guerreiro LP (2019) Preventing and removing wax deposition inside vertical wells: a review. J Pet Explor Prod Technol 9(3):2091–2107. https://doi.org/10.1007/s13202-019-0609-x

8. Chen ZZ, Gang HZ, Liu JF, Mu BZ, Yang SZ (2019) A thermal-stable and salt-tolerant biobased zwitterionic surfactant with ultralow interfacial tension between crude oil and formation brine. J Pet Sci Eng 181(April):106181. https://doi.org/10.1016/j.petrol.2019.06.045
9. Lim ZH, Al Salim HS, Hasiholan B, Ridzuan N (2019) Evaluation of silane-based surfactant nanohybrid as flow improver in a Malaysian light crude oil. IOP Conf Ser Mater Sci Eng 469(1). https://doi.org/10.1088/1757-899X/469/1/012001
10. Ruwoldt J, Humborstad Sørland G, Simon S, Oschmann HJ, Sjöblom J (2019) Inhibitor-wax interactions and PPD effect on wax crystallization: new approaches for GC/MS and NMR, and comparison with DSC, CPM, and rheometry. J Pet Sci Eng 177(November 2018):53–68. https://doi.org/10.1016/j.petrol.2019.02.046
11. Adeyanju OA, Oyekunle LO (2019) Experimental study of water-in-oil emulsion flow on wax deposition in subsea pipelines. J Pet Sci Eng 182(July):106294. https://doi.org/10.1016/j.petrol.2019.106294
12. Anisuzzaman SM, Abang S, Bono A, Krishnaiah D, Karali R, Safuan MK (2017) Wax inhibitor based on ethylene vinyl acetate with methyl methacrylate and diethanolamine for crude oil pipeline. IOP Conf Ser Mater Sci Eng 206(1). https://doi.org/10.1088/1757-899X/206/1/012074
13. Patel MR, Chitte PS, Bharambe DP (2017) Oleic acid based polymeric flow improvers for Langhnaj (North Gujarat, India) crude oil. Egypt J Pet 26(4):895–903. https://doi.org/10.1016/j.ejpe.2015.04.006
14. Lei S, Xu L, Qu C, Jiao H (2016) Influence of Gemini surfactant with modified TiO_2 nanoparticles on the interfacial tension of oil/water. J Dispers Sci Technol 37(10):1494–1501. https://doi.org/10.1080/01932691.2015.1015076
15. Fereidooni Moghadam T, Azizian S, Wettig S (2017) Effect of spacer length on the interfacial behavior of N,N′-bis(dimethylalkyl)-α,ω-alkanediammonium dibromide gemini surfactants in the absence and presence of ZnO nanoparticles. J Colloid Interface Sci 486:204–210. https://doi.org/10.1016/j.jcis.2016.09.069
16. Sheng SS et al (2020) Structure-activity relationship of anionic-nonionic surfactant for reducing interfacial tension of crude oil. J Mol Liq 313:112772. https://doi.org/10.1016/j.molliq.2020.112772
17. Pal N, Kumar N, Saw RK, Mandal A (2019) Gemini surfactant/polymer/silica stabilized oil-in-water nanoemulsions: design and physicochemical characterization for enhanced oil recovery. J Pet Sci Eng 183(September):106464. https://doi.org/10.1016/j.petrol.2019.106464
18. Akinyemi OP, Udonne JD, Oyedeko KF (2018) Study of effects of blend of plant seed oils on wax deposition tendencies of Nigerian waxy crude oil. J Pet Sci Eng 161(December 2017):551–558. https://doi.org/10.1016/j.petrol.2017.12.017
19. Kamal MS, Adewunmi AA, Sultan AS, Al-Hamad MF, Mehmood U (2017) Recent advances in nanoparticles enhanced oil recovery: rheology, interfacial tension, oil recovery, and wettability alteration. J Nanomater 2017. https://doi.org/10.1155/2017/2473175
20. Hao LZ, Al-Salim HS, Ridzuan N (2019) A review of the mechanism and role of wax inhibitors in the wax deposition and precipitation. Pertanika J Sci Technol 27(1):499–526
21. Alnaimat F, Ziauddin M (2020) Wax deposition and prediction in petroleum pipelines. J Pet Sci Eng 184(August):106385. https://doi.org/10.1016/j.petrol.2019.106385
22. Ridzuan N, Adam F, Yaacob Z (2016) Screening of factor influencing wax deposition using full factorial experimental design. Pet Sci Technol 34(1):84–90. https://doi.org/10.1080/10916466.2015.1122625
23. Li P, Guo Y, Lu Z, Zhang W, Hou L (2019) Syntheses, surface activities and aggregation morphologies of a series of novel itaconic acid based asymmetrical gemini surfactants. J Mol Liq 290:111218. https://doi.org/10.1016/j.molliq.2019.111218

Design of Industrial Processes

Industrial Case Study: Cyclone Analysis for Large Scale Calcium Carbonate Looping Process

Muhammad Afiq Zubir, Abreeza Manap, Nurfanizan Afandi, Rabi'atul' Adawiyah Zainuddin, and Mohd Kamaruddin Abd Hamid

Abstract The Calcium Carbonate Looping (CCL) process has the potential to increase net power output as well as reduce CO_2 emissions. Even though there have been studies on CCL, most of them were focused on the reaction system of this process. Unlike previous studies, this work deals with the analysis of variables that may affect the separation efficiency of the cyclone separator in a CCL process. The analyses were divided according to five variables: (1) particle solid distribution (PSD), (2) diameter of cyclone, (3) number of cyclone, (4) inlet temperature, and (5) inlet pressure. The output of this work will help users to identify the variables that affect the efficiency of cyclone separators, thus allowing the users to decide on the best operation condition for high-efficiency cyclone separator designs. The findings showed that increasing PSD and inlet temperature could improve the efficiency of the cyclone. Meanwhile, increasing inlet pressure and cyclone diameter could reduce the effectiveness of the cyclone.

Keywords Cyclone Analysis · Calcium Carbonate Looping · 750 MW_{el} Coal-fired Power Plant

---

M. A. Zubir · M. K. A. Hamid
Process Systems Engineering Centre (PROSPECT), Research Institute for Sustainable Environment (RISE), Universiti Teknologi Malaysia, UTM Johor Bahru, 81310 Johor, Malaysia

A. Manap (✉) · N. Afandi
College of Engineering, Universiti Tenaga Nasional, Jalan IKRAM-UNITEN, Kajang, 43000 Selangor, Malaysia
e-mail: abreeza@uniten.edu.my

M. A. Zubir · A. Manap · R. A. Zainuddin
Institute of Sustainable Energy, Universiti Tenaga Nasional, Jalan IKRAM-UNITEN, Kajang, 43000 Selangor, Malaysia

M. A. A. Zaini et al. (eds.), *Proceedings of the 3rd International Conference on Separation Technology*, Lecture Notes in Mechanical Engineering,
https://doi.org/10.1007/978-981-16-0742-4_20

1 Introduction

The increasing release of CO_2 to the atmosphere has been a major focus recently. The increase in CO_2 emission will cause an overabundance of greenhouse gases that trap additional heat. Consequently, it will cause Earth's temperature to increase which will melt the ice at the poles and lead to the rise of sea level. One of the primary sources of CO_2 gas emission is from the combustion of fossil fuel at coal-fired power plants [1]. Because of this, many studies have been done on the capturing of CO_2 emitted by the coal-fired power plants. Compared to other technologies, Calcium Carbonate Looping (CCL) is one of the promising technologies for large-scale CO_2 capture [2]. CCL offers the possibility to generate extra power through steam generation by using tremendous amount of heat excess from high-temperature carbonation and calcination reactions [3].

The CCL technology was initially proposed by Shimizu et al. [4], which utilized a reversible reaction: the carbonation and calcination of limestone/calcium carbonate ($CaCO_3$), as shown in Eq. 1. Over the years, many studies have been carried out to improve the performance of this technology. Example improvements include the use of water as a reactivation agent for sorbent [5, 6], utilizing indirect heat from the combustor to the calciner [7], and improving reaction rate through various operation temperatures [8]. Currently, there are several CCL pilot plants with different scales and configurations, varying from 3 kWth up to 1.9 MWth [9–12].

$$CaO(s) + CO_2(g) \leftrightarrow CaCO_3(s)\ \Delta H\ 25\,^\circ C = \pm 178.2\ \text{kJ/mol} \quad (1)$$

Many intensive studies have been done on CCL technology; however, less attention has been given on the efficiency of the cyclone separator used in this technology. Hence, this paper aims to analyze variables that may affect the cyclone separator's efficiency, by using Aspen Plus. The efficiency of the cyclone is essential since it will affect the amounts of CO_2 gas recovered and $CaCO_3$ make-up required by the system. A large-scale coal-fired power plant from Malaysia was used as a reference for designing the CCL process.

2 Process Model Description

A 750 MW_{el} coal-fired power plant from Malaysia desired to reduce waste carbon dioxide released in the flue gas using calcium carbonate looping (CCL) technology. The flue gas produced by the plant flowed at 933.43 kg/s with temperature 108.4 °C. The flue gas content is displayed in Table 1.

According to Ströhle et al. [13], a large-scale calcium carbonate looping plant may have two carbonators. The use of two carbonators can reduce the size required by the carbonator reactors and other process units after the reactor. Hence, the flue gas produced by the coal-fired power plant was reduced by half, representing one

Table 1 Flue gas content from the coal-fired power plant

Components	Mass %
Carbon dioxide (CO_2)	18.94
Nitrogen (N_2)	59.67
Nitrogen Oxde (NO)	0.05
Oxygen (O_2)	4.76
Sulfur dioxide (SO_2)	0.02
Water vapor (H_2O)	16.56

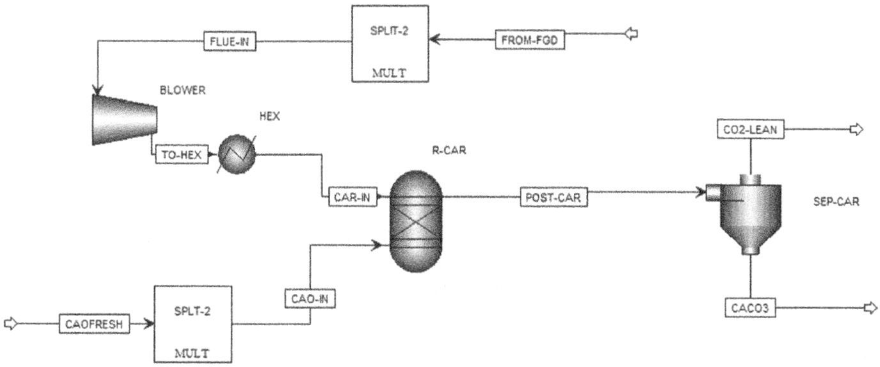

Fig. 1 Part of process flow diagram (PFD) of CCL system

half of the two process lines. Figure 1 shows the process flow of a single process line.

The flue gas is controlled at 650 °C to fulfill the desired operating condition of the carbonation process. In the carbonator reactor, calcium oxide (CaO) recycled from the calciner process is fed into the carbonator reactor to capture carbon dioxide in the flue gas and convert it to calcium carbonate ($CaCO_3$). Then, the carbonator's product is sent to a cyclone separator to remove lean CO_2 flue gas through stack. The combined $CaCO_3$ and unconverted CaO is transferred to the calciner reactor. The inclusion of a blower and heat exchanger before the carbonator is to control the temperature and pressure in the carbonator reactor and cyclone separator. The initial number of cyclones used is similar to a previous study [13], which is six.

3 Cyclone Separator Analysis

In this study, the analyzed variables were inlet temperature and pressure, particle size distribution (PSD) of the feed, cyclone diameter, and number of cyclone used. The analyses were carried on the cyclone separator of the carbonator's products. Several existing cyclone design considerations, as listed below, were taken into account in this study.

1. The pressure drop of the cyclone should not exceed 10 in. of H_2O (0.025 bar).
2. The common diameter of the cyclone should not exceed 1.5 m except for Stairmand high throughput (HT), which is up to 5 m.
3. The inlet gas velocity should not exceed 30 m/s.
4. The operating pressure should not exceed atmospheric pressure (if possible).

For this study, the Leith-Licht correlation was used since the correlation is accurate for inlet velocities approximately 25 m/s. In addition, Stairmand HT cyclone type was used since the flowrate of the flue gas was huge. Other variables such as density, viscosity, and the number of turns in the cyclone were kept constant throughout all analyses. The desired separation efficiency is $\geq$99.5%.

4 Results and Discussion

4.1 The Effect of Particle Size Distribution

The purpose of this analysis was to determine the effect of particle size distribution (PSD) of solid particles on the efficiency of the cyclone separator used for the separation of the carbonator's product. The analysis was carried out using CaO with three different types of PSD ranges, as shown in Table 2. During the analysis, variables such as cyclone diameter, inlet temperature, and inlet pressure were kept constant. The cyclone diameter was kept at the possible maximum (5 m) whilst the temperature and pressure were set at values similar to the ones for the carbonator reactor (650 °C, 1 bar). Based on Fig. 2, the efficiency of the cyclone separator increased with increasing PSD of solid particles. This is because the bigger the particles than cut point diameter (d_{pc}), the higher the efficiency of the cyclone [14].

Table 2 Composition of PSD for CaO

PSD (μm)	Mass %		
	Type A	Type B	Type C
<25	2.0	0.3	–
25–50	23.7	1.7	0.3
50–100	45.8	23.7	1.7
100–150	26.8	45.8	23.7
150–200	1.7	26.8	45.8
200–350	–	1.7	26.8
350–500	–	–	1.7

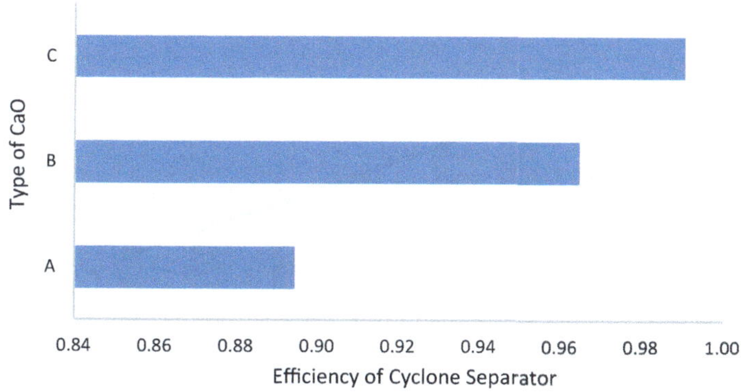

Fig. 2 The effect of PSD on the efficiency of the cyclone separator

4.2 The Effect of Cyclone Diameter

The purpose of this analysis was to determine the effect of cyclone diameter on the efficiency of the cyclone separator. The cyclone diameter was varied from 3.5 to 5 m. During the analysis, variables such as PSD of solid particles, inlet temperature, inlet pressure of the cyclone, and number of cyclones were kept constant. The inlet pressure and temperature of the cyclone used were taken from previous analysis whilst the PSD of type C was used due to its highest cyclone efficiency operation. Based on Fig. 3, the reduction of the cyclone's diameter causes the inlet velocity to increase and generated stronger centrifugal forces, which leads to the increase in efficiency of the cyclone [15, 16]. However, it was found that the inlet velocity exceeded the maximum feasible range (30 m/s) when the diameter was reduced to 4.5 m or lower.

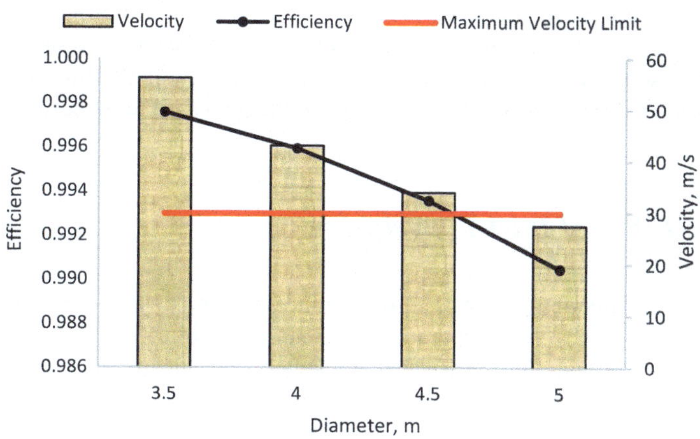

Fig. 3 The effect of cyclone diameter on the velocity and efficiency of the cyclone separator

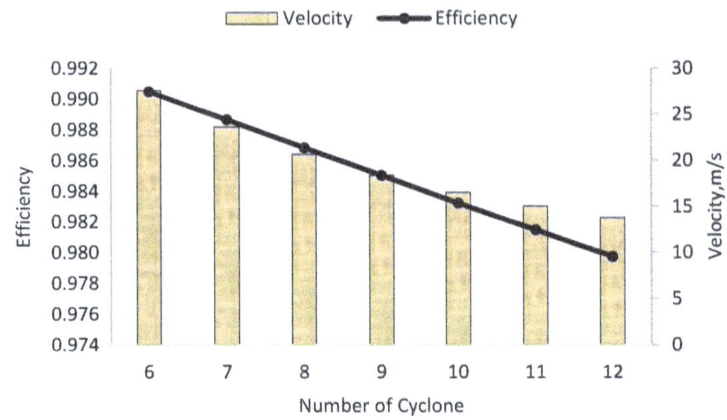

Fig. 4 The effect of number of cyclones on the velocity and efficiency of the cyclone separator

4.3 The Effect of Number of Cyclone

The purpose of this analysis was to determine the effect of number of cyclones on the efficiency of the cyclone separation. The number of cyclones were varied from 6 up to 12 cyclones. Numbers of cyclones lower than 6 were not possible since the maximum feasible velocity limit into the cyclones would be exceeded. During the analysis, variables such as PSD of solid particles, inlet temperature, inlet pressure, and diameter of cyclone were kept constant. The PSD, inlet pressure, and inlet temperature used were taken from previous analysis whilst the diameter was set at 5 m to ensure feasible cyclone operation. Based on Fig. 4, increasing the number of cyclones causes the velocity to drop, which leads to lower cyclone efficiency [17].

Since Fig. 4 shows that no cyclone design has a separation efficiency higher than 99.5%, the simultaneous effect of number of cyclones used and cyclone diameter was analyzed and the results are shown in Fig. 5. Based on the figure, the 99.5% efficiency is obtained at a diameter of 3.4 m and 12 number of cyclones. Diameters lower than 3.4 m were not considered since the velocity of the fluid entering the cyclone exceeded the feasible velocity range at the analyzed temperature range.

4.4 The Effect of Inlet Temperature

The purpose of this analysis was to determine the effect of inlet temperature on the efficiency of the cyclone separator. The inlet temperature was varied from 550 to 650 °C. This temperature range was considered because most carbonator reactors' operation temperatures are within this range, as discussed in the literature [8]. During the analysis, variables such as PSD of solid particles, inlet pressure, cyclone diameter, and number of cyclones were kept constant. The cyclone diameter was

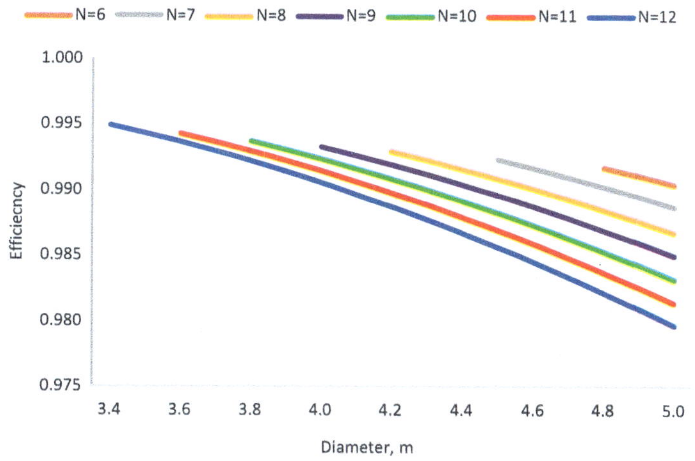

Fig. 5 The simultaneous effect of cyclone diameter and number of cyclones on the efficiency of the cyclone separator

set at 3.4 m whilst the number of cyclones was set at 12 as proposed in previous analysis. Based on Fig. 6, the reduction of inlet temperature causes the efficiency of the cyclone to decrease. As the temperature decrease, the velocity of the fluid entering the cyclone also decreases [18], which consequently reduces the efficiency of the cyclone. Hence, inlet temperature was not reduced to prevent reduction in efficiency and avoid additional cost due to purchasing a heat exchanger to reduce the inlet temperature.

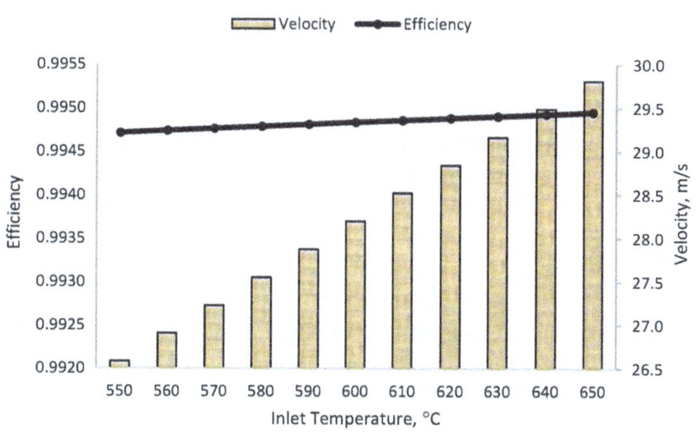

Fig. 6 The effect of inlet temperature on the velocity and efficiency of the cyclone separator

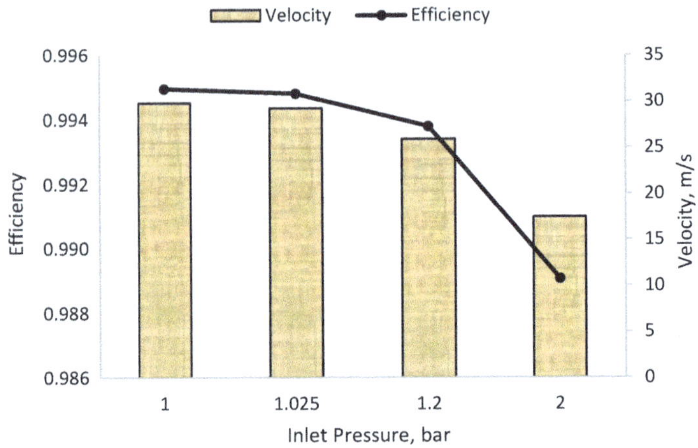

Fig. 7 The effect of inlet pressure on the velocity and efficiency of the cyclone separator

4.5 The Effect of Inlet Pressure

The purpose of this analysis was to determine the effect of inlet pressure on the efficiency of the cyclone separator. In this study, the pressure was set at 1, 1.025, 1.2, and 2 bar. During the analysis, variables such as PSD of solid particles, inlet temperature, diameter of cyclone, and number of cyclones were kept constant. The PSD, number of cyclones, and diameter of the cyclone used were taken from previous analysis whilst the temperature was maintained at 650 °C. Based on Fig. 7, increasing the cyclone's inlet pressure reduces the inlet velocity [19], thus reducing the efficiency of the cyclone. At 2 bar, the efficiency is below the targeted efficiency (99.5%). It was desired to have a cyclone design with a lower pressure, thus, the pressure of the carbonator's cyclone was kept at 1 bar as in the carbonator reactor. It should be noted that the actual inlet pressure of the cyclone might be slightly lower than desired pressure because of the pressure drop across the carbonator reactor.

5 Conclusion

Analyses on variables which may affect the efficiency of a cyclone separator for the carbonator's product in a Calcium Carbonate Looping (CCL) system was successfully carried out. Based on the findings, increasing the cyclone's Particle Size Distribution (PSD) of solid and inlet temperature would lead to the increase of the cyclone's efficiency. However, increasing the pressure, diameter and number of the cyclone would show opposite results. Hence, cyclones with large PSDs, high inlet temperatures, low inlet pressures, small diameters, and lower number of cyclones are more favorable as they would improve the efficiency of the cyclones and reduce the make-up $CaCO_3$ required by the CCL system. Further studies on these variables in relation to the operation and capital costs of the CCL system should be carried out.

Acknowledgements The authors acknowledge Universiti Tenaga Nasional, UNITEN R&D Sdn Bhd and Institute of Sustainable Energy of UNITEN for the lab facilities. This work was financially supported by the UNITEN R&D Sdn. Bhd. under the Seeding Fund (U-TG-RD-18-28) and Bold Grant (10436494/B/2019141).

References

1. Moazzem S, Rasul MG, Khan MMK (2012) A review on technologies for reducing CO_2 emission from coal fired power plants. Thermal Power Plants 11:227–254. https://doi.org/10.5772/31876
2. Zhou L, Duan L, Anthony EJ (2019) A calcium looping process for simultaneous CO_2 capture and peak shaving in a coal-fired power plant. Appl Energy 235:480–486. https://doi.org/10.1016/j.apenergy.2018.10.138
3. Hilz J, Haaf M, Helbig M, Lindqvist N, Ströhle J, Epple B (2019) Scale-up of the carbonate looping process to a 20 MWth pilot plant based on long-term pilot tests. Int J Greenhouse Gas Control 88:332–341. https://doi.org/10.1016/j.ijggc.2019.04.026
4. Shimizu T, Hirama T, Hosoda H, Kitano K, Inagaki M, Tejima K (1999) A twin fluid-bed reactor for removal of CO_2 from combustion processes. Chem Eng Res Des 77:62–68. https://doi.org/10.1205/026387699525882
5. Ramkumar S, Fan LS (2010) Thermodynamic and experimental analyses of the three-stage calcium looping process. Ind Eng Chem Res 49(16):7563–7573. https://doi.org/10.1021/ie100846u
6. Mutch GA, Anderson JA, Vega-Maza D (2017) Surface and bulk carbonate formation in calcium oxide during CO_2 capture. Appl Energy 202:365–376. https://doi.org/10.1016/j.apenergy.2017.05.130
7. Junk M, Reitz M, Ströhle J, Epple B (2013) Thermodynamic evaluation and cold flow model testing of an indirectly heated carbonate looping process. Chem Eng Technol 36(9):1479–1487. https://doi.org/10.1002/ceat.201300019
8. Ramezani M, Tremain P, Doroodchi E, Moghtaderi B (2017) Determination of carbonation/calcination reaction kinetics of a limestone sorbent in low CO_2 partial pressures using TGA experiments. Energy Procedia 114:259–270. https://doi.org/10.1016/j.egypro.2017.03.1168
9. Charitos A, Hawthorne C, Bidwe AR, Sivalingam S, Schuster A, Spliethoff H, Scheffknecht G (2010) Parametric investigation of the calcium looping process for CO_2 capture in a 10 kW_{th} dual fluidized bed. Int J Greenhouse Gas Control 4:776–784. https://doi.org/10.1016/j.ijggc.2010.04.009
10. Arias B, Diego ME, Abanades JC, Lorenzo M, Diaz L, Martinez D, Alvarez J, Sanchez-Biezma A (2013) Demonstration of steady state CO_2 capture in a 1.7 MW_{th} calcium looping pilot. Int J Greenhouse Gas Control 18:237–245. https://doi.org/10.1016/j.ijggc.2013.07.014
11. Chang MH, Huang CM, Liu WH, Chen WC, Cheng JY, Chen W, Wen TW, Ouyang S, Shen CH, Hsu HW (2013) Design and experimental investigation of calcium looping process for 3-kW_{th} and 1.9-MW_{th} facilities. Chem Eng Technol 36:1525–1532. https://doi.org/10.1002/ceat.201300081
12. Ströhle J, Junk M, Kremer J, Galloy A, Epple B (2014) Carbonate looping experiments in a 1 MWth pilot plant and model validation. Fuel 127:13–22. https://doi.org/10.1016/j.fuel.2013.12.043

13. Ströhle J, Galloy A, Epple B (2009) Feasibility study on the carbonate looping process for post-combustion CO_2 capture from coal-fired power plants. Energy Procedia 1(1):1313–1320. https://doi.org/10.1016/j.egypro.2009.01.172
14. Norelyza H, Rashid M, Hajar S, Nurnadia A (2014) MR-deDuster: a dust emission separator in air pollution control. Jurnal Teknologi 58:85–88. https://doi.org/10.11113/jt.v68.3025
15. Gimbun J, Choong TS, Fakhru'l–Razi A, Chuah TG (2004) Prediction of the effect of dimension, particle density, temperature, and inlet velocity on cyclone collection efficiency. Jurnal Teknologi, 40(1):37–50.https://doi.org/10.11113/jt.v40.421
16. Azadi M, Azadi M (2012) An analytical study of the effect of inlet velocity on the cyclone performance using mathematical models. Powder Technol 217:121–127. https://doi.org/10.1016/j.powtec.2011.10.017
17. Miller BG (2017) Particulate Formation and Control Technologies. In: Clean coal engineering technology. Butterworth-Heinemann, pp 419–465. https://doi.org/10.1016/B978-0-12-811365-3.00008-9
18. Jitschin W (2016) Gas laws and kinetic theory of gases. Handb Vac Technol, 29–82. https://doi.org/10.1002/9783527688265.ch3
19. Qin R, Duan C (2017) The principle and applications of Bernoulli equation. J Phys: Conf Ser 916:012038. https://doi.org/10.1088/1742-6596/916/1/012038

Optimization of Anthocyanins Extracts from Roselle (*Hibiscus sabdarifa*) Petals Using Ultrasonic-Assisted Extraction Method

Syafiqah Redzuan, Chai Yee Ho, Zuhaili Idham, Suzana Yusuf, Nicky Rahman Putra, Mohd Azizi Che Yunus, Syamila Mansor, and Muhammad Syafiq Hazwan Ruslan

Abstract Roselle petals contain high anthocyanins, a good antioxidant, and natural dark red colorants. Anthocyanins help in reducing risks of cardiovascular disease and colon cancer. But anthocyanins are easily affected by extraction process. Therefore, this study proposed a new method of extracting anthocyanins using ultrasonic-assisted extraction (UAE) from roselle petals with three extraction parameters. The extraction parameters are sample's particle size (0.125, 0.375 and 0.625 mm), solvent to solid ratio (10:1, 15:1 and 20:1 mL solvent/g solid) and extraction time (5, 10 and 15 min). The optimization of process parameters aims to achieve the highest extraction yield of extract, Total Anthocyanins Content (TAC), and Antioxidant Activity (AA). The results show that 0.125 mm of particle size, 10:1 mL solvent/g solid, 15 min of extraction obtained the best percent mass yield (64.72%), TAC (70.97 mg/L), and AA (90.05%). Method-wise, this study showed that the ultrasonic-assisted extraction gives better quality of roselle petals extracts than the maceration extraction.

Keywords Anthocyanins · *Hibiscus sabdarifa* · Roselle petals · Ultrasonic-assisted extraction · Antioxidants

S. Redzuan · C. Y. Ho · S. Yusuf
Department of Chemical Engineering, Universiti Teknologi PETRONAS, 32610 Seri Iskandar, Perak, Malaysia

C. Y. Ho · S. Yusuf
Centre of Biofuel and Biochemical Research, Universiti Teknologi PETRONAS, 32610 Seri Iskandar, Perak, Malaysia

Z. Idham · N. R. Putra · M. A. C. Yunus · M. S. H. Ruslan (✉)
Centre of Lipids Engineering and Applied Research, Universiti Teknologi Malaysia, 81310 Johor Bahru, Johor, Malaysia
e-mail: syafiqhazwan@uitm.edu.my

S. Mansor
Department of Food Biotechnology, Faculty of Science and Technology, Universiti Sains Islam Malaysia, 71800 Nilai, Negeri Sembilan, Malaysia

M. S. H. Ruslan
Faculty of Chemical Engineering, Universiti Teknologi MARA, 40450 Shah Alam, Selangor, Malaysia

M. A. A. Zaini et al. (eds.), *Proceedings of the 3rd International Conference on Separation Technology*, Lecture Notes in Mechanical Engineering,
https://doi.org/10.1007/978-981-16-0742-4_21

1 Introduction

Anthocyanins are generally accepted as the most critical group of water-soluble pigments, which occur naturally in plant foods [1]. More than 6,000 members of the polyphenol phytochemicals flavonoid family are found in various types of plant foods, fruits (eg: roselles, blackberries, raspberries), and vegetables [2]. These polyphenols are responsible for the purple, blue, orange, and red colors in many vegetables and fruits. Anthocyanin is very sensitive to temperature. Extraction temperature exceeding 40 °C can cause degradation of the active component and turning the extracts brown. Thus, there is a need to apply a more thermal sensitive extraction method to preserve the compound.

Anthocyanins is an excellent source of natural antioxidants that act as radical stabilizer and anti-inflammatory properties. Antioxidants also can prevent and reduce the cardiovascular and cancer diseases. Thus, these are the main reasons for their increasing popularity in the current population diet [3]. A clinical test on roselle extract has been done on a human prostate cancer cell and suggested that the extract could reduce the cancer cells' growth via intrinsic and extrinsic pathway [4]. Other than that, roselle petals also contain high anthocyanins, a good natural antioxidant source, and natural colorant [5]. However, they can quickly degrade due to the stability of anthocyanins is affected by the extraction parameters [6, 7].

Several extractions methods have been implemented to extract anthocyanins over the years [8]. However, most conventional extraction methods, such as hydrodistillation and maceration, have serious drawbacks, such as high solvent use and prolonged extraction time. A newly introduced extraction methods such as the Ultrasonic-Assisted Extraction (UAE) can solve the drawbacks aforementioned [8, 9].

UAE is a non-thermal technology that can serve as an alternative to traditional extracts [7]. Ultrasonic oscillators generate waves that propagate through the medium. The oscillation of rarefaction and compression causes bubble nuclei to form. As the bubbles' size increases, it will burst and generate a shear force that causes macro-turbulence in the medium. This phenomenon is due to cavitation, which can enhance the solvent's penetration deep into the sample matrix and increases the extraction efficiency for temperature sensitive materials. UAE method also contributes to environmental preservation by reducing the use of solvents, fossil energy, and the generation of hazardous substances. The utilization of ultrasonic radiation for extraction gives high reproducibility, simple operations, lower operating temperature, and lower loss of bioactive mixes contrast with the conventional method. Maceration and soxhlet have bound working conditions, especially temperature, limiting the extraction efficiency since temperature impacts the target solutes' kinetic and diffusivity.

Thus, this study aims to optimize the UAE's parameters for the anthocyanin's extraction from roselle petals. The optimization of process parameters aims to achieve the highest extraction yield of extract, Total Anthocyanins Content (TAC), and

Antioxidant Activity (AA). This study also compares the extraction quality between maceration extraction as the baseline of the extraction method with the UAE.

2 Methodology

2.1 *Sample Preparation*

A 2 kg of dried roselle petals were purchased from a local supplier, ground and sieved to achieve 0.125, 0.375, and 0.625 mm mean particle sizes. The process was conducted by utilizing blender and sieving instruments accessible in Block P Laboratory, Universiti Teknologi PETRONAS.

2.2 *Ultrasonic-Assisted Extraction*

A jacketed beaker was first set-up and loaded up with 50 mL of distilled water as the solvent used for the extraction. A 2.5 g of ground roselle petals were then inserted into the jacketed beaker. Tap water-line was fitted to the jacketed beaker to keep the extraction temperature below 40 °C. The ultrasonic generator and run clock were controlled at 20 kHz and 5 min, respectively. After the extraction was completed, the solvent was filtered using a Whatman-40 filter paper. The filtrate was transferred into a round bottom flask to remove the excess solvent using a rotary evaporator unit (BUCHI Rotavapor R-215). The concentrated extract was oven-dried (40 °C, 3–5 h) until no measurable weight loss is observed. The extract was then kept in a freezer (−20 °C) for further analysis. The design of experiment is shown in Table 1. A Box-Behnken design was chosen due to the limitation of samples and the simplicity of the design (3 factors and 3 levels, 5 replication of mid-point).

2.3 *Analysis*

2.3.1 Extraction Yield

Extraction yield was determined by weighing the samples after the drying process and comparing it to the sample's initial weight.

Table 1 Design of experiment for UAE extraction

Run	X_1: Solvent to solid ratio	X_2: Particle size	X_3: Time extraction
	mL/g	mm	min
1	10.00	0.125	10.00
2	20.00	0.125	10.00
3	10.00	0.625	10.00
4	20.00	0.625	10.00
5	10.00	0.375	5.00
6	20.00	0.375	5.00
7	10.00	0.375	15.00
8	20.00	0.375	15.00
9	15.00	0.125	5.00
10	15.00	0.625	5.00
11	15.00	0.125	15.00
12	15.00	0.625	15.00
13	15.00	0.375	10.00
14	15.00	0.375	10.00
15	15.00	0.375	10.00
16	15.00	0.375	10.00
17	15.00	0.375	10.00

2.3.2 Total Anthocyanins Content (TAC)

The determination of anthocyanins was done using the AOAC standard [10]. Anthocyanins were determined using a UV–visible spectrophotometer (UV-1800, Shimadzu, Japan) at 520 nm. A test portion should not exceed 10 mL due to the capacity of the cuvette used. A mixture of 1-part test portion and 4-part buffer was prepared for the analysis. The appropriate dilution factor was determined by diluting the test portion with pH 1.0 buffer until the absorbance at 520 nm is within the spectrophotometer's linear range. This dilution factor was used to prepared two dilutions of the test sample with one in h pH 1.0 buffer and the other with a pH 4.5 buffer. The absorbances were measured at a wavelength of 520 and 700 nm for pH 1.0 buffer and pH 4.5 buffer, respectively. The diluted test portions were recorded versus a blank cell filled with distilled water. Anthocyanin concentration was calculated and expressed as cyanidin-3-glucoside equivalents, as in Eq. 1.

$$Anthocyanin\ pigment\ (\mathrm{mg\,CGE/L}) = \frac{A \times MW \times DF \times 10^3}{\varepsilon \times l} \quad (1)$$

where A is absorbance value $(A_{520\mathrm{nm}} - A_{700\mathrm{nm}})$ of pH 1.0 – $(A_{520\mathrm{nm}} - A_{700\mathrm{nm}})$ pH 4.5, MW is molecular weight for cyanidin-3-glucoside (cyd-3-glu) (449.2 g/mol), DF is dilution factor, l is path length (cm), ε is 26 900 molar extinction coefficient, for cyanidin-3-glucoside (L mol^{-1} cm^{-1}), 10^3 is factor for conversion from g to mg.

2.3.3 Antioxidant Activity (AA)

The determination of the Antioxidant Activity (AA) was done using 2-diphenyl-1-picrylhydrazyl (DPPH) radical scavenging assay. The methodology was adopted from the referred literature [11]. A 5 mg/L methanolic DPPH stock solution was prepared and kept within the fridge until utilized. Next, 2.5 mL of extract was blended with a (1) mL of the methanolic DPPH solution and kept within the darkroom for 30 min. The absorbance was measured at 518 nm using a UV–Vis spectrophotometer (model Shimadzu UV-1800). The DPPH scavenging activity was determined using Eq. 2. A blank reading replaces the extracts with distilled water.

$$DPPH\ scavenging\ activity\ (\%) = \frac{A_{blank} - A_{sample}}{A_{blank}} \times 100\% \qquad (2)$$

2.4 Statistical Analysis

Experimental design and analysis via response surface methodology (RSM) was conducted using Design Expert V6.0.4 (Stat-Ease Inc., Minneapolis, USA) software. The software developed the model equation, plotting the response graphs, and predicting the optimum parameters over all three responses to obtain maximum yield, TAC, and the AA. The analysis was conducted on the regression fitness measured by the value of the coefficient of determination (R^2), analysis of variance (ANOVA), and p-value.

3 Results and Discussion

This work discussed the effect of each parameter (sample particle size, solid to solvent ration and extraction time) and their statistical fitness towards the responses (extraction yield, total anthocyanins content and antioxidant activity). The study also compares the quality of extracts obtained from the UAE to the maceration extraction method to measure the proposed UAE-method's efficiency.

3.1 *Effect of Sample Particle Size*

Particle size has been known to affect the composition and mass fraction of antioxidants in the extract [12]. Figure 1 illustrates that the maximum yield of 64.72% was obtained for the lowest particle size of 0.125 mm. The extraction yield decreases as the sample particle size increases from 0.125 to 0.625 mm. Lower particle size provides a higher total surface area for the extraction. This allows a higher mass transfer rate from the sample to the solvent. A study shows that larger particle size of wheat results in higher antioxidant properties compared with the smaller particle size [13]. This result is coherent with the surfaces plot in Fig. 2 that shows the TAC concentration.

In contrast, the highest TAC concentration of 65.32 mg CGE/L was observed at 0.375 mm. This high response can be explained by the total surface area available for mass transfer. Generally, by lowering the particle size, the total surface area accessible by the solvent is increased. This phenomenon promotes the extraction process and reduces the mass transfer resistance due to the distance traveled by the solvent. However, particle sizes that is too small can cause a negative impact on the results due to the loss of precious material in the grinding process. Mechanical shearing and tearing of the sample matrices cause the exposed solute loss to surrounding cause less amount being collected in the extraction process [13].

Figure 3 shows the highest percentages of DPPH radical scavenging activities were observed at a particle size of 0.125 mm. The result shows the highest antioxidant activity observed was 90.05%. This outcome corresponds with the research conducted by [13], whereby smaller particle sizes gave higher oil yield. Sample particle size can affect and regulate the mass transfer kinetics and solvent access into the soluble component. The reduction of sample particle size can increase the intrinsic capacity for the solvent's diffusion into the target component, which improves the

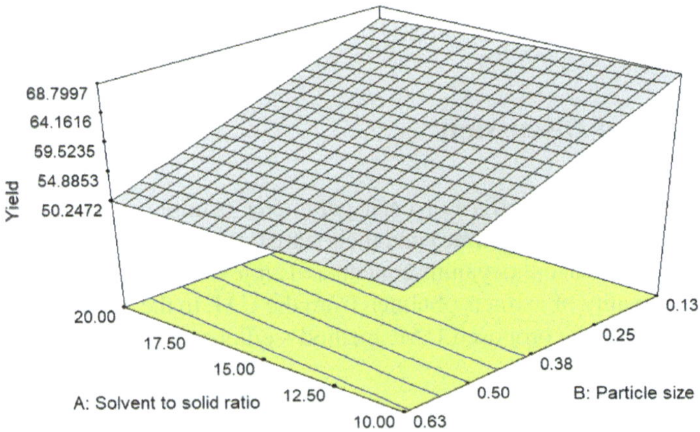

Fig. 1 Surface plots for the effect of mean particle size on extraction yield (0.125 mm, 15 min extraction time)

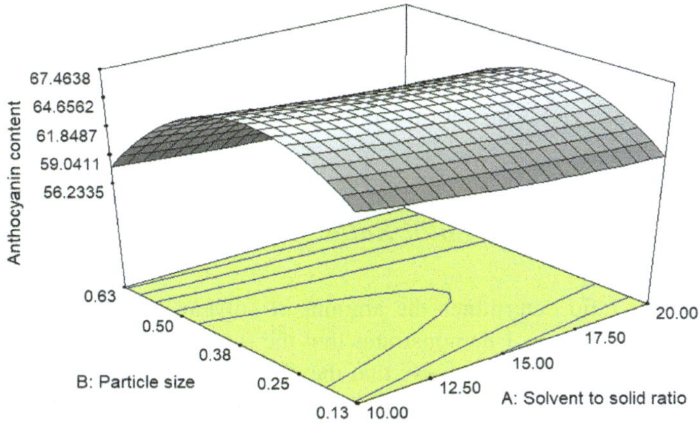

Fig. 2 Surface plots for the effect of mean particle size and solvent on solid ratio on total anthocyanins content at 10 min of extraction time

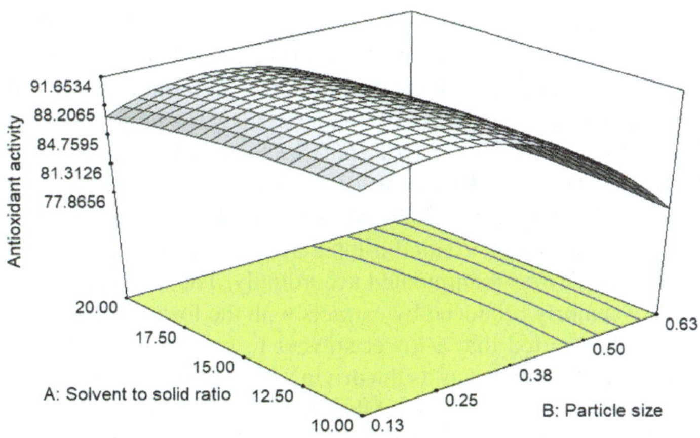

Fig. 3 Surface plots for the effect of mean particle size on antioxidant activity at 10 min of extraction time

extraction efficiency. Compared with the previous results, the AA corresponds to the yield extracted rather than the TAC amount obtained. This can be explained by the fact that TAC is the measure of anthocyanins extractable from the sample, whereby AA reflects on the extract activity. AA is a synergistic effect of the whole extract rather than the anthocyanin content obtained from the extract. Some other unknown component may contribute to the overall activity has yet to be determined.

Previous research showed that 52% of extract yield bioactive compounds from hazelnut shells with particle size 0.5–1.0 mm compared to 28.6% of yields for particle size between 1.00 and 2.0 mm [14]. Even so, as claimed by Makanjuola (2017), the smallest particle size may not always yield the highest antioxidant property.

Although particle size reduction corroborates an increase in extraction efficiency, a critical particle size might be reached, which will not significantly differ in the extraction efficiency [12]. Compared to anthocyanins extraction, results did prove smaller particle size yields higher oil yield.

3.2 *Effect of Solvent to Solid Ratio*

Solid to solvent ratio determines the amount of solvent need to be used for the extraction process. Figure 1 demonstrates that the ratio of 10:1 yielded the highest yield of 68.80%. Figure 1 also shows that the ratio did not significantly impact the curvature or shift of trend of the extraction across the particle size. The solvent to solid ratio for ultrasonic has a less effect on the percent mass yield of extraction since different ratios gave similar results. Each extraction method requires a different set of operating parameters to improve extracts and recovery of the bioactive compounds [9]. The saturation of the solvent causes this has not been achieved even at a ratio of 10:1. Saturation and solubility can be the factors affecting the yield of extracts. However, since the temperature and pressure of the process were kept constant, the solubility of the solute into the solvent should not change significantly. Like a study conducted by Giannoccaro et al. (2006), the lowest sugar amounts were obtained at 80 °C with solvent to solid ratio of 10:1 and 15:1 than with 1:5 [15].

The lowest solvent to solid ratio of 10:1 for ultrasonic extraction showed the highest TAC amount, which was 70.97 mg CGE/L. The solvent was kept fixed at 50 ml for ultrasonic extraction to suit the apparatus's available size capacity, whereas the mass of the sample was manipulated accordingly. The results indicated that the highest TAC was majorly produced by extracts with the lowest solvent to solid ratio (10:1). A research asserted that a lower solvent to solid ratio results in a greater concentration gradient, which acts as the driving force for the mass transfer between solute and solvent [7]. It is also reported that TAC was increased drastically with the increase of solvent to solid ratio. The increase of phenolic compounds may result from a higher amount of solvent that can penetrate plant cells [16].

For AA, an indistinct pattern was acquired in Fig. 3 contrasting and the yield rate pattern in Fig. 1. No critical contrast of AA observed was the solvent to solid ratio increases or decreases. The highest AA observed was 90.05% at a 10:1 ratio. It can be supported by the cumulative impact of present radical free searching exercises [17]. The antioxidant activity measured in this experiment considers the cumulative effect of all antioxidants present in the extracts that were also contributed from other extraction parameters.

A recent study by Mancini et al. (2018) confirmed that antioxidant capacity is an index derived from providing a single estimate of antioxidant activity from all bioactive compounds. The total bioactive compounds in the extracts from most natural product extracts were correlated with DPPH antioxidant assay. Hence, the high percentages of AA at the lowest solvent to solid can be deduced to be the effect of

DPPH scavenging activity other present bioactive compounds in the extracts other than anthocyanin content [18].

3.3 *Effect of Extraction Time*

Figure 4 shows the UAE produces the highest yield of 67.81% at the longest extraction time of 15 min and the smallest particle size of 0.125 mm. This indicates that longer extraction time gives a longer extraction process to take place, which allows the diffusion of solvent deep inside the sample matrix to extract the solvent from the core of the sample.

Figures 5 and 6 show the effect of extraction time on the TAC and AA, respectively. The highest TAC was produced at the extraction of 15 min, which yielded 70.97 mg CGE/L anthocyanins. There is no significant difference in the highest AA that was noticed across the extraction time tested. The highest percentage of AA was observed at 90.95% for ultrasonic at 15 min. The prolonged time of extraction potentially increases the loss of solvent with bioactive compounds by vaporization [19, 20]. Thus, 15 min at 40 °C as practices in the study is recommended to be performed for the anthocyanin's extraction.

Medina-Torres et al. (2017) mentioned that extraction efficiency is highly influenced by extraction time, specifically in UAE, where it is asserted that ultrasound increases as a function of time with two main stages; washing and diffusion extraction [21]. Washing occurs in the first 10 to 20 min of extraction, where approximately 90% of soluble components on the surfaces of the matrix dissipate, indicating rapid recovery of the phenolic compounds. Meanwhile, slow extraction is a stage where mass transfer from solute to solvent occurs only via diffusion, which could last up

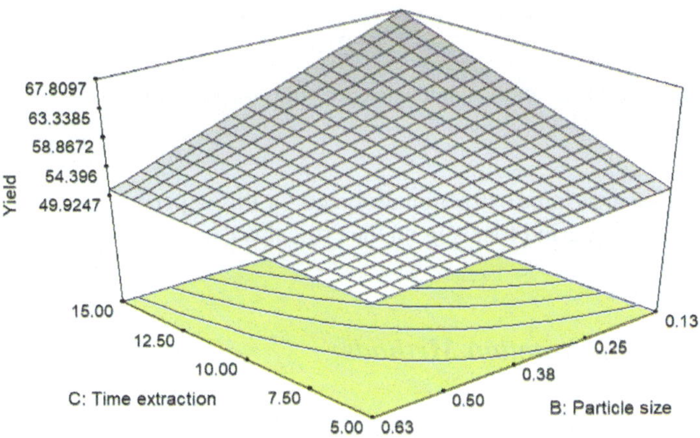

Fig. 4 Surface plots for the effect of time extraction ratio on yield at solvent to solid ratio on 15:1

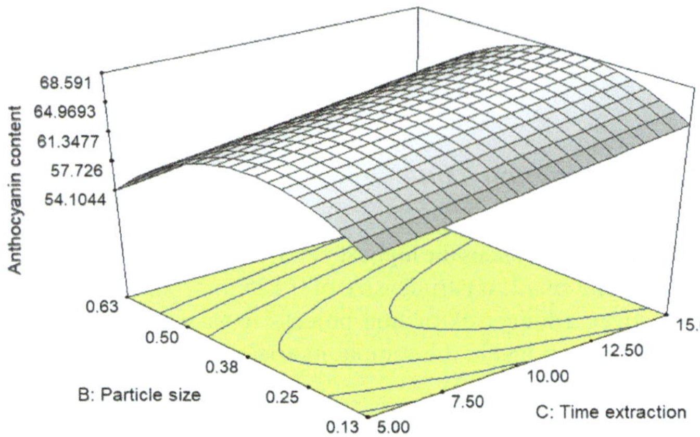

Fig. 5 Surface plots for effect of time extraction ratio on anthocyanin content at solvent to solid ratio on 15:1

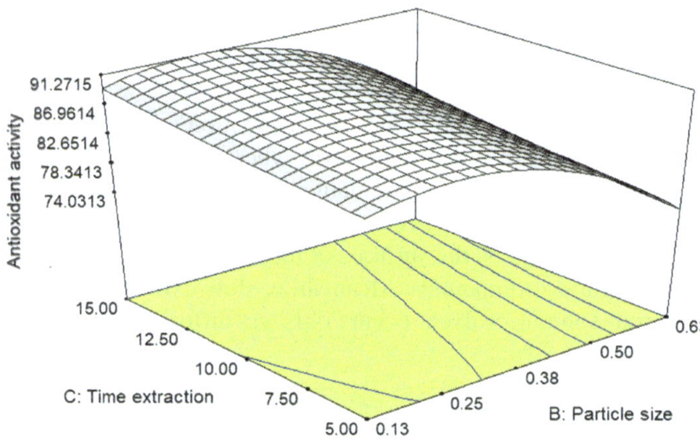

Fig. 6 Surface plots for the effect of time extraction ratio on antioxidant activity at solvent to solid ratio on 15:1

to 100 min. The ultrasonic extraction designed for this experiment was conducted in a relatively shorter period focusing on the critical washing stage.

3.4 *Effect of Extraction Methods*

This section of the study compares the two extractions method's performance, namely maceration and ultrasonic-assisted extraction. Maceration extraction was selected to be the control as it is the most basic method of extraction. The maceration extraction

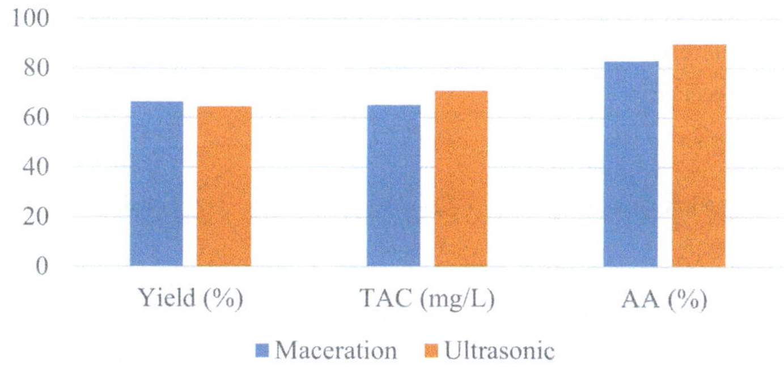

Fig. 7 Maximum yield, TAC, and AA for maceration and ultrasonic assisted extraction method

was conducted at room temperature while maintaining the UAE method's parameters except for extraction time that was increased from 1 to 3 h.

Figure 7 demonstrated that the maceration extraction method results in a slightly higher measure in the percentage of extraction yield (66.67%) compared to the UAE method (64.72%). This may be due to the substantial difference of sample size for both methods, which were 50.0 g for maceration extraction and 5.0 g of sample for UAE. The considerable distinction on this physical perspective was one of the contributing elements towards the higher percent mass yield created by maceration extraction [11].

However, for TAC, extraction using the UAE method produced 70.97 mg CGE/L compared to maceration extraction, which yielded 65.29 mg CGE/L. The UAE method utilizes acoustic cavitation assistance, making the development of bubbles rise due to the wave oscillation to the molecules. The collision between the molecules and bubbles results in a shear force, causing a decrease in particle size, thus encourages mass exchange [21].

The AA analysis of extracts shows that ultrasonic extraction obtained 90.05% DPPH scavenging activities, whereas the maceration extraction method showed 83.20% DPPH scavenging activity. These results demonstrated no significant difference, which are supported by previous studies [22–24]. The main difference between these two methods is the introduction of an ultrasonic probe to the system. Maceration extraction was maintained at room temperature (about 26 °C), while the introduction of the ultrasonic waves to the extraction calls for a need to keep the extraction at 40 °C. A slight disparity in operating temperature may cause the extraction content to be different. As temperature increases, the solubility of the solute in the solvent also increases. A higher amount of soluble solute extracted may lead to higher TAC and AA observed in the analysis.

Table 2 The proposed actual coefficient for the optimized equation for yield, TAC and AA for maceration and ultrasonic assisted extraction method

Coefficient	Maceration			UAE		
	Yield (%)	TAC (mg CGE/L)	AA (%)	Yield (%)	TAC (mg CGE/L)	AA (%)
Model	Quadratic	Quadratic	Quadratic	2FI	Quadratic	Quadratic
Mean	+80.50	+55.05	+96.72	+46.29	+53.24	+81.84
X_1	−3.13	−1.41	−1.037	−0.17	−0.39	+0.35
X_2	+54.90	+67.51	−0.380	+6.77	+63.02	+38.90
X_3	−13.34	+4.421	−1.21	+1.94	+0.91	−0.14
X_1^2	+0.06	+0.04	+0.04	–	+0.01	−0.03
X_2^2	−66.99	−98.59	−34.41	–	−103.40	−97.85
X_3^2	+3.40	−0.53	+0.22	–	+0.002	+0.02
X_1X_2	+0.70	−0.07	−0.66	+0.16	+0.27	+0.67
X_1X_3	+0.45	+0.03	−0.02	−0.003	−0.03	+0.01
X_2X_3	−7.40	+0.84	+6.98	−2.87	+0.07	+0.23
R^2	0.9196	0.9662	0.9930	0.8880	0.9917	0.9996
Model *p*-value	0.0044	0.0002	<0.0001	0.0003	<0.0001	<0.0001
Standard deviation	2.07	1.14	0.57	2.19	0.59	0.17

3.5 Optimization of Extraction Condition

For optimization purposes, all three parameters; solvent to solid ratio (X_1), mean particle size (X_2), and extraction time (X_3) factors, and three responses were examined using Design-Expert Version 6.0.4. A second-order polynomial with interaction model was selected, and the proposed equation for the optimization purposes is shown in Table 2.

Table 2 showed that all the models are significant, with the coefficient of determination (R^2) at least 0.888. Tested with significant levels of 0.05, the predicted data shows a good agreement with the experimental runs with a 95% confidence level. It also can be seen that the most influential parameters in all responses are particle size (X_2 and X_2^2) as both are having the highest coefficient magnitude, as shown in Table 2. This can be supported as the particle size getting lower, the total accessible surface area increased, thus increasing the interaction between the solvent and solute. Meanwhile, the lowest influence on the extraction regime is dominated by the interaction between solvent to solid ratio and extraction time (X_1X_2). This shows that the extraction process still runs as lower solvent to solute saturation. Thus, increasing the solvent does not impact the result much.

The optimization of parameters and responses was conducted for both extraction methods. The parameters were set to 'in range' while the responses were set to

Table 3 Optimized condition for maceration and ultrasonic extraction method

Condition	Specification	Maceration	UAE
X_1	In range	10	10
X_2	In range	0.35 mm	0.24 mm
X_3	In range	3.00 h	15 min
Yield (%)	Maximum	66.38557	64.72
TAC (mg CGE/L)	Maximum	66.19	76.22
AA (%)	Maximum	83.38	91.60
Desirability		0.995	0.983

maximum. This is to ensure that the optimized condition could yield the maximum amount with the best quality of extracts. The results are tabulated in Table 3.

Maceration extraction acquired high desirability compared to the UAE method. However, the UAE method achieved the main objective of maximizing the anthocyanin content. Moreover, the UAE method only takes 15 min compared with maceration, which takes 3 h of extraction time. From an economic point of view, the UAE method is preferable because it is faster with better antioxidant quality. Even though the yield is 2% less than the maceration method, the quality of extract makes up for the lacking. The desirability function of both results is satisfactory as it is more than 95%. It can be concluded that the optimum condition for extraction of roselle petals with regard to maximizing extracts yield, TAC, and AA was 15 min of extraction time, the mean particle size of 0.24 mm, solvent to solid ratio of 10:1, and using the UAE method.

4 Conclusion

In this investigation, the examination of extraction techniques on roselle petals was conducted using the UAE method and proven to achieve a better result of extraction yield and total anthocyanin content when compared with the maceration extraction method. The results show that 0.125 mm of particle size, 10:1 mL solvent/g solid, 15 min of extraction obtained the best percent mass yield (64.72%), TAC (70.97 mg/L), and AA (90.05%). Method-wise, this study showed that the ultrasonic-assisted extraction gives better quality of roselle petals extracts than the maceration extraction. It is concluded that UAE method is a new sustainable, innovative green technology that can increase the extraction efficiency, reduce time and energy-consuming procedures. This study suggests the development of a kinetic and solubility model that takes into account the ultrasonics amplitude and frequency factors into consideration of the models.

Acknowledgements This research would like to acknowledge Centre of Biofuel and Biochemical Research (CBBR), Universiti Teknologi PETRONAS, Centre for Research on Ionic Liquid

(CORIL), Universiti Teknologi PETRONAS, and Centre of Lipids Engineering and Applied Research (CLEAR), Universiti Teknologi Malaysia for the success of the study.

References

1. Khoo H, Azlan A, Tang S, Lim S (2017) Anthocyanidins and anthocyanins: colored pigments as food, pharmaceutical ingredients, and the potential health benefits. Food Nutr Res 61(1):1361779. Clerk Maxwell J (1892) A treatise on electricity and magnetism, vol 2, 3rd edn. Clarendon, Oxford, pp 68–73
2. Densie W (2014) Anthocyanins. Today's Diet 16(3):20
3. Bowen-Forbes CS, Zhang Y, Nair MG (2010) Anthocyanin content, antioxidant, anti-inflammatory and anticancer properties of blackberry and raspberry fruits. J Food Compos Anal 23(6):554–560. https://doi.org/10.1016/j.jfca.2009.08.012
4. Da-Costa-Rocha I, Bonnlaender B, Sievers H, Pischel I, Heinrich M (2014) *Hibiscus sabdariffa L.*—a phytochemical and pharmacological review. Food Chem 165:424–443
5. Tsai P-J, Mcintosh J, Pearce P, Camden B, Jordan BR (2002) Anthocyanin and antioxidant capacity in Roselle (*Hibiscus Sabdariffa L.*) extract. Food Res Int 35(4):351–356. https://doi.org/10.1016/s0963-9969(01)00129-6
6. Tiwai B, Patras A, Brunton N, Cullen P, O'Donnell C (2010) Effect of ultrasound processing on anthocyanins and color of red grape juice. Ultrason Sonochem 17(3):598–604. https://doi.org/10.1016/j.ultsonch.2009.10.009
7. Golmohamadi A, Möller G, Powers J, Nindo C (2013) Effect of ultrasound frequency on antioxidant activity, total phenolic and anthocyanin content of red raspberry puree. Ultrason Sonochem 20(5):1316–1323. https://doi.org/10.1016/j.ultsonch.2013.01.020
8. Azmir J, Zaidul ISM, Rahman MM, Sharif KM, Mohamed A, Sahena F, Jahurul MHA, Ghafoor K, Norulaini NAN, Omar AKM (2013) Techniques for extraction of bioactive compounds from plant material: a review. J Food Eng 117(1):426–436
9. Garcia-Salas P, Morales-Soto A, Segura-Carretero A, Fernández-Gutiérrez A (2010) Phenolic-compound-extraction systems for fruit and vegetable samples. Molecules 15(12):8813–8826
10. Lee J, Durst R, Wrolstad R (2005) AOAC 2005.02: total monomeric anthocyanin pigment content of fruit juices, beverages, natural colorants, and wines—pH differential method, pp 37–39
11. Rithwan F, Zhari S, Yunus MAC, Hadzri HM (2014) Efficacy of biological activity of *Andrographis paniculata* extracted by using supercritical carbon dioxide (Sc-CO_2) extraction
12. Makanjuola SA (2017) Influence of particle size and extraction solvent on antioxidant properties of extracts of tea, ginger, and tea–ginger blend. Food Sci Nutr
13. Manyele IF (2008) Investigation of the effect of particle size on groundnut-oil solvent extraction. Tanzan J Eng Technol 2(2):1–13
14. Yuan B, Lu M, Eskridge KM, Isom LD, Hanna MA (2018) Extraction, identification, and quantification of antioxidant phenolics from hazelnut (Corylus avellana L.) shells. Food Chem 244:7–15
15. Giannoccaro E, Wang YJ, Chen P (2006). Effects of solvent, temperature, time, solvent-to-sample ratio, sample size, and defatting on the extraction of soluble sugars in soybean. J Food Sci 71(1)
16. Živković J, Šavikin K, Janković T, Ćujić N, Menković N (2018) Optimization of ultrasound-assisted extraction of polyphenolic compounds from pomegranate peel using response surface methodology. Sep Purif Technol 194:40–47
17. Rani AJ, Mythili SV (2014) Study on total antioxidant status in relation to oxidative stress in type 2 diabetes mellitus. J Clin Diagn Res: JCDR 8(3):108

18. Mancini FR, Affret A, Dow C, Balkau B, Bonnet F, Boutron-Ruault MC, Fagherazzi G (2018) Dietary antioxidant capacity and risk of type 2 diabetes in the large prospective E3N-EPIC cohort. Diabetologia 61(2):308–316
19. Dent M, Kovačević DB, Bosiljkov T, Dragović-Uzelac V (2017) Polyphenolic composition and antioxidant capacity of indigenous wild Dalmatian sage (*Salvia officinalis L.*). Croat Chem Acta 90(3). https://doi.org/10.5562/cca3231
20. Durling N, Catchpole O, Grey J, Webby R, Mitchell K, Foo L, Perry N (2007) Extraction of phenolics and essential oil from dried sage (*Salvia officinalis*) using ethanol–water mixtures. Food Chem 101(4):1417–1424. https://doi.org/10.1016/j.foodchem.2006.03.050
21. Medina-Torres N, Ayora-Talavera T, Espinosa-Andrews H, Sánchez-Contreras A, Pacheco N (2017) Ultrasound assisted extraction for the recovery of phenolic compounds from vegetable sources. Agronomy 7(3):47
22. Thoo YY, Ho SK, Liang JY, Ho CW, Tan CP (2010) Effects of binary solvent extraction system, extraction time and extraction temperature on phenolic antioxidants and antioxidant capacity from mengkudu (*Morinda citrifolia*). Food Chem 120(1):290–295
23. Ghosh M, Sinha BN, Seijas JA, Vázquez-Tato MP, Feás X (2014) Flavonoids and phenolic compounds from Litsea polyantha Juss. bark. In: The 18th international electronic conference on synthetic organic chemistry. Multidisciplinary Digital Publishing Institute
24. Othman A, Mukhtar NJ, Ismail NS, Chang SK (2014) Phenolics, flavonoids content and antioxidant activities of 4 Malaysian herbal plants. Int Food Res J 21(2)

Effect of Irradiation Time, Solvent to Solid Ratio and Power on Microwave-Assisted Extraction of *Colubrina asiatica*

Nurul Nadrah Mohd Zabidi, Eraricar Salleh, Norhayati Pa'e, Yanti Maslina Mohd Jusoh, and Khairul Azly Zahan

Abstract In the past decades, herbal plants had been used widely by indigenous or even urban people for disease treatment or taken orally as appetizers, including *Colubrina asiatica* (*C. asiatica*) or known as Peria Pantai. Previous studies reported that this plant contains a high amount of active compounds that are responsible for foaming ability and various bioactivities. However, these active compounds are sensitive towards many factors; temperature, pH, light, and others in the production line. Thus, this study was conducted to optimize the extraction conditions using microwave-assisted extraction (MAE) using Response Surface Methodology (RSM) and the Box-Behnken Design method was selected. The parameters used in RSM were irradiation time (3, 5, and 7 min), solvent to solid ratio (10, 20, and 30 mL/g), and power (300, 400, and 500 W). The optimal extraction for the extract was 6.06 min irradiation time, solvent to solid ratio 28.31 mL/g, and 445 W of microwave power provide 42.63 w/w %. The actual yield value from the experimental procedure was 43.1 ± 0.05 w/w %. The total saponin content of the extract was determined using Folin-Ciocaltaeu's colorimetric method with a reading of 15.36 ± 0.001 mg ESE/mg.

Keywords *Colubrina asiatica* · Foaming agent · Microwave-assisted extraction · Solvent extraction · Ultrasonic-assisted extraction

1 Introduction

Nowadays, the majority of the world's population uses various cleansing agents in their daily lives to maintain proper hygiene and avoid germs or bacterial infections that can cause undesired diseases, especially the millions of people in the vast rural

N. N. M. Zabidi · E. Salleh (✉) · N. Pa'e · Y. M. M. Jusoh
Faculty of Engineering, School of Chemical & Energy Engineering, Universiti Teknologi Malaysia (UTM), 81310 Johor Bahru, Johor, Malaysia
e-mail: eraricar@utm.my

K. A. Zahan
Faculty of Engineering Technology, Universiti Tun Hussein Onn Malaysia, 86400 Parit Raja, Batu Pahat, Johor, Malaysia

M. A. A. Zaini et al. (eds.), *Proceedings of the 3rd International Conference on Separation Technology*, Lecture Notes in Mechanical Engineering,
https://doi.org/10.1007/978-981-16-0742-4_22

areas of developing countries. In the formulation of either cosmetic or pharmaceutical products, a surfactant is often present, which is usually used as detergent and emulsifier. The major concern is that most of the ingredients that comprise these cleansing agents are chemical-based, including synthetic surfactant that is harmful to human health. Examples of those chemical-based ingredients are sodium lauryl sulphate (SLS) and ammonium laurel sulphate (ALS). Chemical-based foaming agents that are extensively applied worldwide in food, cosmetics, and pharmaceutical preparations can be harmful to human health and the environment. It may also lead to skin reactions such as inflammation and can cause irritant contact dermatitis [10]. Generally, the inclusion of a natural foaming agent as a substitute for the chemical-based foaming agent will result in a significant improvement to its consumer-friendly and environmentally friendly qualities. In addition, the presence of a natural foaming agent would lead to an increase in ecosystem conservation effort and a decrease in chemical pollution. Thus, the next logical approach is to replace chemical-based ingredients with natural ones.

Aside from the effectiveness of cleansing agents towards dirt and stains, the antimicrobial activity of the cleansing agent needs to be considered as well in order to maintain its hygiene. Some medicinal plants, such as *C. asiatica*, can act as antimicrobial agents [13]. *C. asiatica* was chosen in this study because it believes to contain high saponin content that will possess foaming properties. Chen et al. [3] reported that crude saponin content in the seed of *C. oleifera* was 8.34% and its total saponin content in the crude saponin extract was 39.5% (w/w). Besides that, the potential of *C. asiatica* as a natural foaming agent has yet to be widely explored. Therefore, this research could present a new approach in the selection of a natural foaming agent and cleansing agent production studies in the sense that it will produce a product with good and safer properties for human consumption. Also, some natural foaming agents that have been studied previously have less stability of foam and possess weak foam properties. According to Somdee et al. [20], *C. asiatica* has exceptional advantages in phytomedicine because of the diversity in biological activities, including antimicrobial, anti-inflammatory, and antioxidant effects.

Microwave-assisted extraction (MAE) is a green extraction technique with shorter irradiation time, less solvent used, and lower CO_2 emission compared to other conventional extraction methods [19]. Ahmad et al. [1] reported the used of MAE for encapsulation of anthocyanin-rich product are environmentally friendly, low-cost, low energy and require short processing period. Other researcher also reported higher yield of saponin extract from Safed musli using MAE when compared to other extraction technique such as maceration, soxhlet and sonication [4]. Therefore, this study aims to optimize the extraction of *C. asiatica* by using MAE with three main parameters, which were time, microwave power, and solvent to solid ratio, to give the best yield of saponin.

2 Methodology

2.1 *Preparation of* Colubrina asiatica *Leaves Samples*

Colubrina asiatica (*C. asiatica*) leaves were carefully washed with tap water to remove undesired substances or particles. After that, the samples were dried for a day using an oven (Memmert, Germany). The leaves of *C. asiatica* were left in the oven at 52 °C for 2 days to make sure that they were completely dry. The dried samples were ground using a commercial electric blender (Waring® Products Division, U.S.A.). Then, the sample powder was sieved to obtain particles of small size. For further experimental work, the dried sample was placed in a sealed transparent container and stored in a −30 °C freezer to maintain the optimum condition and freshness of the sample.

2.2 *Extraction of* Colubrina asiatica *Leaves Using Microwave-Assisted Extraction*

A method from Kerem et al. [7] was modified and used for the extraction of *C. asiatica* using MAE, in which 95% methanol was used as a solvent instead of water. Extraction solvents have a significant effect on the yield of bioactive compounds, and methanol has been found to be superior to other solvents for the extraction of saponins. The *C. asiatica* leaves powder was weighed according to the solvent to solid ratio at a fixed volume of solvent; 40 mL. For the ratio of solvent to solid 10:1, 4 g of leaves powder were used. The mixture was poured into a 100 mL round bottom flask and was placed inside a microwave. The mixture was irradiated in a closed system microwave extractor using two levels of heating: 1 min of preheating to set temperature and irradiation time based on the experimental design. The extraction procedure was performed via MAE using MAS-II Plus Microwave Synthesis/Extraction Reaction Workstation (SINEO Microwave Chemistry Technology Co. Ltd., China). The temperature was set at 60 °C, and the solvent volume was 40 mL. The mixture was filtered using a vacuum pump (model: DOA-P504-BN) with Whatman qualitative filter paper No 1 and then the solvent was evaporated completely using a rotary evaporator (IKA RN10 Digital, Germany). The extracts were stored in a 4 °C refrigerator for further analysis. The experimental processes were performed in triplicate and the average values were calculated.

Table 1 Independent parameters for *C. asiatica* extraction

Parameters	Coded factor	Factor level		
		−1	0	+1
Irradiation time (min)	*A*	3	5	7
Solvent to solid ratio (mL/g)	*B*	20	30	40
Power (W)	*C*	300	400	500

2.3 *Optimization of* **Colubrina asiatica** *Extract by Using Microwave-Assisted Extraction*

Table 1 shows the independent parameters of Box-Behnken Design (BBD) that include irradiation time (*A*), solvent to solid ratio (*B*), and power (*C*). The range of the parameters was set according to the preliminary results obtained prior to the optimization process. BBD was chosen because it has been frequently applied in analytical method optimizations due to its advantages, which include the reduction of the number of experiments. This advantage resulted in lower reagent consumption and significantly less laboratory work considering its ability to develop mathematical models to evaluate the significance of the relation between the factors [5]. A total of 17 runs were generated using BBD in the Design of Expert version 6 software, as shown in Table 2.

The design included 1 block, 5 center points to assess pure error and the parameters values were fitted according to Eq. 1 as a polynomial equation:

$$Y = \alpha_0 + \alpha_1 A + \alpha_2 B + \alpha_3 C + \alpha_{11} A^2 + \alpha_{22} B^2 + \alpha_{33} C^2 2 + \alpha_{12} AB + \alpha_{13} AC \quad (1)$$

where; *Y* represents the total oil yield in MAE while α_{ij} values represent constant regression coefficients, *A*: Irradiation time, *B*: Solvent to Solid Ratio and *C*: Power

2.4 *The Yield of* **Colubrina asiatica** *Leaves Extract*

After all extractions completed, the extract was filtered to remove undesired particles, and the extracts were evaporated by using a rotary evaporator to separate the solvent and the crude extract. All the extract was stored in the refrigerator under 4 °C until further analysis. The weight of extracts was recorded. Then the percentage of yield was calculated according to the equation shown in Eq. 2.

$$\text{Yield(w/w)}\ (\%) = \frac{Weight\ of\ extract\ produced\ (\text{g})}{Weight\ of\ ground\ sample\ (\text{g})} \times 100 \quad (2)$$

Table 2 Independent parameters for *C. asiatica* extraction

Run	Time, min (*A*)	Solvent to solid ratio, mg/L (*B*)	Power, W (*C*)
1	7	30	400
2	3	20	300
3	7	10	400
4	5	20	400
5	3	30	400
6	7	20	300
7	5	30	500
8	5	30	300
9	3	20	500
10	5	20	400
11	5	20	400
12	5	20	400
13	7	20	500
14	5	10	500
15	5	10	300
16	5	20	400
17	3	10	400

2.5 Determination of Saponin

The saponin content was measured according to the method described by Vuong et al. [21] and Pham et al. [16]. Briefly, 0.25 ml of the extract was sequentially mixed with 0.25 ml of 8% (w/v) vanillin solution and 2.5 ml of 72% (v/v) H_2SO_4, which then was incubated in a water bath (Ratek, Ratek Instruments Pty., Ltd., Victoria, Australia) at 60 °C for 15 min. The solution was then cooled in an ice water bath to room temperature (RT). The absorbance was measured at 560 nm using a UV–VIS spectrophotometer (Cary 50 Bio, Varian Australia Pty., Victoria, Australia). Aescin was used as the standard, and different concentrations of aescin from 100 to 1000 μg/mL were prepared to generate the standard curve. The content of saponin was expressed as mg aescin equivalents per gram of dried sample (mg ESE/g). The total saponin content concentration was calculated by using leaves powder dry weight and extract volume following the Eq. 3.

$$Total\ saponin\ content\ concentration\ (\%) = \frac{\frac{\text{mg ESE}}{volume\ of\ extract}}{g DW\ of\ ground\ sample} \quad (3)$$

3 Results and Discussion

3.1 *Effect of Irradiation Time on Total Yield of* Colubrina asiatica

Figure 1 shows the experimental result for the effect of irradiation time on the extract yield (w/w) %. Based on Fig. 1, the results show that the highest extract yield was obtained at time 5 min with 41.27 (w/w) % followed by irradiation time of 7 min, 20 min, 10 min and 3 min with oil yield of 33.10 (w/w)%, 31.49 (w/w)%, 24.95 (w/w)% and 24.40 (w/w)% respectively. The result shows that the extract yield increased from 3 min to 5 min but decreased for the extraction time of 7 min onwards. The highest extract yield percentage at 5 min was significant ($P < 0.05$) compared to other irradiation times. Extract yield percentage from irradiation time of 10 and 20 min did not show a significant difference with statistical analysis ($P > 0.05$) of extract yield compared to 3, 5 and 7 min.

These results show that if the ground leaf was exposed to irradiation time for a longer time, it could not yield the highest amount of extract. This result is supported by previous research by Ren et al. [17], who discovered that the highest yield of steroid saponins from *Dioscorea zingiberensis* by using microwave-assisted extraction was at 5 min. Koyu et al. [8] reported a significant decrease in anthocyanin content from *Morus nigra* fruits when the extraction was done at 15 min. In contrast, it was at the highest yield when extracted for 10 min.

It shows that longer irradiation time does not yield more targeted bioactive compounds, but it might reduce targeted bioactive metabolites content because of the simultaneous extraction of other compounds. This is because the irradiation time is affected by the dielectric properties of the solvent. Extreme heat up may occur on extensive exposure to microwave heating for solvents like water, ethanol and methanol [11]. Minimum time extraction was desirable to minimize the usage of

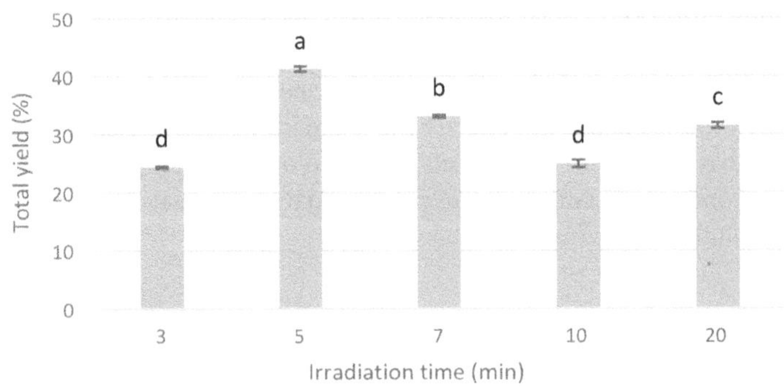

Fig. 1 Percentage of *C. asiatica* total yield based on different irradiation time. Different letters on top column indicate significant difference (P-value < 0.05)

power. Thus, for this study, the temperature range for optimization was set up from 3 to 7 min only because 10 and 20 min did not show significant extract yield extracted ($P > 0.05$).

3.2 *Effect of Solvent to Solid Ratio on Total Yield of* C. asiatica

Figure 2 shows the experimental result for the effect of solvent to solid ratio on the extract yield (w/w) % of *C. asiatica.* It shows that the highest extract yield was obtained at solvent to solid ratio of 30 mL/g, followed by 40 mL/g, 20 mL/g and 10 mL/g with extract yield of 41.51 (w/w) %, 39.42 (w/w) %, 36.70 (w/w) % and 27.72 (w/w) % respectively. The extract yield increased with the increasing of ratio, however, became depleted at 40 mL/g. Extract yield extracted for the solvent to solid ratio of 10–30 mL/g was significant ($P < 0.05$) to each other. It showed that there was no significant difference ($P > 0.05$) in the extract yield was observed between 30 and 40 mL/g.

Based on Fig. 2, it can be deduced that higher solvent to solid ratio did not guarantee more extract yield. This shows that MAE does not require a high quantity of solvent to obtain an optimum yield of extract from *C. asiatica* leaf. Yuan et al. [22] reported a similar pattern of result on the extraction of *Vaccaria segetalis* with different solvent to solid ratio from 10 to 50 mL/g where the extract yield was increased from 10 to 30 mL/g but started to decrease after 40 mL/g.

A higher ratio of solvent volume to solid is commonly desirable for effective extraction process in conventional extraction methods. This is in line with results proposed by Hu et al. [6] which stated higher ratio of solvent volume provide greater contact area for higher yield of total saponins. However, homogeneous and effective heating was ensured by using the optimum solvent to solid ratio. This shows that

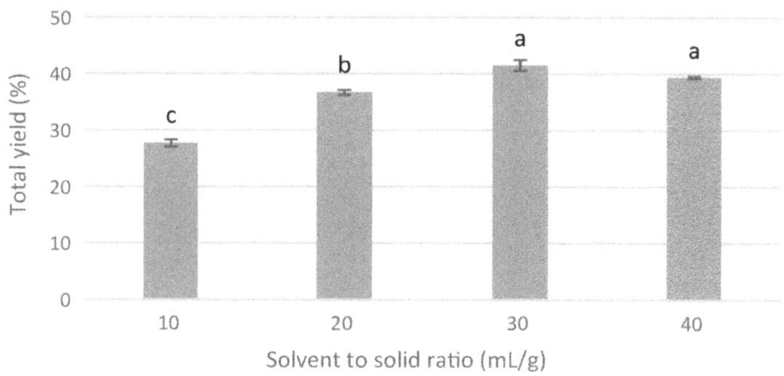

Fig. 2 Percentage of *C. asiatica* total yield based on different solvent to solid ratio. Different letters on top column indicate significant difference (P-value < 0.05)

MAE process required adequate amount of solvent because higher ratio may yield lower recoveries, which may be due to insufficient stirring to distribute the sample thoroughly through the solvent in the MAE [11]. Based on these results, 40 mL/g was excluded in the range of the parameters for optimization due to its effect on the reduction of extract yield.

3.3 Effect of Power on Total Yield of Colubrina asiatica

Figure 3 shows the experimental result for the effect of power on the extract yield (w/w)%. It shows that the extract yield of *C. asiatica* was highest at the power of 500 W. The highest yield was 42.54%, followed by 38.16%, 30.97% and 29.96% at the power of 500 W, 400 W, 300 W and 600 W respectively. The result shows no significant difference ($P > 0.05$) between extract yield at 300 W and 600 W. The extract yield was increased with power from 300 W to 500 W, however, decreased at 600 W. This phenomenon shows low extract yield at both lowest power and highest power used, 300 W and 600 W respectively.

According to Benmoussa et al. [2], the extraction of cumin essential oil resulted in the low oil yield at lowest and highest power used. The same pattern was observed in this study where the extract yield was increased with the increased of power; however, decreased after a certain power level. Research from Olalere et al. [14] on microwave extraction of piperine-oleoresin from black pepper (*Piper nigrum*) also showed the same pattern of extract yield. The extract yield of *Piper nigrum* was low at the power of 300 W, increasing at 400 W but decreasing at 450 W. Similar finding was reported by Ren et al. [18] on the extraction of *Sargassum thunbergii* polysaccharides using

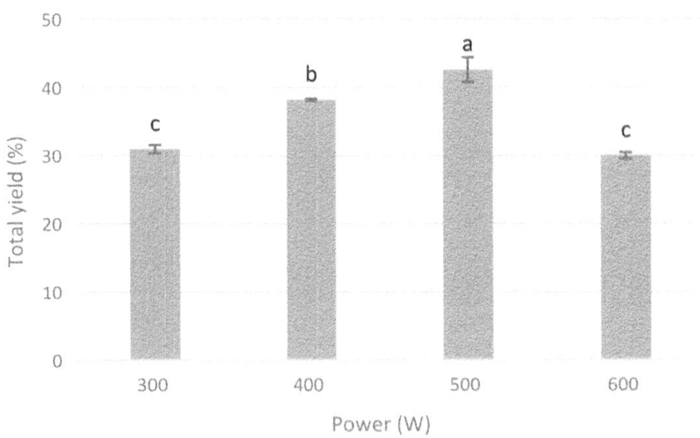

Fig. 3 Percentage of C. *asiatica* total yield based on different irradiated power. Different letters on top column indicate significant difference (P-value < 0.05)

power ranging from 200 to 1000 W showed that extract yield was increasing from 200 to 600 W. After 600 W, the extract yield was slightly decreased due to polysaccharides degradation under excessive power exposure. Based on these results, 600 W was not included in the optimization experiment because the results showed degradation of extract yield at 600 W.

3.4 Optimization of Microwave Assisted Extraction Method of Colubrina asiatica

Various parameters of extraction in this method are conducted and optimized. The optimum condition from selected ranges was obtained to produce the highest yield of extract for further analyses. An empirical quadratic polynomial model corresponding to the BBD was fitted to correlate the relationship between three independent variables (irradiation time, solvent to solid ratio and power) and the response (yield) in order to predict the optimum conditions of extract yield. Figure 4 shows the graph of actual versus predicted values of the response. The determination coefficient (R^2) of the quadratic regression model was 0.9379. High values of R^2 showed that most variation of extract yield could be explained and predicted by the obtained model. Hence, the model can be used to explain the response. The interaction term that gave the largest effect to the yield was AC. The relationship between the response (yield) and independent variables (irradiation time, solvent to solid ratio and power) was given in Table 3.

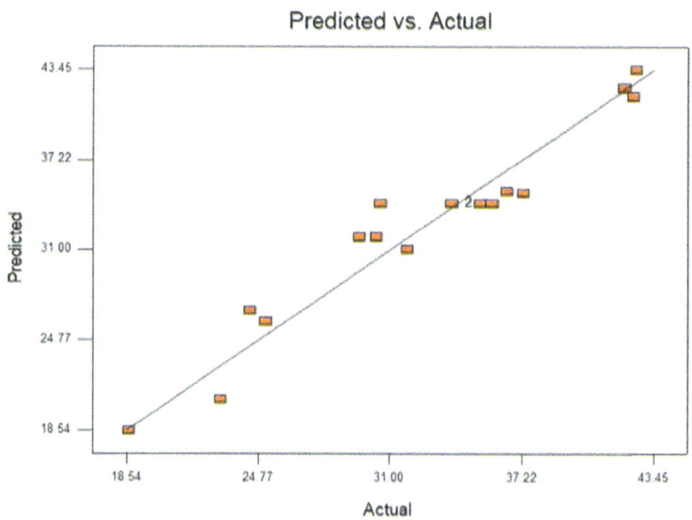

Fig. 4 Predicted values versus experimental data (actual values) for *C. asiatica* extract yield

Table 3 Actual regression coefficient of the quadratic model and the corresponding P-value

Factor	Regression coefficient	df	P-value
Intercept	34.19	1	0.0019
A Time	6.69	1	0.0002
B Solvent to solid ratio	5.12	1	0.0009
C Power	1.09	1	0.2817
A^2	−3.14	1	0.0444
AC	3.77	1	0.0244
BC	3.20	1	0.0456

The regression coefficients of the quadratic model equation were used to describe the extract yield from *C. asiatica* leaf. Table 3 summarized the actual parameters of the extract yield. Its shows that the extract yield was positively affected by the irradiation time, solvent to solid ratio and power. Based on the P-value, irradiation time was the most significant parameter followed by solvent to solid ratio and power. This result indicates that the extract yield increases as the factors increase. The quadratic term of irradiation time was negatively affected the extract yield and significant ($P < 0.05$). The interaction between time and power has significantly ($P < 0.05$) shown a positive effect on the response, together with the interaction between ratio and power. The most significant interaction was observed between irradiation time and power, followed by the interaction between the solvent to solid ratio and power.

The model was verified based on the results obtained from response surface analysis; the optimal extraction for the extract was 6.06 min irradiation time, solvent to solid ratio 28.31 mL/g, and 445 W of microwave power. Based on the optimal conditions, the predicted yield value was 42.63 w/w %. The actual yield value from the experimental procedure was 43.1 ± 0.05 w/w % at 6 min irradiation time, solvent to solid ratio of 28 mL/g, and 445 W microwave power. The optimal conditions are reasonable and adequate as the percentage error was 0.47%.

Table 4 shows the percentage of yield and total saponins content (TSC) of *C. asiatica* leaves extract produced from the optimal condition. The result indicates that methanol is a good solvent for the extraction of saponin. Low alcohols such as methanol and ethanol have usually been used as effective solvents for the extraction of saponins from plant materials [9]. According to Pham et al. [15], *Catharanthus roseus* leaf MAE extraction by using methanol as solvent gave a higher amount of TSC compared to extraction using hexane and ethyl acetate as solvent. This result is because saponin is polar compounds. Thus, the usage of methanol, which has a higher polarity, is useful to increase the polarity of the solvent and extract a higher

Table 4 *C. asiatica* extract yield and total saponin content of *C. asiatica* extract using MAE in optimal condition

Optimum extract yield (%)	43.1 ± 0.05
Total saponin content (mg ESE/mg)	15.36 ± 0.001

yield of the saponin compound [15]. In addition, another supported study by Ngo et al. [12] reported that absolute methanol as solvent had the highest yield of saponin TSC (15.6%) from the root of *Salacia chinensis* followed by absolute ethanol, 50% methanol and 50% acetone (14.3%, 12.3%, and 12.2%, respectively).

4 Conclusion

In conclusion, the extraction of *C. asiatica* using microwave-assisted extraction (MAE) showed that the highest extraction yield obtained at 6 min irradiation time, solvent to solid ratio of 28 mL/g and 445 W microwave power with 43.1 ± 0.05 w/w % extract yield. The experimental data of MAE were analyzed using Box-Behnken Design from response surface methodology (RSM) to study the factor that affects the extract yield. The results show that irradiation time and solvent to solid ratio have a significant effect ($P < 0.05$) on extract yield. The experiment was considered satisfactory as the R^2 was 0.9379, which has a good correlation between experimental data and calculated data.

References

1. Ahmad F, Mohd Jusoh YM, Abang Zaidel DN, Zakaria ZY, Muhamad II (2020) Optimization of microwave-assisted encapsulation of Black Mulberry (Morus nigra) extract. Chem Eng Trans 78:145–150. https://doi.org/10.3303/CET2078025
2. Benmoussa H, Elfalleh W, He S, Romdhane M, Benhamou A, Chawech R (2018) Microwave hydrodiffusion and gravity for rapid extraction of essential oil from Tunisian cumin (Cuminum cyminum L.) seeds: optimization by response surface methodology. Ind Crops Prod 124:633–642. https://doi.org/10.1016/j.indcrop.2018.08.036
3. Chen YF, Yang CH, Chang MS, Ciou YP, Huang YC (2010) Foam properties and detergent abilities of the saponins from Camellia oleifera. Int J Mol Sci 11(11):4417–4425. https://doi.org/10.3390/ijms11114417
4. Deore SL, Baviskar BA, Rangari AS (2015) Rapid and high yield extraction method for saponins from Safed musli. Pharmacogn J 7(4):210–214. https://doi.org/10.5530/pj.2015.4.1
5. Ferreira SLC, Bruns RE, Ferreira HS, Matos GD, David JM, Brandão GC, dos Santos WNL (2007) Box-Behnken design: an alternative for the optimization of analytical methods. Anal Chim Acta 597:179–186. https://doi.org/10.1016/j.aca.2007.07.011
6. Hu T, Guo YY, Zhou QF, Zhong XK, Zhu L, Piao JH, Jiang JG (2012) Optimization of ultrasonic-assisted extraction of total saponins from Eclipta prostrasta L. using response surface methodology. J Food Sci 77(9):975–982. https://doi.org/10.1111/j.1750-3841.2012.02869.x
7. Kerem Z, German-Shashoua H, Yarden O (2005) Microwave-assisted extraction of bioactive saponins from chickpea (Cicer arietinum L). J Sci Food Agric 85(3):406–412. https://doi.org/10.1002/jsfa.1989
8. Koyu H, Kazan A, Demir S, Haznedaroglu MZ, Yesil-Celiktas O (2018) Optimization of microwave assisted extraction of Morus nigra L. fruits maximizing tyrosinase inhibitory activity with isolation of bioactive constituents. Food Chem 248:183–191. https://doi.org/10.1016/j.foodchem.2017.12.049

9. Le AV, Parks SE, Nguyen MH, Roach PD (2018) Effect of solvents and extraction methods on recovery of bioactive compounds from defatted Gac (Momordica cochinchinensis Spreng.) seeds. Separations 5(39):1–13. https://doi.org/10.3390/separations5030039
10. Lemery E, Briançon S, Chevalier Y, Bordes C, Oddos T, Gohier A, Bolzinge MA (2015) Skin toxicity of surfactants: structure/toxicity relationships. Colloids Surf A: Physicochem Eng Asp 469:166–179. https://doi.org/10.1016/j.colsurfa.2015.01.019
11. Mandal V, Mohan Y, Hemalatha S (2007) Microwave assisted extraction—an innovative and promising extraction tool for medicinal PHCOG REV.: review article extraction tool or medicinal plant research. Pharmacogn Rev 1(1):7–18
12. Ngo TV, Scarlett CJ, Bowyer MC, Ngo PD, Vuong QV (2017) Impact of different extraction solvents on bioactive compounds and antioxidant capacity from the root of Salacia chinensis L. J Food Qual 2017:1–8. https://doi.org/10.1155/2017/9305047
13. Nivas D, Dethe UL, Gaikwad DK (2015) In vitro antioxidant activities and antimicrobial efficacy of Asian snakewood; Colubrina asiatica (L.) Brong. Res J Med Plant 9(7):307–320. https://doi.org/10.3923/rjmp.2015.307.320
14. Olalere O, Nour A, Alara O, Akbari S, Yunus R (2018) Evaluation of optimization parameters in microwave reflux extraction of piperine-oleoresin from black pepper (Piper nigrum). Beni-Suef Univ J Basic Appl Sci 7. https://doi.org/10.1016/j.bjbas.2018.07.006
15. Pham HNT, Vuong QV, Bowyer MC, Scarlett CJ (2018) Ultrasound-assisted extraction of Catharanthus roseus (L.) G. Don (Patricia White cultivar) stem for maximizing saponin yield and antioxidant capacity. J Food Process Preserv 42(5):1–12. https://doi.org/10.1111/jfpp.13597
16. Pham HNT, Nguyen VT, Vuong QV, Bowyer MC, Scarlett CJ (2015) Effect of extraction solvents and drying methods on the physicochemical and antioxidant properties of Helicteres hirsuta Lour. Leaves. Technologies 3(4):285–301. https://doi.org/10.3390/technologies3040285
17. Ren Y, Chen Y, Hu B, Wu H, Lai F, Li X (2015) Microwave-assisted extraction and a new determination method for total steroid saponins from Dioscorea zingiberensis C.H. Wright. Steroids 104:145–152. https://doi.org/10.1016/j.steroids.2015.09.008
18. Ren B, Chen C, Li C, Fu X, You L, Liu RH (2017) Optimization of microwave-assisted extraction of Sargassum thunbergii polysaccharides and its antioxidant and hypoglycemic activities. Carbohydr Polym 173:192–201. https://doi.org/10.1016/j.carbpol.2017.05.094
19. Samad NA, Abang Zaidel DN, Salleh E, Mohd Yusof AH, Dailin DJ, Abang Zaidel DN (2019) Optimization of Plectranthus amboinicus (Lour.) Spreng extraction process using microwave-assisted technique. Chem Eng Trans 72:397–402. https://doi.org/10.3303/CET1972067
20. Somdee T, Mahaweerawat U, Phadungkit M, Yangyuen S (2016) Antioxidant compounds and activities in selected fresh and blanched vegetables from northeastern Thailand. Chiang Mai J Sci 43(4):834–844
21. Vuong QV, Hirun S, Roach PD, Bowyer MC, Phillips PA, Scarlett CJ (2013) Effect of extraction conditions on total phenolic compounds and antioxidant activities of Carica papaya leaf aqueous extracts. J Herb Med 3:104–111. https://doi.org/10.1016/j.hermed.2013.04.004
22. Yuan XH, Fu LN, Gu CB, Zhang YD, Fu YJ (2014) Microwave-assisted extraction and antioxidant activity of vaccarin from the seeds of Vaccaria segetalis. Sep Purif Technol 133:91–98. https://doi.org/10.1016/j.seppur.2014.06.002

A Selection Design of Experiment for Optimization of Process Variables for Supercritical Carbon Dioxide Using Response Surface Methodology: A Review

Norlisa Mili, Nor Faadila Mohd Idrus, Zuhaili Idham, Noor Azwani Mohd Rasidek, Azizul Azri Bin Mustaffa, and Mohd Azizi Che Yunus

Abstract Supercritical carbon dioxide (SC-CO_2) which is an alternative and green method for extraction of natural plants. Many parameters in SC-CO_2 including pressure, temperature, extraction time, particle size, CO_2 flowrate and composition of solvent modifiers were optimized to improve the extraction efficiencies by using a response surface methodology (RSM). RSM is a collection of statistical and mathematical techniques useful for developing, improving, and optimizing processes. To adequately resolve the above-mentioned issues, a design of experiment (DOE) is required. Recently, in providing a better decision support analysis, the necessity for a specific experimental design modelling is based on statistical methodologies via response surface. Hence, objective of this study is to review the previous studies on various innovative designs that have been employed for SC-CO_2 extraction using RSM Consequently, the present paper provides an insight into RSM by scrutinizing the design, modelling, predicting, and optimizing SC-CO_2 process with an acceptable accuracy for different designs of experimental work with an appropriate model that fits the desired approach of the researchers.

Keywords Design of experiment · Optimization · Response surface methodology · Supercritical carbon dioxide

---

N. Mili · N. F. M. Idrus · Z. Idham · N. A. M. Rasidek · M. A. C. Yunus (✉)
Centre of Lipids Engineering & Applied Research (CLEAR), Ibnu Sina Institute for Scientific & Industrial Research, Universiti Teknologi Malaysia (UTM), 81310 Johor Bahru, Johor, Malaysia
e-mail: azizi@cheme.utm.my

N. Mili
e-mail: mnorlisa@unimas.my

N. Mili · N. F. M. Idrus · Z. Idham · N. A. M. Rasidek · A. A. B. Mustaffa · M. A. C. Yunus
Faculty of Chemical & Energy Engineering, Universiti Teknologi Malaysia (UTM), 81310 Johor Bahru, Johor, Malaysia

N. Mili
Department of Chemical Engineering and Energy Sustainability, Faculty of Engineering, UNIMAS, 94300 Kota Samarahan, Sarawak, Malaysia

M. A. A. Zaini et al. (eds.), *Proceedings of the 3rd International Conference on Separation Technology*, Lecture Notes in Mechanical Engineering,
https://doi.org/10.1007/978-981-16-0742-4_23

1 Introduction

The factors of a manufacturing production involve time and expenses, which are crucial criteria for the manufacturing sector as they need to make economic profits. Besides, the safety of the process is also considered. Accordingly, as a green method, supercritical fluid carbon dioxide (SC-CO_2) method is an excellent offer for fast extraction, utilizes moderate temperatures, eliminates clean-up steps, and avoids the use of harmful organic solvents. This process is an ideal criterion for industries involving extractions of target compounds from foods and natural products [1, 2]. Mainly, the effectiveness of SC-CO_2 depends on several parameters, including speed of flow, friction, temperature, pressure, time of extraction, and so on. Modelling and optimizing processing parameters by techniques such as Response Surface Methodology (RSM) is an effective way to increase the process efficiency in conditions where several variables can affect the outcome [3].

Over the past decade, there has been a sustained research activity in the optimization of using RSM. Figure 1 shows the total number of documents recently published pertaining to RSM process optimization from 2010 to 2019 (data extracted from Scopus). Intriguingly, the total number of RSM documents published increased significantly from 866 articles published in 2010 to 2857 in 2019. Increased publications indicate that researchers are becoming more interested in this study. Subsequently, the documents are published according to the subject area, as shown in Fig. 2. It can be seen that these published papers contribute to process optimization in engineering (14.6%), chemical engineering (12.3%), chemistry (11%), agricultural and biological sciences (10.8%), Environmental Science (10%), Biochemistry, Genetics and molecular biology (8.4%), and material science (6.3%), respectively.

In recent years, many researchers have shown rigorous interest in optimizing the extraction process of SC-CO_2. One of the known challenges in this approach is to select the appropriate experimental design to be applied. The previous investigations

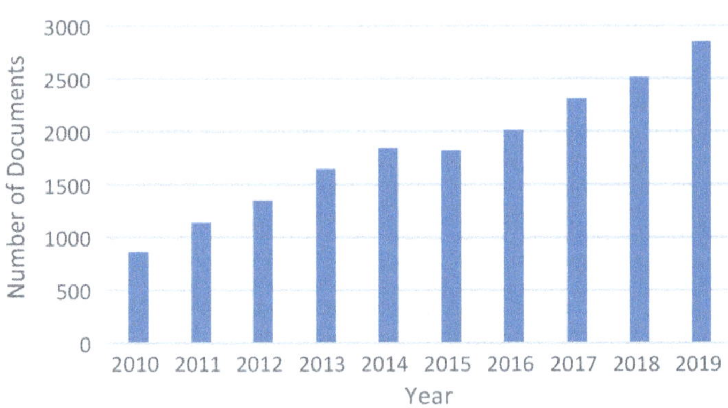

Fig. 1 Documents published from 2010 to 2019 related to RSM based on Year (data extracted from Scopus)

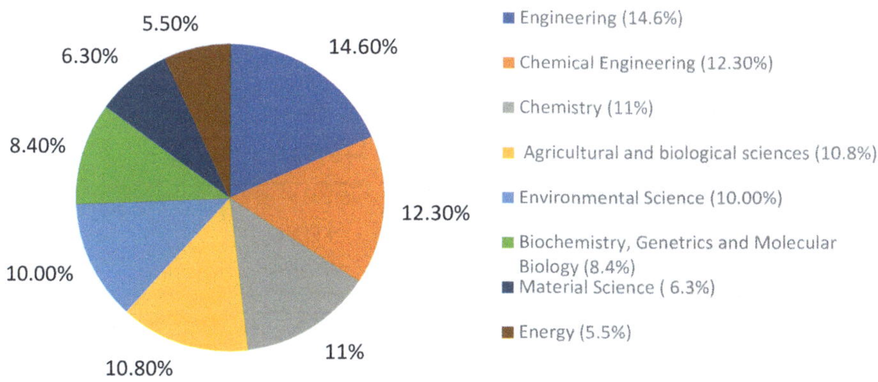

Fig. 2 Documents published from 2010 to 2019 related to RSM based on subject area (data extracted from Scopus)

using RSM in SC-CO_2 have been established for process development, improvement, and optimization [4–6]. In addition to SC-CO_2, process optimization using RSM also applies to the internal combustion engine [7], wastewater treatment processes [8], and analytical chemistry [9]. Furthermore, having a greater insight into the selection of experimental designs would provide useful information on the correlation between the experimental studies, input factors, and output responses [7].

Due to the extensive application of RSM in engineering and sciences, SC-CO_2 are actively employed RSM in their experimental designs. The primary advantage of RSM in innovative design is its ability to reduce the total of experiments amount to be performed and consume a limited amount of chemical [10]. Conversely, it is beneficial for the SC-CO_2 extraction process. In this paper, initially, the core concept is introduced. Then the procedure of optimization is elucidated. Finally, the example of studies, including their design selection, objectives, and output is reviewed. This review paper covers the SC-CO_2 from natural plants. The ultimate objective of this study is to review the previous studies on various innovative designs that have been employed for SC-CO_2 extraction using RSM.

2 Theory and Procedures of RSM

RSM consists of a set of mathematical and statistical techniques in which the response of concern is based on a number of factors, and the purpose of the design is to model and optimize the research design [11]. Experimental variables are indicators that can be adjusted independently of one another, namely a factor or independent variables. On the contrary, the expected outcomes from the experiments are defined by dependent variables or responses.

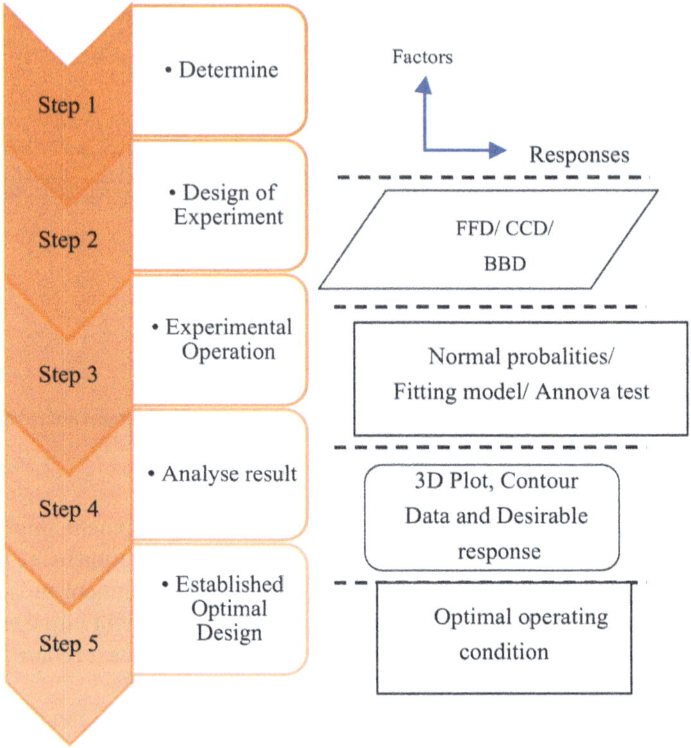

Fig. 3 Flow chart of RSM for SC-CO_2

Optimization from the perspective of industrial and research design is a central concern because it influences the output yield of the product. As a result, RSM contributes to the less energy, resources, and time that needed from the technical procedure and research experimentation. Published research articles on the subject mainly implemented the experimental flow chart, as illustrated in Fig. 3.

2.1 *Determination of Factors and Responses*

Generally, the optimization technique coincides with a selection design to identify essential factors before continuing to optimization design. Multiple factors may have an impact on the process response studied, and the identification of small contributions from each is applied to almost unfeasible. Thus, it is imperative to obtain those factors with significant impacts. In order to determine the vital factors and responses for DOE, the screening design is required. A review made by Sharif et al. [12] reported that two-level full factorial, two- level factorial and Plackett-Burman design are mostly executed in the screening process. Although the screening design for factors and responses determination can be done, an alternative step can be taken. Since optimization is widely used, the researcher also can rigorously define their factors and responses by literature review based on their research aims and objectives.

The results of the responses are considered to enhance and extend the results of the optimization. In addition, it is also necessary to determine the appropriate range (level) for these factors against which the elements are to be evaluated as this affects the accuracy of the model equation. Prior analysis to process the information is crucial in order to narrow the selection of the significant factors.

2.2 Design of Experiment

Design of Experiment (DOE) is a primary step after the determination of factors and responses. The number of factors and responses influences the selection of the DOE. Deciding an adequate design is to rely on the research objectives, the suitability of experiments, cost-effectiveness, time consumption, and other significant aspects [12]. Furthermore, the selected convenient design has a significant impact on the development of a response surface and the accuracy of the expected model. In the optimization of SC-CO_2 extraction process, the most applied design are Box-Behnken design (BBD), central composite design (CCD), and full Factorial design (FFD). Meanwhile, the popular software for applying DOE on SC-CO_2 are Design Expert (State-Ease Inc.), Statistical (Stat Soft), and Minitab (Minitab Inc.). In the optimization analysis, FFD covers all points of experimental design. Whereas two experimental designs which are BBD and CCD reference points are located at various places within the cubic space of experimentation table generated from the appropriate software. The BBD thus incorporates no experimental cubic space vertices, and the CCD assesses borderline areas.

2.3 Experimental Operation

A mathematical model acquired after the function is fitted to the data may not efficiently and adequately describe the experimental behavior. A reliable improvement on the performance of the fitted model is by the use of variance analysis (ANOVA). ANOVA's core concept is to compare the treatment variation (change in variable level combination) to the variation resulting from the errors implicit in the measurement of the response produced. In comparison, The relevance of regression used for the prediction of responses can be assessed, taking into account the components of experimental variance. Furthermore, for the response surface analysis, a simplified second-order polynomial model has been proposed, as Eq. (1).

$$Y = \beta_0 + \sum_{i=1}^{k} \beta_1 + \sum_{i=1}^{k} \beta_{ii} x_i^2 + \sum_{i<y}^{k} \beta_{ij} x_i x_j \tag{1}$$

where Y is the predicted response, β_0 is the intercept term, β_1 is the linear effect, β_{ii} is the squared effect, and β_{ij} is the interaction effect.

2.4 Result Analysis

RSM intended to obtain optimum conditions for the area generated using the linear models that can be used to achieve the research objective. Response surface plots such as 3D contour and surface plots are useful for establishing desirable response values and operating conditions. However, if the test area cannot be relocated for physical or instrumental factors, the study needs to identify the best operational conditions in the evaluated experimental environment.

2.5 Established Optimal Design

The established optimal design significantly provides the best results and data presentation. However, experiment should be conducted to check the reliability in optimum conditions of theoretically determined models. The most used are Chi-Square and T-tests to determine the difference between experimental data and predicted values. The experimental mistake between theoretical and experimental values can be determined by another way to test model validity.

3 Concept of SC-CO_2 Process

Figure 4 shows schematically the fundamental process of SC-CO_2 extraction process. In the late 1980s, few researchers explored the use of supercritical liquids (SCF) in analytical applications before the development of commercial Supercritical Fluid Extraction (SFE) instruments. In the early 1990s, increased automation and instrument design developments resulted in an increase in SFE work in a variety of applications. SFE is based on the supercritical fluid (SF) solvation properties that occurred from pressure and temperature above the critical point of a substance. This technique is frequently performed using carbon dioxide (CO_2) for many factors such as CO_2 has an approximately low critical temperature (7.38 MPa and Tc = 31.05 °C) which are non-flammable, highly pure at reasonably cheap, and can be extracted effortlessly. Therefore, carbon dioxide (CO_2) is the best and favourable solvent among researchers in SC-CO_2. Accordingly, it already becomes apparent that SC-CO_2 is

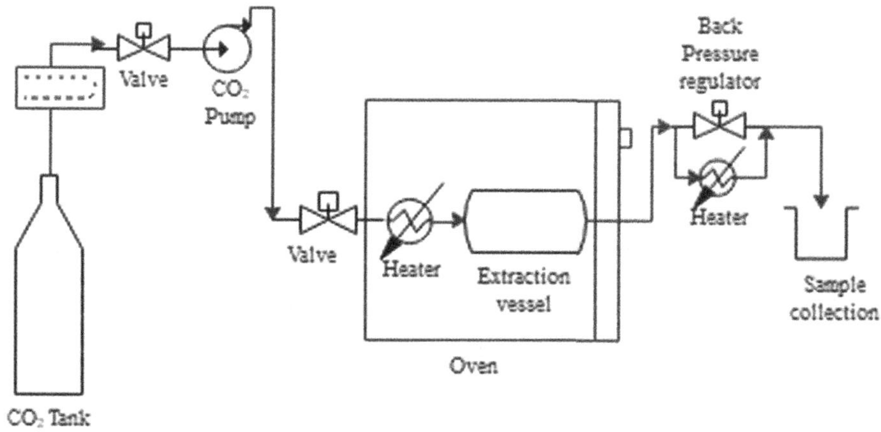

Fig. 4 Schematic diagram of SC-CO_2 extraction unit

an appropriate, environmentally-friendly alternative to the production of essential oils, taking into account the lipophilic characteristics of plant essential oils [13]. The primary limitation of CO_2, however, is that the polarity needed for the extraction of polar substances is not present. Nevertheless, the presence of co-solvents changes CO_2 polarity by increasing the solubility of the extracted components.

Theoretically, when the temperature and the pressure of a pure substance surpass a critical value, it reaches the supercritical point. The critical value or point of the substance is defined by their critical temperature (T_c) and pressure (P_c). Any substance below the critical point could emerge in the state of liquid, solid, or gas and above it or in the state of supercritical if adequate pressure is operated. If the properties are between liquid and gas, the density of the fluid is very similar to liquid. Certainly, viscosity close to gas, and the diffusivity of the fluid is intermediate, under supercritical conditions [14]. As such, SC-CO_2 has several operational benefits over conventional solvents. Therefore, it is a better transportation than fluids because of its low viscosity and relatively high diffusiveness that promotes faster extraction process. In addition, supercritical fluids increase the solvation power with high densities. The tremendous benefit of using SC-CO_2 lies in the ability, by changes of its temperature and/or its pressure will directly modify the density and solubility of the fluids. Furthermore, SC-CO_2 extraction also has been employed as the reaction medium during biocatalytic processes that enhances the conveyance characteristics of substrates at the active location of enzymes, thus increases the conversion of the reaction [15]. Wrona et al. [16] reported the trends and examples of SC-CO_2 application to separate medical and nutritional compound groups like flavonoids, saponins, carotenoids, and phenols.

Overall, the efficiencies of SC-CO_2 are based on number of parameters or factors, including flow rate, pressure, temperature, extraction time, particle size, modifier, extraction time, CO_2 flowrate and other important factors. In situations like this, where several variables affect the outcome or responses, processing parameters are effectively modeled and optimized through techniques such as RSM. Among the most studied areas by many researchers is optimization of SC-CO_2 using RSM.

4 Application of RSM in SC-CO_2

The aims of optimization mainly to maximize output that can be obtained for a given amount of experimental data. Several researchers have successfully implemented this tool for optimizing. Most researchers and engineers in this field tried to use RSM to solve the problem with the output responses between single and combined factors of the experimental variables. This review only covers the SC-CO_2 of natural plants. The DOE selection also has to consider the limitations of the equipment used in the research, such as limitations of temperature and pressure range. A list of experimental designs with their applications on SC-CO_2 is presented in Table 1. The implications of the reviewed observed that most researchers are favour BBD and CCD as their DOE.

5 Conclusion

These findings show that RSM's role in optimizing process variables is to predict the best condition for the input and output range. To use SC-CO_2, DOE must be chosen to suit an appropriate mathematical model and determine the consistency of the factors with the data obtained. The interaction between the factors and responses are based on experimental designs and the statistical results. Consequently, by increasing the weightings given in accordance with the importance of the responses, the prediction model can be improved. However, the disadvantage of RSM is to provide optimum results for multiple responses. The challenge is to obtain the best optimum condition that can satisfy more than one response. In the case of SC-CO_2, the use of FFD is not frequent compared to BBD and CCD since the desirability value for higher numbers of variables, however, based on three factorial design levels, provide a more efficient and revised method. However, the BBD and CCD the Box–Behnken based on FFD present a more efficient metric and based on according to the reviewed papers done in this research.

Table 1 Published article related to optimization of SC-CO_2 using RSM

Material	Extract	DOE	Software	Optimum conditions	Optimum response	Aim of study	References
Hibiscus sabdariffa L. (Roselle)	Roselle extract	Historical data	Design-expert 6.4	P = 27.5 MPa T = 50.8 °C	Yield = 163.26 mg/	Obtain the highest yield of Roselle extract using SC-CO_2	[17]
Pogostemon cablin	Essential oil	BBD	Design Expert	P = 24.5 MPa T = 47 °C t = 119 min	Oil yield = 2.4356%	Influence of the extraction variables on the extraction yield of patchouli oil	[4]
Cymbopogon citronella leaves	Essential oil	BBD	Design-expert V8.0.6	P = 25 MPa T = 35 °C t = 120 min *f* = 18L/h	Oil yield = 4.40%	To obtain the highest essential oil yield from C. citronella leaves	[18]
Camelina sativa seed	Astaxanthin-rich Camelina seed oil	CCRD	Design Expert	P = 41.6 MPa T = 36.6 °C ec = 42.0% ethanol	Astaxanthin concentration = 4 37 μg/g oil	(i) To study the effects of ethanol modified-SC-CO_2 extraction parameters, and concentration of astaxanthin (ii) To optimize the ethanol-modified SC-CO2 extraction conditions	[19]

(continued)

Table 1 (continued)

Material	Extract	DOE	Software	Optimum conditions	Optimum response	Aim of study	References
Echium vulgare seed	Echium vulgare seed oil	CCD	Modde 5.0 (Umetrics, Sweden)	P = 400 bar T = 62 °C	Oil yield = 4.40% 14.4% Oleic acid = 16.2% and polyunsaturated fatty acids = 72.3%	(i) To investigate the effect of extraction temperature and pressure on the yield of echium oil obtained (ii) To determine the physicochemical properties and fatty acid profile of the oil samples obtained at optimum conditions.	[20]
Swietenia mahagoni seed	Seed oil	Three-level factorial design	Statistica version 7.0 (STatSoft, EUA)	Oil yield: P = 30 MPa T = 60 °C β-sitosterol content P = 30 MPa T = 40 °C	Oil yield = 14.45% and β-sitosterol content = 0.9204%	To maximum oil yield and β-sitosterol content from Swietenia mahagoni seeds	[21]
Melaleuca cajuputi leaves	Volatile oil	CCRD	Design Expert 8.0.4	P = 24.91 MPa T = 43.1 °C f = 5.88 ml/min	Oil yield = 1.24 wt%	To maximize highest yield of volatile oil	[22]

(continued)

Table 1 (continued)

Material	Extract	DOE	Software	Optimum conditions	Optimum response	Aim of study	References
Patchouli leaves	Patchouli oil	BBD	Design-Expert 6.0.8	Oil yield: P = 146.35 atm T = 38 °C t = 74 min Alcohol content: P = 135.17 atm T = 41.45 °C t = 252.62 min	Oil yield = 6.41% and Alcohol content = 25.34%	To improve the quality of patchouli oil	[23]
Eryngium billardieri	Essential oil	CCD	Design Expert 7.0	P = 30 MPa T = 34.85 °C ps = 0.75 mm t = 130 min	Oil yield = 0.8522%	To optimize the essential oil yield	[24]
Tomato	Lycopene rich extracts (oleoresins)	CCD	Stat Ease Design-Expert 8.0.6.1	Oleoresin yield: P = 53.7 MPa T = 73.9 °C t = 155 min Cis-lycopene isomers: P = 55 MPa T = 52 °C t = 180 min	Oleoresin yield Cis-lycopene isomers 62% (of total lycopene)	To optimize the isolation of the non-polar tomato by-product fraction (oleoresin) and cis-lycopene isomers	[25]

(continued)

Table 1 (continued)

Material	Extract	DOE	Software	Optimum conditions	Optimum response	Aim of study	References
Crocus sativus petals	Polar and non-polar compound	BBD	Design-Expert v.7	P = 16.4 MPa T = 62 °C t = 47 min	Phenolic content = 1423 mg/100 g, total flavonoid content = 180 mg/100 g, total anthocyanin content = 103.4 mg/100 g, 2,2-diphenyl-1 picrylhydrazyl free radical-scavenging activity = 74.5% and Ferric reducing/antioxidant power value = 3.9 mM)	To investigate the effect of extraction conditions i.e. temperature, time, and extraction pressure on total phenolic content (TPC), total flavonoid content (TFC), total anthocyanin content (TAC), and antioxidant activity which was measured with two in vitro methodologies of: ferric reducing/antioxidant power (FRAP) and 2,2-diphenyl-1-picrylhydrazyl (DPPH) free radical-scavenging activity	[26]
Cleome coluteoides Boiss	Essential oil	CCD	Design Expert 7.0	P = 22 MPa T = 34.85 °C ps = 0.58 mm t = 135 min	Oil yield = 0.6589% P = 220 bar, T = 308 K, dp = 0.58 mm and t = 135 min, under which conditions, the maximum yield of 0.6589 was obtained.	To optimize the essential oil yield	[27]

(continued)

Table 1 (continued)

Material	Extract	DOE	Software	Optimum conditions	Optimum response	Aim of study	References
Gynostemma pentaphyllum	Seed oil	BBD	Design-Expert 8.0.6	P = 32 MPa T = 43 °C t = 160 min	Oil yield = 35.96 ± 0.21%	To optimize extraction parameters of oil yield extraction to achieve the highest yield	[28]
Pomegranate (Punica granatum L.) peel	Essential oil	CCD	Design-Expert 7.0	P = 35.46 MPa T = 55 °C mc = 150 μL t = 30 min	Oil yield = 1.18% (w/w)	To optimize the essential oil yield	[29]
Microalgae	Essential oil	CCD	Design-Expert Software, version 6.0.1	P = 64 MPa T = 49.3 °C	Temperature had a greater effect on response variables than CO_2 density	To evaluate the effect of extraction conditions on oil and carotenoids recovery from microalgae	[30]
Castor (Ricinus communis L) seed	Castor oil	BBD	Design-Expert	P = 29.90 MPa T = 63.72 °C f = 4.15 ml/min	9.29% oil yield	To optimize the oil yield	[31]
Quinoa seeds	Essential oil	CCD	Design software	P = 18.5 MPa T = 130 °C t = 180 min	Tocopherol concentration of 336.0 mg/100 g of oil	To obtain tocopherol-enriched oils from Quinoa	[32]
Licorice plant root	Glycyrrhizic acid	CCRD	Minitab software version 7	P = 29.60 MPa T = 68 °C f = 2 ml/min	GA recovery = 54.4%	To extract Glycyrrhizic acid (GA)	[33]
Anoectochilus roxburghii using	Essential oil	CCD	Design-Expert software version 7.0.0	P = 35 MPa T = 37.6 °C t = 150 min f = 8 L/h	1.677% oil yield	To optimize the SFE conditions to obtain the highest possible yield of essential oils from A. roxburghii	[34]

(continued)

Table 1 (continued)

Material	Extract	DOE	Software	Optimum conditions	Optimum response	Aim of study	References
Phaleria macrocarpa seed	Unsaturated oil	CCD	Minitab software version 16	P = 42 MPa T = 72 °C f = 4.5 ml/min	52.9% oil yield	To optimize oil yield	[35]

Acknowledgements This work was financially supported by the Universiti Teknologi Malaysia under the Fundamental Research University Grant Scheme (FRGS), FRGS/1/2018/STG07/UTM/02/13 (UTM Vote:4F997) from Ministry of Higher Education Malaysia.

References

1. Mohd Idrus NF, Yian LN, Idham Z, Aris NA, Putra NR, Abdul Aziz AH, Che Yunus MA (2018) Mini review: application of supercritical carbon dioxide in extraction of propolis extract. Malays J Fundam Appl Sci 14:387–396. https://doi.org/10.11113/mjfas.v14n4.1088
2. Abdul Aziz AH, Putra NR, Kong H, Che Yunus MA (2020) Supercritical carbon dioxide extraction of sinensetin, isosinensetin, and rosmarinic acid from Orthosiphon stamineus leaves: optimization and modeling. Arab J Sci Eng 45:7467–7476. https://doi.org/10.1007/s13369-020-04584-6
3. Yousefi M, Rahimi-Nasrabadi M, Pourmortazavi SM, Wysokowski M, Jesionowski T, Ehrlich H, Mirsadeghi S (2019) Supercritical fluid extraction of essential oils. TrAC Trends Anal Chem 118:182–193. https://doi.org/10.1016/J.TRAC.2019.05.038
4. Xiong K, Chen Y, Shen S (2019) Experimental optimization and mathematical modeling of supercritical carbon dioxide extraction of essential oil from Pogostemon cablin. Chin J Chem Eng 27:2407–2417. https://doi.org/10.1016/j.cjche.2019.03.004
5. Ndayishimiye J, Chun BS (2017) Optimization of carotenoids and antioxidant activity of oils obtained from a co-extraction of citrus (Yuzu ichandrin) by-products using supercritical carbon dioxide. Biomass Bioenergy 106:1–7. https://doi.org/10.1016/j.biombioe.2017.08.014
6. Krakowska A, Rafińska K, Walczak J, Buszewski B (2018) Enzyme-assisted optimized supercritical fluid extraction to improve Medicago sativa polyphenolics isolation. Ind Crops Prod 124:931–940. https://doi.org/10.1016/j.indcrop.2018.08.004
7. Yusri IM, Abdul Majeed APP, Mamat R, Ghazali MF, Awad OI, Azmi WH (2018) A review on the application of response surface method and artificial neural network in engine performance and exhaust emissions characteristics in alternative fuel. Renew Sustain Energy Rev 90:665–686. https://doi.org/10.1016/j.rser.2018.03.095
8. Bashir MJK, Amr SSA, Aziz SQ, Aun NC, Sethupathi S, Technology G, Tunku U, Rahman A (2015) Wastewater treatment processes optimization using response surface methodology (RSM) compared with conventional methods: review and comparative study. Middle-East J Sci Res 23:244–252. https://doi.org/10.5829/idosi.mejsr.2015.23.02.52
9. Bezerra MA, Santelli RE, Oliveira EP, Villar LS, Escaleira LA (2008) Response surface methodology (RSM) as a tool for optimization in analytical chemistry. https://doi.org/10.1016/j.talanta.2008.05.019
10. Karimifard S, Alavi Moghaddam MR (2018) Application of response surface methodology in physicochemical removal of dyes from wastewater: a critical review. Sci Total Environ 640–641:772–797. https://doi.org/10.1016/j.scitotenv.2018.05.355
11. Myers RH, Montgomery DC, Geoffrey Vining G, Borror CM, Kowalski SM (2004) Response surface methodology: a retrospective and literature survey. J Qual Technol 36:53–78. https://doi.org/10.1080/00224065.2004.11980252
12. Sharif KM, Rahman MM, Azmir J, Mohamed A, Jahurul MHA, Sahena F, Zaidul ISM (2014) Experimental design of supercritical fluid extraction—a review. https://doi.org/10.1016/j.jfoodeng.2013.10.003
13. Fornari T, Vicente G, Vázquez E, García-Risco MR, Reglero G (2012) Isolation of essential oil from different plants and herbs by supercritical fluid extraction. J Chromatogr A 1250:34–48. https://doi.org/10.1016/j.chroma.2012.04.051

14. Brunner G (1994) An introduction to fundamentals of supercritical fluids and the application to separation processes. Springer Science & Business Media, New York, NY
15. Dias ALB, dos Santos P, Martínez J (2018) Supercritical CO2 technology applied to the production of flavor ester compounds through lipase-catalyzed reaction: a review. J CO_2 Util 23:159–178. https://doi.org/10.1016/j.jcou.2017.11.011
16. Wrona O, Rafińska K, Możeński C, Buszewski B (2017) Supercritical fluid extraction of bioactive compounds from plant materials. J AOAC Int 100:1624–1635. https://doi.org/10.5740/jaoacint.17-0232
17. Lee W, Hamidah S, Setapar M, Mohd H (2020) Process optimization of supercritical CO_2 extraction of Roselle using response surface methodology 16:30–33
18. Wu H, Li J, Jia Y, Xiao Z, Li P, Xie Y, Zhang A, Liu R, Ren Z, Zhao M, Zeng C, Li C (2019) Essential oil extracted from Cymbopogon citronella leaves by supercritical carbon dioxide: antioxidant and antimicrobial activities. J Anal Methods Chem 2019. https://doi.org/10.1155/2019/8192439
19. Xie L, Cahoon E, Zhang Y, Ciftci ON (2019) Extraction of astaxanthin from engineered Camelina sativa seed using ethanol-modified supercritical carbon dioxide. J Supercrit Fluids 143:171–178. https://doi.org/10.1016/j.supflu.2018.08.013
20. Bilgiç-Keleş S, Şahin-Yeşilçubuk N, Barla-Demirkoz A, Karakaş M (2019) Response surface optimization and modelling for supercritical carbon dioxide extraction of Echium vulgare seed oil. J Supercrit Fluids 143:365–369. https://doi.org/10.1016/J.SUPFLU.2018.09.008
21. Md Norodin NS, Md Salleh L, Machmudah S, Mustafa NM, Hartati H, Ismail R (2018) Extraction of β-sitosterol from Swietenia mahagoni seeds by using supercritical carbon dioxide (SC-CO_2) extraction. Malays J Fundam Appl Sci 14:411–417. https://doi.org/10.11113/mjfas.v14n3.1082
22. Kueh BWB, Yusup S, Osman N (2018) Supercritical carbon dioxide extraction of Melaleuca cajuputi leaves for herbicides allelopathy: optimization and kinetics modelling. J CO_2 Util 24:220–227. https://doi.org/10.1016/j.jcou.2018.01.005
23. Utomo EP, Marina, Warsito, Agustian E (2018) Optimization of supercritical CO_2 extraction process to improve the quality of patchouli oil by response surface methodology approach. Indones J Chem 18:235–241. https://doi.org/10.22146/ijc.26605
24. Sodeifian G, Sajadian SA, Saadati Ardestani N (2017) Experimental optimization and mathematical modeling of the supercritical fluid extraction of essential oil from Eryngium billardieri: application of simulated annealing (SA) algorithm. J Supercrit Fluids 127:146–157. https://doi.org/10.1016/j.supflu.2017.04.007
25. Urbonaviciene D, Viskelis P (2017) The cis-lycopene isomers composition in supercritical CO_2 extracted tomato by-products. LWT—Food Sci Technol 85:517–523. https://doi.org/10.1016/j.lwt.2017.03.034
26. Ahmadian-Kouchaksaraie Z, Niazmand R (2017) Supercritical carbon dioxide extraction of antioxidants from Crocus sativus petals of saffron industry residues: optimization using response surface methodology. J Supercrit Fluids 121:19–31. https://doi.org/10.1016/j.supflu.2016.11.008
27. Sodeifian G, Saadati Ardestani N, Sajadian SA, Ghorbandoost S (2016) Application of supercritical carbon dioxide to extract essential oil from Cleome coluteoides Boiss: experimental, response surface and grey wolf optimization methodology. J Supercrit Fluids 114:55–63. https://doi.org/10.1016/j.supflu.2016.04.006
28. Wang L, Wang X, Wang P, Xiao Y, Liu Q (2016) Optimization of supercritical carbon dioxide extraction, physicochemical and cytotoxicity properties of Gynostemma pentaphyllum seed oil: a potential source of conjugated linolenic acids. Sep Purif Technol 159:147–156. https://doi.org/10.1016/j.seppur.2016.01.007
29. Ara KM, Raofie F (2016) Application of response surface methodology for the optimization of supercritical fluid extraction of essential oil from pomegranate (Punica granatum L.) peel. J Food Sci Technol 53:3113–3121. https://doi.org/10.1007/s13197-016-2284-y
30. Millao S, Uquiche E (2016) Extraction of oil and carotenoids from pelletized microalgae using supercritical carbon dioxide. J Supercrit Fluids 116:223–231. https://doi.org/10.1016/j.supflu.2016.05.049

31. Danlami JM, Zaini MAA, Arsad A, Yunus MAC (2015) A parametric investigation of castor oil (Ricinus comminis L) extraction using supercritical carbon dioxide via response surface optimization. J Taiwan Inst Chem Eng 53:32–39. https://doi.org/10.1016/j.jtice.2015.02.033
32. Przygoda K, Wejnerowska G (2015) Extraction of tocopherol-enriched oils from Quinoa seeds by supercritical fluid extraction. Ind Crops Prod 63:41–47. https://doi.org/10.1016/j.indcrop.2014.09.038
33. Hedayati A, Ghoreishi SM (2015) Supercritical carbon dioxide extraction of glycyrrhizic acid from licorice plant root using binary entrainer: experimental optimization via response surface methodology. J Supercrit Fluids 100:209–217. https://doi.org/10.1016/j.supflu.2015.03.005
34. Shao Q, Deng Y, Liu H, Zhang A, Huang Y, Xu G, Li M (2014) Essential oils extraction from Anoectochilus roxburghii using supercritical carbon dioxide and their antioxidant activity. Ind Crops Prod 60:104–112. https://doi.org/10.1016/j.indcrop.2014.06.009
35. Azmir J, Zaidul ISM, Sharif KM, Uddin MS, Jahurul MHA, Jinap S, Hajeb P, Mohamed A (2014) Supercritical carbon dioxide extraction of highly unsaturated oil from Phaleria macrocarpa seed. Food Res Int 65:394–400. https://doi.org/10.1016/j.foodres.2014.06.049

Assessing the Validity of Torsional Motion on Dynamic Analysis of Rotor Bearing System Using Finite Element

Abdul Malek Abdul Wahab, Z. A. Rasid, Ahmad Khushairy Makhtar, and Jamaluddin Mahmud

Abstract The rotor-bearing system has been used extensively in machinery industry. The precise prediction of dynamic behaviour is vital for preventing any catastrophic occurrence from happening due to a vibration phenomenon. This work is based on Nelson's shaft model that contains two translation components and three rotary inertia components. Besides the bending and transverse shear deformation elements, additional torsional motions are included in the model. The objective is to investigate the effect of different level stiffness of the isotropic bearing and the torsional element towards the whirling phenomenon. The Campbell diagram obtained by applying Bolotin method to the finite element equation of motion is employed here to predict the whirling frequency. It was discovered that the torsional motion has different substantial effect on whirling speed for various levels of bearing stiffness. The highest impact of 8.57% difference can be seen for soft bearing that occurs in the 1st mode corresponds to forward frequency direction.

Keywords Bearing stiffness · Finite element · Torsional motion · Whirling speed

1 Introduction

Rotor-bearing systems have broad applications in industrial types of machinery. Thus, research on the dynamic characteristic of these systems is essential at the design stage, use, and maintenance procedure [1]. Besides analysing real physical, researchers for a long time have formed mathematical models or computational processes that are substantial confidence to represent the actual machinery [2, 3]. The study on predicting the whirling motion has been conducted for a long time and continuing

A. M. A. Wahab (✉) · A. K. Makhtar · J. Mahmud
Faculty of Mechanical Engineering, Universiti Teknologi MARA, 40450 Shah Alam, Selangor, Malaysia
e-mail: abdulmalekabdulwahab@gmail.com; abdmalek@uitm.edu.my

Z. A. Rasid
Malaysia-Japan International Institute of Technology (MJIIT), Universiti Teknologi Malaysia Kuala Lumpur, Jalan Sultan Yahya Petra, 54100 Kuala Lumpur, Malaysia

M. A. A. Zaini et al. (eds.), *Proceedings of the 3rd International Conference on Separation Technology*, Lecture Notes in Mechanical Engineering,
https://doi.org/10.1007/978-981-16-0742-4_24

due to its contribution to vibration failure in the rotating shaft system. Shahgholi et al. [4] found that the values of nonlinear forward and backward frequencies were lower in the second mode compared to the first mode for free nonlinear vibrations of the nonlinear slender rotating shaft.

Seshendra Kumar and Sundara Siva Rao [5] demonstrated the lateral motions interact substantially with torsional degrees of freedom if the resonance frequencies were similar for whirling characteristics of geared rotor system for combined torsional-lateral vibrations. Lien-Wen and Der-Ming [6] employed the Finite Element (FE) method to demonstrate the natural whirl speeds for Timoshenko shaft were noticeably affected by the end conditions of rotating shafts and the effects of both shear deformation and rotary inertia.

Whalley and Abdul-Ameer [7] contoured the shaft profile to compute the whirling speed and rotational frequency of shaft- rotor systems. Taplak and Parlak [8] attained critical speed, Campbell diagram, and the response of the rotor in evaluating the dynamic characteristics of a small gas turbine rotor. The Campbell diagram was employed to predict the critical speed of the rotating shaft. For high-speed rotating machinery, the estimation of critical speed is even more essential since the system is able of rotating at very close to critical speed.

Based on the study by Nelson [9], this work investigated the dynamic characteristics of a rotating system under the undamped isotropic bearing condition. The effect of different levels of stiffness of isotropic bearing and the torsional element towards the whirling phenomenon was investigated by FE model. Bolotin's approach [10] was employed to establish Campbell's diagram from the FE matrix.

2 Materials and Methods

2.1 The Finite Element Formulation

In this study, the governing equation of motion for vibration response for the shaft is based on energy approaches. Energy-based approaches provide a means to model flexible bodies with distributed mass. The FE formulation is based on the Timoshenko beam with five degrees of freedom (DOF) per node element that was suggested by Nelson [9] and Rao [11]. The governed equation considered the axial torsional deformation as one of the DOF. For this paper, this FE model is called the 5DOF model, while the formulation that is developed based on the four DOF per node element that excluded the torsional deformation is called the 4DOF model.

Figure 1 shows the element used to establish the 5DOF FE model. The model involves two translation components and three rotary inertia components denoted by (v, w) and $(\theta_y, \theta_z, \theta_x)$, respectively. The translation components are due to bending, transverse shear deformation and torsional motions. While rotary inertia components are due to bending and torsional motions. The shaft is discretized into elements with

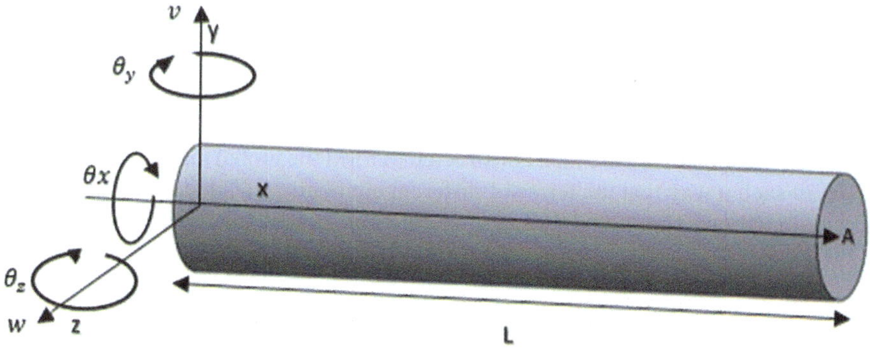

Fig. 1 The torsional-bending element (5DOF)

two nodes which the translation and rotatory element are represented in term of a nodal displacement vector $\{q^{(e)}\} = (v^1, w^1, \theta_y^1, \theta_z^1, \theta_x^1, v^2, w^2, \theta_y^2, \theta_z^2, \theta_x^2)^T$.

Based on energy approaches, the formulation of the potential energy equation in terms of nodal displacement vector is given as;

$$U^{(e)} = \frac{1}{2}\{q^{(e)}\}^T\left(\left[K_b^{(e)}\right] + \left[K_s^{(e)}\right] + \left[K_t^{(e)}\right]\right)\{q^{(e)}\} - \frac{1}{2}\{q^{(e)}\}^T\left[K_g^{(e)}\right]\{q^{(e)}\} \quad (1)$$

where $[K_b^{(e)}]$, $[K_s^{(e)}]$, $[K_t^{(e)}]$ and $[K_g^{(e)}]$ are the elemental bending, shear, torsional and geometric stiffness matrices, respectively.

The elemental kinetic energy is

$$T^{(e)} = \frac{1}{2}\{\dot{q}^{(e)}\}^T\left(\left[M_d^{(e)}\right] + \left[M_r^{(e)}\right] + \left[M_t^{(e)}\right]\right)\{\dot{q}^{(e)}\} + \frac{1}{2}\{\dot{q}^{(e)}\}^T\left[G^{(e)}\right]\Omega\{q^{(e)}\} \quad (2)$$

where the matrices $\left[M_d^{(e)}\right]$, $\left[M_r^{(e)}\right]$ and $\left[M_t^{(e)}\right]$ are the element translational, rotary inertia and torsional mass matrix, respectively. The gyroscopic element matrix $[G^{(e)}]$ is derived from Euler angle transformation to make the beam in rotating condition [12]. The shaft is considered to rotate at constant angular velocity Ω about the *x*-axis and small deformation is neglected.

The bearing model used in this work is based on a linear relationship between forces, displacements, and velocities. This bearing model produces radial support only and does not provide any moment or longitudinal load on the shaft. The force–displacement and the force–velocity relationship of the bearing element are based on the model by Lalanne [13]. The equation of bearing element can be expressed as;

$$F_B = \left[D_B^{(e)}\right]\left\{\dot{q}_B^{(e)}\right\} + \left[K_B^{(e)}\right]\left\{q_B^{(e)}\right\} = 0 \quad (3)$$

where F_B is the force applied towards the damping element and the stiffness element of the bearing that are represented by $\left[D_B{}^{(e)}\right]$ and $\left[K_B{}^{(e)}\right]$, respectively. $\left\{q_B^e\right\}^T = [u_B \ w_B \ \theta_{By} \ \theta_{Bz}]^T$ are the generalized nodal displacement vectors of the bearing element.

The element matrices are then constructed to form the global matrices of the rotor model in such a way that the terms of different elements related with the same node are combined up. The global U and T of rotating Timoshenko shaft are as shown below.

$$U = \frac{1}{2}\{q\}^T[K]\{q\} - \frac{1}{2}\{q\}^T\left[K_g\right]\{q\} \tag{4}$$

$$T = \frac{1}{2}\{\dot{q}\}^T[M]\{\dot{q}\} + \frac{1}{2}\{\dot{q}\}^T\Omega[G]\{q\} \tag{5}$$

where the global matrices $[K]$, $[M]$, $[G]$ and $\left[K_g\right]$ of the shaft, system corresponds to the elastic, mass, gyroscopic and geometric stiffness matrices, respectively. Applying the Lagrange's equation, the matrix form for governing equation of motion for the rotating shaft is

$$[M]\{\ddot{q}\} + \Omega[G]\{\dot{q}\} + \left([K] - \left[K_g\right]\right)\{q\} = 0 \tag{6}$$

For elements with lumped properties such as bearings, their features were added to those nodes as shown in Fig. 2. It is, therefore, essential to discretize the rotor shaft in such a way that these elements lie on a node. Finally, the global matrices are employed to define the dynamic properties of the rotor.

The equation of motion is formulated into Mathieu-Hill's equation and is solved using Bolotin's theory [14, 15]. As $\omega_0 = \phi/2$ and $\Omega \neq 0$, the equation for whirling analysis becomes;

$$\left(\phi^2[M_E] + \phi[G_E] + [K_E]\right)q = 0 \tag{7}$$

where ω_0 is the non-rotating natural frequency of shaft.

2.2 Materials

Figure 3a shows the shaft supported by the rigid bearing, which is used in this paper. L is the length of the shaft, while d is the diameter of the shaft. Ω is the rotating speed of the shaft. The bearing is based on a linear relationship between forces, displacements, and velocities. Figure 3b shows the bearing and shaft cross-section with the corresponding stiffness and damping properties of the bearing where C_{by} and C_{bz} are damping bearing at y and z-axis, respectively while K_{by} and K_{bz} are the

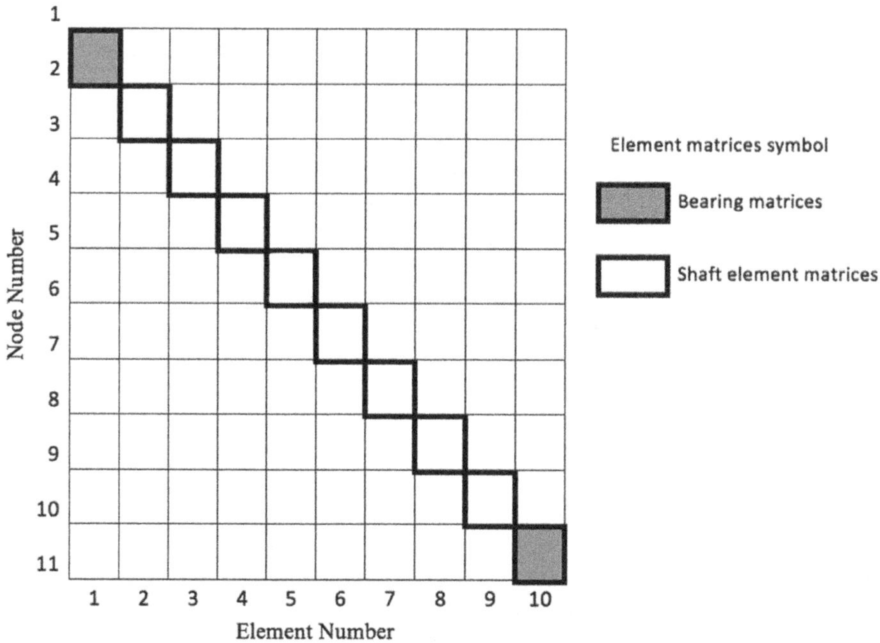

Fig. 2 The structure of the assembled global matrix of the rotor model shown

stiffness of the bearing in y and z-axis, respectively. For this study, the damping for the bearing is neglected. This bearing model produces radial support only and does not provide any moment or longitudinal load on the shaft. The force–displacement and the force–velocity relationship of the bearing elements are based on the model by Lalanne [13]. The geometry and mechanical properties of the shaft are given in Table 1.

3 Results and Discussion

3.1 Model Validation

The formulation of 5DOF FE model was validated by comparing the results with the previous work by Souza and Paulista [16]. The validation was made on the natural frequency of a fixed-free shaft system. The length (*L*) of the shafts is 1 and 0.1 m in radius (*r*). The material is a steel with Young's modulus E $=$ 210 GPa, density $\rho =$ 7850 kg/m^3 and Poisson's ratio $\nu = 0.3$. The shafts are discretized into ten elements.

The natural frequency obtained by the 5DOF model was validated. Four modes of vibration are recorded as shown in Table 2. As can be seen in the table, the 5DOF gives entirely accurate results for both static and rotating shafts. The maximum relative

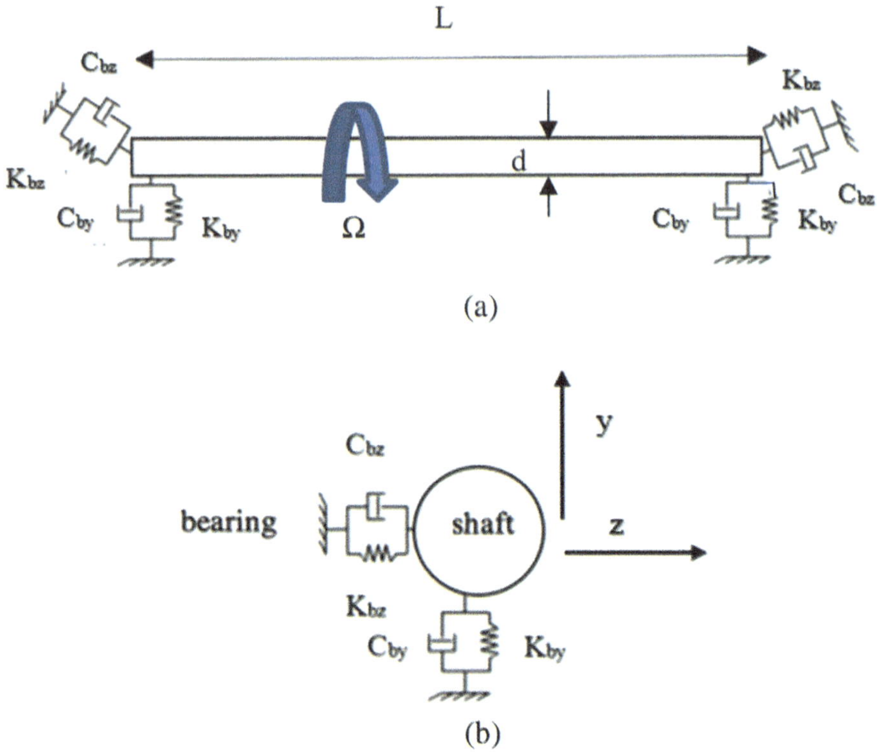

Fig. 3 **a** The rotating shaft understudy. **b** The bearing model

Table 1 Geometric and material properties for the rotating shaft

Young's modulus, E	207 GPa
Modulus of rigidity, G	79.6 GPa
Poisson's ratio, v	0.303
Density, ρ	7833 kg/m^3
Radius, r	Round tube with $r_{inner} = 0$ m, $r_{outer} =$ 0.0508 m
Length, L	1.27 m
Shear factor, κ	0.9
Yield strength, S_y	400 MPa

error of the natural frequency at static condition is 0.079%, and that due to rotating condition is less than 31.9%. Then, the 5DOF model is compared with the 4DOF model to investigate the effect bearing stiffness and torsional element on the whirling speed of the shaft-bearing systems in the following section.

Table 2 The validation of the natural frequencies of a fixed-free disk shaft system corresponds to the 5DOF model

	Mode	1	2	3	4
Non-rotating shaft	Ref. [16] (rad/s)	4.55	28.49	79.78	156.34
	5DOF (rad/s)	4.55	28.48	79.72	156.22
	Error (%)	0.00	0.04	0.08	0.08
Rotating shaft (10.47 rad/s)	Ref. [16] (Forward) (rad/s)	4.86	28.87	80.09	156.68
	5DOF (Forward) (rad/s)	4.49	28.42	79.66	156.15
	Error (%)	7.57	1.56	0.54	0.34
	Ref. [16] (Backward) (rad/s)	6.76	29.30	80.24	156.77
	5DOF (Backward)(rad/s)	4.60	28.54	79.79	156.29
	Error (%)	31.92	2.60	0.57	0.31

3.2 *The Effect of Bearing on Dynamic Behaviour of Shaft*

Static Condition

In this section, the effects of bearing stiffness level and torsional element towards the natural frequency of static shaft are discussed. The natural frequencies and mode shapes were recorded. Results from 5 and 4DOF models were compared to investigate the significance of torsional motion. Three cases of undamped bearing with different levels of stiffness were employed, with values of 2.49×10^7, 1.1×10^7 and 4.5×10^5 N/m. These values correspond to stiff, intermediate, and soft bearing condition. The bearing is assumed to have equal stiffness in all radial directions K_{by} and K_{bz}.

First, the natural frequencies of static shaft with different bearing stiffness were recorded. Tables 3 and 4 shows the comparison of natural frequencies between 5 and 4DOF model for different bearing condition. In general, different stiffness of bearing produces different values of natural frequency, although the shaft systems have the same geometry and properties. It can be said that for both mode 1 and mode 2, the natural frequencies of the shaft with stiff bearing are higher compared to the shaft with soft bearing at static condition as agreed by past research [17].

Table 3 The natural frequencies of the static shaft system correspond to the 4DOF and 5DOF models for different bearing levels at 1st mode

Bearing condition	Natural frequency (Hz) 4DOF	Natural frequency (Hz) 5DOF	Differences (%)
Stiff	91.15	91.21	0.1
Intermediate	70.93	70.94	0.0
Soft	16.69	16.69	0.0

Table 4 The natural frequencies of the static shaft system correspond to the 4DOF and 5DOF models for different bearing levels at 2nd mode

Bearing condition	Natural frequency (Hz) 4DOF	Natural frequency (Hz) 5DOF	Differences (%)
Stiff	203.63	204.17	0.3
Intermediate	139.83	140.24	0.3
Soft	28.99	29.08	0.3

For example, as shown in Table 3 for the first mode in the static condition, the natural frequency for shaft with stiff bearing is 91. 21 Hz while the natural frequency for shaft with soft bearing is 16.69 Hz. For the second mode, as shown in Table 4 the natural frequency in static condition is 204.2 Hz for stiff bearing and 29.09 Hz for soft bearing. For both conditions of the 1st and 2nd mode, the natural frequencies obtained for stiff bearing is 81.7% and 85.7% more than those of the soft bearing, respectively. With these big differences of natural frequency, it shows that bearing with different stiffness can alter the natural frequency of the shaft system. The softer the bearing the less natural frequency of the shaft system will become. This condition is very similar behavior with mass-spring system where the natural frequency is proportional with square root of stiffness as mass constant.

Rotating Conditions

In this section, the effects of torsional element and bearing stiffness on whirling speed were discussed. The whirling speed is a condition where whirl motion occurs due to the speed of the rotating shaft equals to the natural frequency of shaft. This whirling motion can be in similar direction or opposite direction of rotating shaft [17]. The Campbell diagram was used to estimate whirling speed. Figure 4 and Fig. 5 illustrated the Campbell diagram presenting frequency response as rotating speed varies for 1st and 2nd modes, respectively. The term BF is the backward frequency and FF is the forward frequency that are corresponding to the backward and forward mode [18].

The vibration frequencies correspond to the 5DOF model changes rapidly with speed compared to that of the 4DOF model, especially for shafts with intermediate and stiff bearings for both modes. However, the vibration frequencies for shaft with soft bearing slightly changes with the increase of speed. This result is in line with the work of Fischer and Strackeljan [19] and Vance et al. [20].

The changes in frequency for the rotating shaft system are due to gyroscopic effect. The gyroscopic effect splits vibration frequency into the FF and BF components as the speed of rotation increases. The smallest changes in frequency for the soft bearing are due to the shaft rotates in cylindrical mode in a planar motion. In this condition, the effect of gyroscopic is minimum.

The vibration frequency corresponds to the 4DOF model does not drastically change compare to the 5DOF model as the rotating speed increases. This difference occurs due to the ability of the 5DOF model to provide the torsional effect in the rotating shaft, which significantly affects the gyroscopic term.

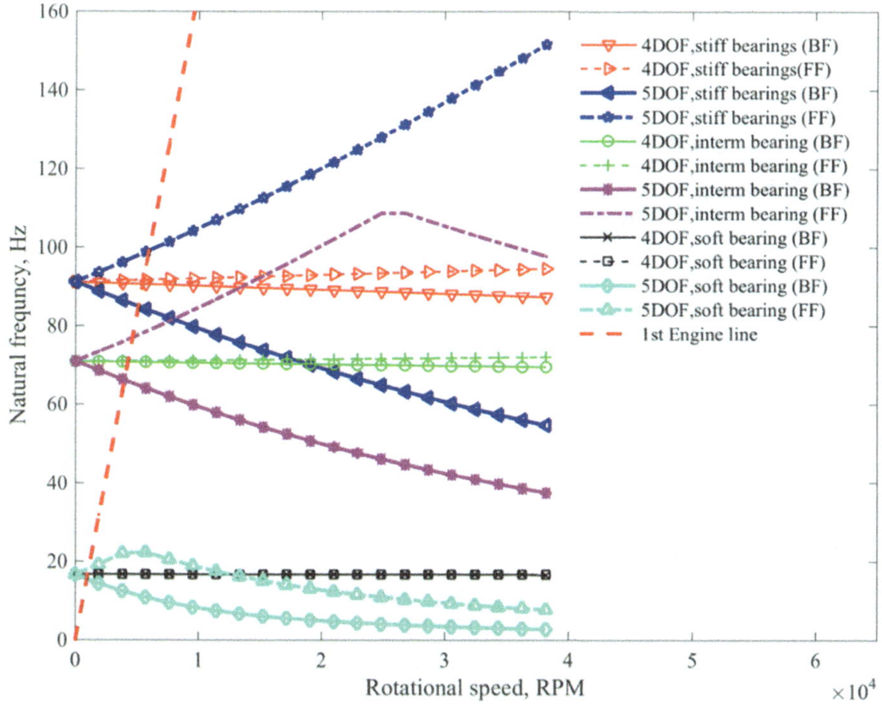

Fig. 4 The effect of three levels of bearing stiffness on whirling frequency of shaft at 1st mode of frequency

Tables 5 and 6 show the whirling speed for different types of bearing stiffness. The whirling speed is the reading at the intersection between 1st Engine line and frequency line in the Campbell diagram. The stiff bearing condition has the highest whirling speed, followed by intermediate and soft bearing condition for both modes. The significant effect of torsional motion can be observed by the differences between 5 and 4DOF models for both the 1st and 2nd modes.

The highest effect is for the soft bearing in FF direction with an 8.6% difference. While the lowest impact is for stiff bearing where in BF direction the difference is 6.3%, this case is for the 1st mode condition. In the 2nd mode, the highest difference is 6.8% for stiff and intermediate bearings in FF direction. The lowest difference is for soft bearing that gives 3.9% difference. It shows that different levels of bearing stiffness can affect the significance of torsional motion, and the effect is different per level of stiffness. Thus, it is safe to say that the additional torsional motion is vital to be considered in modeling the bearing-shaft system for accurate prediction of whirling speed.

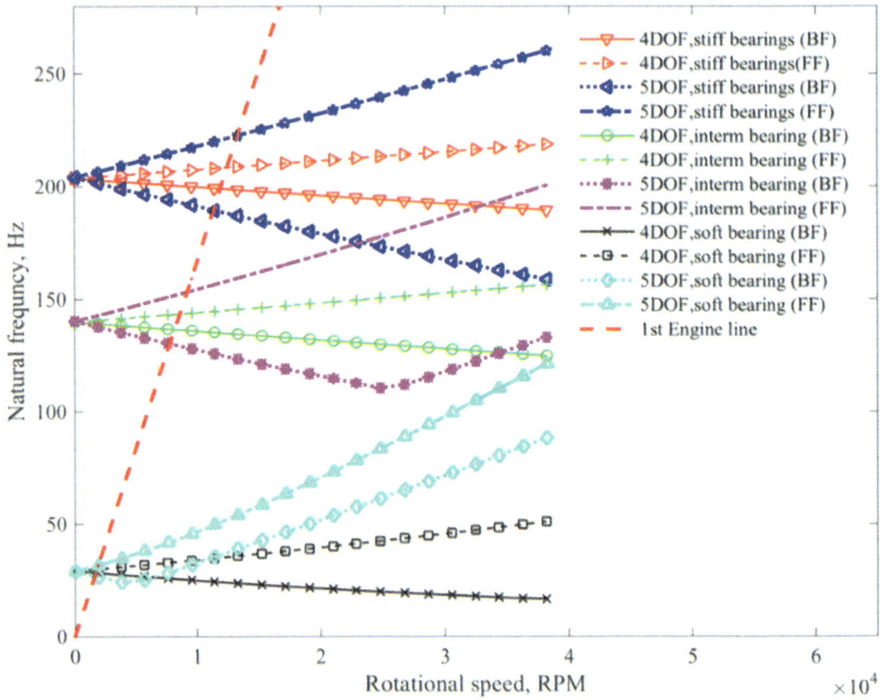

Fig. 5 The effect of three levels of bearing stiffness on whirling frequency of shaft at 2nd mode of frequency

Table 5 The comparison of whirling speed between 4 and 5DOF model for different bearing stiffness at 1st mode

Bearing condition	Frequency direction	Whirling speed (RPM) 4DOF	Whirling speed (RPM) 5DOF	Differences (%)
Stiff	BF	5443	5099	6.32
	FF	5510	5949	7.97
Intermediate	BF	4249	3973	6.50
	FF	4269	4622	8.27
Soft	BF	1003	935.8	6.70
	FF	1003	1089	8.57

Table 6 The comparison of whirling speed between 4 and 5DOF model for different bearing stiffness at 2nd mode

Bearing condition	Frequency direction	Whirling speed (RPM) 4DOF	Whirling speed (RPM) 5DOF	Differences (%)
Stiff	BF	11950	11370	−4.85
	FF	12520	13370	6.79
Intermediate	BF	8203	7811	−4.78
	FF	8613	9196	6.77
Soft	BF	1690	1623	−3.96
	FF	1795	1910	6.41

4 Conclusion

The dynamic behavior of a rotating system under the undamped isotropic bearing condition has been successfully modeled and analyzed using FE method. The stiffness level of bearing contributes into forming different natural frequency and vibration frequency at static and rotating conditions. The torsional element shows significant impact on whirling frequency for rotating shaft and increasing as bearing stiffness reduce. Thus, torsional element needs to be considered in predicting unstable operating condition of rotating shaft-bearing system especially at high speed. Further, the method used in this study can be employed for predicting instability condition of rotating shaft system at different parameters.

Acknowledgements Acknowledgement is gratefully given to the Malaysia-Japan Institute of Technology and Universiti Teknologi MARA for supporting the work through the provision of the FRGS RACER Grant of 600-IRMI/FRGS-RACER 5/3 (092/2019).

References

1. Sternlicht B, Lewis P (1968) Vibration problems with high-speed turbomachinery. J Manuf Sci Eng 90:174–185. https://doi.org/10.1115/1.3604595
2. Wenhui X, Yougang T, Yushu C (2008) Analysis of motion stability of the flexible rotor-bearing system with two unbalanced disks. J Sound Vib 310:381–393. https://doi.org/10.1016/j.jsv.2007.08.001
3. Zhao S, Xu H, Meng G, Zhu J (2005) Stability and response analysis of symmetrical single-disk flexible rotor-bearing system. Tribol Int 38:749–756. https://doi.org/10.1016/j.triboint.2004.11.004
4. Shahgholi M, Khadem SE, Bab S (2014) Free vibration analysis of a nonlinear slender rotating shaft with simply support conditions. Mech Mach Theory 82:128–140. https://doi.org/10.1016/j.mechmachtheory.2014.08.005
5. Seshendra Kumar Karri VS, SundaraSiva Rao BSK (2012) An experimental study of whirling characteristics of gear-pinion rotor system. Int J Mech Appl 2:10–13. https://doi.org/10.5923/j.mechanics.20120202.03

6. Lien-Wen C, Der-Ming K (1991) Finite element analysis of natural whirl speeds of rotating shafts. Comput Struct 40:741–747. https://doi.org/10.1016/0045-7949(91)90241-D
7. Whalley R, Ebrahimi M, Abdul-Ameer A (2007) High-speed rotor-shaft systems and whirling identification. Proc Inst Mech Eng Part C-J Mech Eng Sci 221:661–676. https://doi.org/10.1243/0954406JMES314
8. Taplak H, Parlak M (2012) Evaluation of gas turbine rotor dynamic analysis using the finite element method. Measurement 45:1089–1097. https://doi.org/10.1016/j.measurement.2012.01.032
9. Nelson HD (1980) A finite rotating shaft element using Timoshenko beam theory. J Mech Des 102:793. https://doi.org/10.1115/1.3254824
10. Bolotin VVV (1962) The dynamic stability of elastic systems, vol 2. DTIC Document
11. Rao JS (2011) Finite element methods for rotor dynamics. In: History of rotating machinery dynamics. SE-16. Springer Netherlands, pp 269–297
12. Raffa F, Vatta F (1999) Gyroscopic effects analysis in the Lagrangian formulation of rotating beams. Meccanica 357–366
13. Lalanne M, Ferraris G (1998) Rotordynamics prediction in engineering. John Wiley
14. Bolotin VV (1965) The dynamic stability of elastic systems. Am J Phys 33:752. https://doi.org/10.1119/1.1972245
15. Wahab AMA, Yusof Z, Rasid ZA, Abu A, Rudin NFMN (2018) Dynamic instability of high-speed rotating shaft with torsional effect. Int J Automot Mech Eng 15:6034–6051. https://doi.org/10.15282/ijame.15.4.2018.23.0460
16. Souza C, Paulista UE (2010) Complex modal analysis of a vertical rotor by finite elements method. In: 9th Brazilian conference on dynamics control and their applications, pp 449–457
17. Swanson E, Powell CD, Weissman S (2005) A practical review of rotating machinery critical speeds and modes. Sound Vib 39:10–17
18. Hota M, Vakharia D (2014) A review of rotating machinery critical speeds and modes of vibrations. Int Rev Appl Eng Res 4:241–250
19. Fischer J, Strackeljan J (2006) FEM-Simulation and stability analyses of high speed rotor systems. In: 7th international conference on rotor dynamics. Conference proceedings, Institute of Mechanics and Mechatronics, Vienna University of Technology, pp 25–28
20. Vance J, Zeidan F, Murphy B (2010) Bearings and their effect on rotordynamics. Vibration and rotordynamics. Wiley, Hoboken, NJ, USA, NJ, USA, pp 171–269

Comparing Different Pre-processing Techniques and Machine Learning Models to Predict PM_{10} and $PM_{2.5}$ Concentration in Malaysia

Danny Hartanto Djarum, Zainal Ahmad, and Jie Zhang

Abstract A recent report from world health organization (WHO) showed that there are in total of 3.7 million deaths that were attributed with outdoor ambient pollution. Particulate matter (PM) has been considered by many as the major contributor of air pollution, it poses serious health risks towards human that are mostly associated with cardiovascular and respiratory disease. Even though various studies have been carried out to predict the concentration of PM_{10} and $PM_{2.5}$, there are only a handful of papers that focused on the pre-processing aspect of the prediction. In this study, the importance of data pre-processing steps were assessed for PM_{10} and $PM_{2.5}$ prediction in Malaysia, these cover: Data cleaning (handling missing values), data transformation (standardization, normalization, robust scaling, and min–max scaling), features selection (univariate linear regression and mutual info regression), and dimensionality reduction (principle component analysis) techniques. Four different machine learning models were utilized for the predictions: Multiple linear regression (MLR), random forest regression (RFR), extra tree regression (ETR), and Decision tree regression with AdaBoost (BTR). The result showed that the best PM_{10} and $PM_{2.5}$ prediction accuracy ($R^2 = 0.97, 0.92$) were achieved by using BTR model coupled with normalization and PCA techniques, in which the missing data were imputed with median values.

Keywords Extra tree regression · Decision tree regression with AdaBoost · Pre-processing · Principle component analysis

D. H. Djarum · Z. Ahmad (✉)
School of Chemical Engineering, Universiti Sains Malaysia, Engineering Campus, 14300 Nibong Tebal, Penang, Malaysia
e-mail: chzahmad@usm.my

J. Zhang
School of Engineering, Merz Court, Newcastle University, Newcastle upon Tyne N1 7RU, UK

M. A. A. Zaini et al. (eds.), *Proceedings of the 3rd International Conference on Separation Technology*, Lecture Notes in Mechanical Engineering,
https://doi.org/10.1007/978-981-16-0742-4_25

1 Introduction

In 2019, As much as 90% of the world population still suffers from poor air quality that constantly exceeded world health organization (WHO) guidelines, whereby the majority are from Asian countries [1]. WHO reported that more than 7 million deaths are due to air pollution exposure, in which 4.3 million of the deaths were associated with indoor pollutions, and 3.7 million of the deaths were attributed to outdoor ambient air pollution [2]. In fact, air pollution ranked fifth as the highest mortality contributors in the world [3].

Particulate matter (PM) is considered as the major contributors of health-related air pollutants in the atmosphere along with SO_2, NO_2, and ground level ozone (O_3) [4]. PM can be classified into two different categories, PM_{10} (breathable particulate matter with aerodynamic diameter of less than 10 μm) and $PM_{2.5}$ (respirable particulate matter with aerodynamic diameter of less than 2.5 μm) [5]. Numerous studies have shown that particulate matter (PM) poses serious health risk towards human, which mostly are associated with cardiovascular and respiratory diseases [4]. Majority of the inorganic and organic substances found in particulate matter are toxic and carcinogenic in nature [6]. The size of PM allows it to efficiently transport toxic substances into the body through inhalation and absorption [7]. Zanobetti et al. [8] developed a model which shows that for a 10 $\mu g/m^3$ increase in PM_{10} concentration, the risk of contracting chronic obstructive pulmonary diseases, pneumonia, and cardiovascular disease increases by 2.5%, 1.95%, and 1.27% respectively.

In Malaysia, PM_{10} and $PM_{2.5}$ emerge from various sources such as biomass burning, industry, transport vehicles and domestic cooking. According to Mutalib et al. [9], transport vehicle is the most significant contributor towards the increase of PM in Malaysia. It is estimated that up to 82% of the air emission load in Peninsular Malaysia is due to transport vehicle [10]. Large biomass burning is also another contributor towards the sudden increase of PM in Malaysia, especially due to transboundary pollution from illegal deforestation in Indonesia (the worst occurrences were recorded in year 1997, 2005, and 2015) [11].

A 2015 study indicated that the 24 h mean concentration of $PM_{2.5}$ in Bangi, Selangor, Malaysia is 44.5 $\mu g/m^3$ which exceeded the national air quality standard at that time of 35 $\mu g/m^3$ and WHO maximum recommended concentration of 10 $\mu g/m^3$ [12]. Another study which analyze $PM_{2.5}$ concentration continuously for 14 days in Pahang, shows that $PM_{2.5}$ concentration exceeded the WHO standards 56.85% of the time with the highest recorded $PM_{2.5}$ concentration at 98 $\mu g/m^3$ [13]. During the 2015 transboundary haze period, the 24 h average $PM_{2.5}$ concentration in Malaysia was 136 $\mu g/m^3$, which is significantly higher than the national and WHO standard [14].

Over the years, Malaysia department of environment (DOE) has established multiple air quality monitoring stations which include PM_{10} and more recently $PM_{2.5}$ measurements, which are automated through the Continuous Air Quality Monitoring Stations (CAQMS). By utilizing these measurements, DOE has successfully notified people of dangerous air pollutants concentration in multiple occasions. However, these measurements have also resulted in an immense amount of historical environment quality data which has not been fully utilized in term of analysis and predictions. With the latest advancement in big data analytics and machine learning technology, it is now possible to develop a reliable and accurate data-driven environmental quality prediction model. Specifically, PM_{10} and $PM_{2.5}$ prediction is extremely important in order to plan and mitigate the risk of these toxic containing particulate matters on the public health.

Various studies have been carried out using different type of prediction models to forecast the concentration of PM_{10} and $PM_{2.5}$. Shi et al. [15] proposed a PM_{10} prediction model in Changsha, China, by utilizing back-propagation (BP) artificial neural network. Through the historical PM_{10} and meteorological data they were able to achieve an R^2 values of 0.62. Alam and McNabola [16] compared the performance of different prediction models to estimate PM_{10} concentration, the model studied includes: Multiple linear regression (MLR), nonparametric regression (NPR), and Artificial neural network (ANN). The study shows that ANN has the best performance amongst the three with an R^2 values of 0.66. A viability land use regression (LUR) in predicting $PM_{2.5}$ concentration in Shanghai (China) were assessed by Liu et al. [17] in their study, their model scored an adjusted R^2 values of 0.88. In Malaysia, Ramli et al. [18] developed two different models (PCA-MLR and PCA-ANN) for the next day PM_{10} concentration prediction. They performed principle component analysis (PCA) to reduce the dimensionality of the input features before training the prediction model, they found that PCA improves the model accuracy by 12.9% for PCA-MLR and 13.3% for PCA-ANN. The best performing model was PCA-ANN with an R^2 values of 0.78.

Even though multiple studies have been carried out to develop PM_{10} and $PM_{2.5}$ prediction models, there are only a handful of papers which cover the importance of data pre-processing aspects of the prediction. Pre-processing steps (data cleaning, transformation and reductions) are extremely crucial as real world data often involves noise, missing values, inconsistent data, and superfluous data which in turn could lead to poor model performance [19]. In Tehran, capital city of Iran, Zamani et al. [20] studied the importance of feature selections in predicting $PM_{2.5}$ concentration using the XGBoost approach. Their work shows that, even though many studies have included Aerosol Optical Depth (AOD) data to predict $PM_{2.5}$ concentration, model trained without AOD data fairs better in term of prediction performance scoring R^2 values between 0.77 and 0.81.

In this study, different type of data pre-processing steps were compared and analyzed in order to find the most optimum approach for PM_{10} and $PM_{2.5}$ predictions, these include: Data cleaning (handling missing data), transformations (standard scaler, normalizer, min–max scaler, and robust scaler), features selection (univariate linear regression and univariate mutual info regression) and dimensionality reductions (principle component analysis/PCA). Four machine learning models were investigated to predict PM_{10} and $PM_{2.5}$ concentrations, which are: Multiple Linear Regression (MLR), Random Forest Regression (RFR), Extra Tree Regression (ETR), and Decision Tree Regression with AdaBoost (BTR). R^2 and root mean square error (RMSE) values were employed to evaluate and compare the prediction performance. Lastly, the original dataset was randomly distributed into 80% training set and 20% test set.

2 Methodology

2.1 Study Area

There are in total of 18 air quality monitoring stations within five Malaysian states (Penang, Perak, Perlis, Kedah, and Selangor) are being covered in this study. The total number of data points collected in each state varies, due to different number of monitoring sites have been established between the different states (as shown in Table 1). We could observe from Table 1 that Selangor and Perak have the most monitoring sites which resulted in highest number of data points being collected, whereas Perlis has only one monitoring site which explained the lowest number of data points collected.

As the monitoring sites are in different locations, this may lead to variation in the seasonal trend of the meteorological data, which makes the location attribute an important feature for PM_{10} and $PM_{2.5}$ predictions. Since the location attributes is a non-numerical attribute, this may cause some difficulty when developing the prediction model as most machine learning algorithms prefer to work with numerical

Table 1 List of monitoring sites in five different Malaysian states

States	Site ID	Locations
Penang	CA06P	Seberang Jaya, Pulau Pinang
	CA07P	Seberang Perai, Pulau Pinang
	CA08P	Minden, Pulau Pinang
	CA09P	Balik Pulau, Pulau Pinang
Kedah	CA02K	Langkawi, Kedah
	CA03K	Alor Setar, Kedah
	CA04K	Sungai Petani, Kedah
	CA05K	Kulim Hi-Tech, Kedah
Perak	CA10A	Taiping, Perak
	CA11A	Tasek Ipoh, Perak
	CA12A	Pegoh Ipoh, Perak
	CA13A	Seri Manjung, Perak
Perlis	CA01R	Kangar, Perlis
Selangor	CA18B	Kuala Selangor, Selangor
	CA19B	Petaling Jaya, Selangor
	CA20B	Shah Alam, Selangor
	CA21B	Klang, Selangor
	CA22B	Banting, Selangor

Table 2 List of input and output features for PM_{10} and $PM_{2.5}$ predictions

Features	PM_{10} prediction	$PM_{2.5}$ prediction
Input	Temperature Relative humidity O_3 concentration CO concentration Measuring site Days Months Years Wind direction API	Temperature, Relative humidity CO concentration O_3 concentration Wind speed Measurement site Days Months Years Wind direction API
Output	PM_{10} concentration	$PM_{2.5}$ concentration

values. In order to resolve this issue, the location attribute is converted into numerical binary attribute per category (one-hot encoding) [21]. Through this method, attribute with a true value will be equal to 1 (hot) while all the other attributes are 0 (cold). This method also converts the location attributes from a single variable feature into four set of features whilst explaining the same degree of variance.

2.2 *Data Collection*

The air quality data was obtained from Malaysia's DOE (Department of Environment), the data were collected as a result of continuous daily monitoring in five different Malaysian states from 2014 to 2018. Even though most of the monitoring sites gathers the same number of measuring parameters, Penang state appears to have four more air quality parameters compared to the other states (NO_X, NO, SO_2, NO_2). Since both PM_{10} and $PM_{2.5}$ predictions are carried out independent of each state, it is unnecessary to standardize the parameters of the five states. The complete set of input and output features for PM_{10} and $PM_{2.5}$ prediction can be observed in Table 2.

2.3 *Handling Missing Values*

Missing value is essentially just an attribute in a parameter that were lost during the measuring process. Missing value could be a result of human error during manual data entry process, instrument error, or error in measurement. Presence of missing values could lead to bias and can cause invalid conclusion to be drawn from the results [22]. Thus, it is necessary to carry out a pre-processing stage in which such dataset is prepared and cleaned [23].

Table 3 Percentage of missing values and total data counts in the dataset for each state

States	Missing values (%)		Data counts (rows)	
	$PM_{2.5}$	PM_{10}	PM_{10}	$PM_{2.5}$
Penang	3.7	4.0	3597	2003
Kedah	19.4	28.4	3597	2184
Perak	30.0	22.6	5995	2730
Perlis	37.7	4.0	1199	546
Selangor	24.2	23.4	5995	2730

There is a varying degree of missing values in the air quality dataset obtained from the monitoring sites for different states (as shown in Table 3). In the PM_{10} prediction dataset, Perlis appeared to have the largest number of missing values in the air quality dataset with at 37.7%. It is then followed by the state of Perak, Selangor and Kedah with 30.0%, 24.2%, and 19.4% of missing values respectively. The smallest number of missing values could be found in Penang state at only 3.7%. Whereas in the $PM_{2.5}$ prediction dataset, Kedah appeared to have the largest fraction of missing values at 28.4%, followed by Selangor and Perak at 23.4% and 22.6% respectively. Penang and Perlis, both have the lowest number of missing values at 4.0% each. The default format of the air quality data collected assigned the value of the missing data as 'zero' values. The values displayed in Table 3 depicted the largest missing value from a single feature in each state.

Two different methods were carried out and compared to handle the missing values in the dataset (three if we consider the default zero values). The first approach is imputation with median values, whereby the missing values were substituted by the median value of that parameter. The second approach was removing the row with missing values completely from the dataset, theoretically this approach will be less favorable especially for dataset with a large degree of missing values [24].

2.4 *Feature Scaling*

Feature scaling is one of the most important aspect when it comes to developing data driven prediction models, especially when the dataset involves features with highly varying magnitude, unit, and range. Feature scaling may vary in results when used for different machine learning algorithms, it may either improve the performance significantly or sometimes have no effect at all. For example, PCA (principle component analysis) have high dependencies on normalization in order to yield optimum results [25], whereas algorithms like LDA (linear discriminant analysis and Naive Bayes) may not be much benefited from features scaling as they are designed to give weights on features accordingly. Four different feature scaling approaches were compared in this section: Standard scaler, normalizer, min–max scaler, and robust scaler.

Standard scaler or more commonly known as standardization techniques transform the dataset to have a mean value of zero with a standard deviation of one. The transformed value or often called z-score is calculated by [26]:

$$z = \frac{x - \mu}{\sigma} \tag{1}$$

where, x is the original value, μ is the mean value, and σ is the standard deviation. Standardization could be useful as most machine learning estimator prefer to work with features that looks like a normally distributed data.

The min–max scaler converts the numerical values v in the dataset of a numerical attributes A to a given range R of $[\text{new} - \text{min}_A, \text{new} - \text{max}_A]$. The transformed values for this feature scaling technique will falls between the interval of 0 and 1 [25]. The formula to calculate the transformed value v' is depicted as follow:

$$v' = \text{new} - \text{min}_A + R \cdot \frac{v - \text{min}_A}{\text{max}_A - \text{min}_A} \tag{2}$$

where max_A and min_A are the original maximum and minimum range of the numerical attributes respectively.

Robust scaler eliminates the median value and scale the data based on the interquartile range that has a span between 1st quartile (25th quantile) and the 3rd quartile (75th quantile). The transformed values for robust scaler falls between –2 and 3, which are relatively higher than the other scaling methods. The formula to calculate the transformed values for robust scaler is [26]:

$$v' = \frac{v - Q1_A}{Q3_A - Q1_A} \tag{3}$$

where $\boldsymbol{v}$ is the original value, A is the numerical attribute, and $\boldsymbol{Q}1_A - \boldsymbol{Q}3_A$ is the interquartile range.

Normalizer scales each sample in the dataset instead of the whole parameter into a unit of normalized value. For instance, if the feature has a value of x, y, and z then the transformed values v' can be calculated by this formula [26]:

$$v' = \frac{x_i}{\sqrt{{x_i}^2 + {y_i}^2 + {z_i}^2}} \tag{4}$$

2.5 Feature Selection and Dimensionality Reduction

The features/parameters of the dataset that we use to train the model have a large impact on the model performance, thus having the right features is extremely crucial

[20]. Feature selection could be essentially defined as the process of selecting relevant parameters or attributes that contribute most to the performance of the prediction model. Since features selection essentially reduces the number of parameters, thus the dimensions of the data, the machine learning algorithm should be able to work faster and results in better prediction performance [27]. The main advantages of performing feature selection includes: Increasing prediction accuracy, removing irrelevant data, improving training efficiency, reduce computational cost, and reducing the complexity of the resulting model [28].

Two different type of feature selection techniques were carried out and compared: univariate linear regression and mutual info regression. Univariate linear regression is essentially an f-test for a regression test to asses if any of the independent variables in a multiple linear regression model is significant by utilizing the ratio of variance obtained from the mean squared values. Mutual info regression on the other hand asses the mutual information and measure the dependencies between two random variables that evaluate the information obtained in one variable based on the other random variable [29]. The four least correlated parameters as a result of the features selection technique were dropped from the datasets prior to training the prediction model.

Additionally, principle component analysis (PCA) which is a dimensionality reduction technique was also carried out and the resulting performance was compared with the feature selection techniques. Generally, PCA identify a set of linear transformation of the original features whilst maintaining as much information (variance) in the original data by utilizing fewer number of variables [25]. For each dataset, the optimum dimensions were selected based on the principle components that could explain 95% variance of the original dataset, principle components with lower variance were removed.

2.6 Prediction Model

Four different regression models were compared to predict PM_{10} and $PM_{2.5}$ concentration: (i) Multiple Linear Regression (MLR), (ii) Random Forest Regression (RFR), (iii) Extra Tree Regression (ETR), and (iv) Decision Tree Regression with AdaBoost (BTR). These models were extracted from Python's scikit-learn library. The performance of each of these models were assessed by using the R^2 values and root mean square error (RMSE) metrics. 80% of the dataset were utilized to train the prediction model, whereas 20% of the dataset were utilized to test the prediction model. Air pollution index (API) is dropped from the input features as the values of API itself is a derivation from other air quality parameters.

Multiple linear regression (MLR) essentially is just a statistical technique to model linear relationships between multiple independent variables to predict a single dependent variable. The general formula of MLR could be explained by:

$$Y_i = \beta_0 + \beta_1 X_{1i} + \beta_2 X_{2i} + \cdots + \beta_p X_{pi} + \varepsilon_i \quad (5)$$

where Y is the dependent variable, β_0 is the constant coefficient, $\beta_1, \beta_2, \ldots, \beta_p$ is the regression coefficient of the independent variable $X_1, X_2, \ldots, X_P$, and ε is the residual error (difference between the predicted value and actual measured value). MLR requires two assumptions to be satisfied: (i) The predictor must be an independent variable, and (ii) the residual error should be independent and can be explained by normal distribution, whilst having a vanishing error and constant variance [30].

Random forest regression (RFR) is an ensemble learning method that is based on decision tree technique [20]. Random forest regression has several advantages over other machine learning models such as: Robustness of the model towards parameters specification and ability of the model to handle non-linear properties and strong interactions amongst input variables [31].

The concept of Extra tree regression (ETR) is similar to random forest regression, as both are based on decision tree techniques. ETR utilizes the extremely randomized tree algorithms in which a substantial number of untouched decision trees are created from the training dataset. The final prediction result is then obtained from the average prediction of the decision trees. In a 2006 study, Geurts et al. [32] concluded that ETR yield better accuracy and higher computational efficiency when compared to other prediction models.

Decision tree regression with AdaBoost (BTR) is a regression model that is based on decision tree techniques alike ETR and RFR but boosted using AdaBoost method. Drucker [33] defines AdaBoost as a method that is developed to consecutively build series of regressors based on the weight of sample that was adjusted according to the past prediction errors. This simply means that sample with higher degree of errors will receive larger training weight in the next iterations and the opposite applies for sample will lower degree of errors. Eventually, this will decrease the overall prediction errors as sample with high degree of errors received more attention. The resulting model is a linear combination between decision tree estimator with boosted regressor resulting in better prediction performance.

3 Results and Discussion

3.1 Prediction Model Without Pre-processing

Initially, the four different regression models were trained with the original air quality data without any pre-processing steps. The accuracy of the resulting PM_{10} and $PM_{2.5}$ prediction model on different states can be observed from Tables 4 and 5, where the results displayed are representative of the test accuracy. Note that the default value for the missing data are set to zero.

Table 4 R^2 and RMSE values for different PM_{10} regression models within different states

PM_{10} prediction model	Penang		Kedah		Perak		Perlis		Selangor	
	R^2	RMSE	R^2	RMSE	R^2	RMSE	R^2	RMSE	R^2	RMSE
MLR	0.42	22.11	0.51	18.89	0.25	28.39	0.42	17.64	0.40	34.74
FR	0.65	17.07	0.78	12.46	0.67	18.94	0.71	12.56	0.71	24.32
ETR	0.74	13.58	0.79	12.45	0.75	16.50	0.77	11.22	0.77	21.41
BTR	0.77	12.80	0.79	12.13	0.72	17.18	0.72	12.17	0.78	21.25

Table 5 R^2 and RMSE values for different $PM_{2.5}$ regression models within different states

$PM_{2.5}$ prediction model	Penang		Kedah		Perak		Perlis		Selangor	
	R^2	RMSE	R^2	RMSE	R^2	RMSE	R^2	RMSE	R^2	RMSE
MLR	0.31	14.94	0.23	14.76	0.34	20.21	0.15	13.10	0.20	23.40
FR	0.44	13.95	0.29	14.02	0.23	21.30	0.59	10.79	0.25	22.62
ETR	0.58	11.84	0.39	12.74	0.37	18.03	0.64	7.96	0.31	18.49
BTR	0.58	11.90	0.34	13.24	0.38	17.79	0.58	8.53	0.26	22.45

As shown in Table 4, regression models developed for PM_{10} predictions performed considerably well considering that no pre-processing methods were carried out. Models trained with dataset from Kedah state scored an overall highest value of R^2 and lowest RMSE values, with both ETR and BTR models scoring an R^2 values of 0.79. Between the four different models, Extra-Regression model appears to have an overall better performance compared to other models followed by AdaBoosted-Tree regression model.

Whereas, for $PM_{2.5}$ prediction (shown in Table 5), the performance is not as good as the PM_{10} prediction models. One main reason for this observation is because $PM_{2.5}$ data has a much lower number of data counts compared to PM_{10} data (shown in Table 3). Smaller datasets tend to have higher variance which can lead to poor model performance, and the regression models selected are sensitive towards dataset size. Similar to PM_{10} predictions, Extra-Tree-Regression model has the best overall performance compared to other models followed by Boosted-Tree-regression. The models trained with Perlis dataset also appeared to have the best performance with R^2 values of at 0.64, this could be because Perlis $PM_{2.5}$ dataset has the lowest percentage of missing data as shown in Table 3.

3.2 Handling Missing Values

The first pre-processing steps performed is dataset cleaning, in which various degree of missing values in the dataset were handled by comparing two different methods: (i) Imputing the missing values with the median value of said parameter and (ii) removing the entire row of missing values from the dataset.

As can be observed from Tables 6 and 7, the performance of the prediction models generated after implementing the median value substitution seems to be consistently improved, especially for MLR and BTR. However, not all regression models benefited from this method, for example the performance of extra tree regression model appears to remain the same compared to the out-of-the-box regression models found in Tables 4 and 5. On the other hand, removing the entire row of data when one or more data points is missing seems to degrade the performance of the regression model significantly. One of the reasons behind this is the large percentage of dataset loss once the entire row of data is removed, up to 66.56% as can be observed from Table 6. Large dataset loss eventually leads to smaller dataset with high variance that eventually lead to poor model performance or even worse lead to biased in observation. However, some dataset (Penang PM_{10} and Perak $PM_{2.5}$) appears to be benefited from this method, in which the R^2 values of the regression models seems to be improved. Overall, median value substitution method generated more consistent improvement throughout all the regression models and dataset, thus it is selected to handle the missing data.

Table 6 Performance for different PM_{10} regression models after handling missing values

PM_{10} prediction model	Penang		Kedah		Perak		Perlis		Selangor	
	R^2	RMSE	R^2	RMSE	R^2	RMSE	R^2	RMSE	R^2	RMSE
Median value substitution										
MLR	0.46	21.43	0.51	19.08	0.27	28.01	0.44	16.89	0.40	34.78
FR	0.68	17.18	0.78	13.01	0.65	19.38	0.71	12.52	0.72	23.52
ETR	0.74	13.56	0.79	12.27	0.75	16.33	0.77	11.05	0.79	20.54
BTR	0.77	12.83	0.80	12.12	0.74	16.59	0.74	11.60	0.79	20.27
Removing the entire row										
Data loss	9.95%		39.84%		66.56%		40.70%		40.00%	
MLR	0.50	19.54	0.55	21.03	0.40	20.75	0.38	17.26	0.50	29.13
FR	0.77	14.57	0.73	16.21	0.54	18.29	0.55	14.76	0.65	24.53
ETR	0.79	14.15	0.78	14.56	0.66	15.82	0.66	12.79	0.75	20.62
BTR	0.80	13.76	0.74	15.96	0.57	17.58	0.72	11.63	0.74	21.17

Table 7 Performance for different $PM_{2.5}$ regression models after handling missing values

$PM_{2.5}$ prediction model	Penang		Kedah		Perak		Perlis		Selangor	
	R^2	RMSE	R^2	RMSE	R^2	RMSE	R^2	RMSE	R^2	RMSE
Median value substitution										
MLR	0.32	14.71	0.26	14.04	0.38	17.19	0.44	9.89	0.22	23.02
FR	0.45	13.23	0.30	13.59	0.28	19.12	0.59	8.59	0.25	22.70
ETR	0.56	11.76	0.40	12.63	0.35	18.27	0.64	7.92	0.32	18.30
BTR	0.58	11.70	0.38	12.89	0.42	17.26	0.59	8.43	0.27	22.31
Removing the entire row										
Data loss	4.14%		28.52%		23.44%		4.76%		24.76%	
MLR	0.15	14.53	0.18	17.74	0.48	13.35	0.33	8.55	0.10	20.25
FR	0.33	12.92	0.19	17.65	0.30	15.42	0.34	8.52	0.00	21.35
ETR	0.43	11.97	0.26	16.78	0.44	13.83	0.26	8.99	0.10	20.24
BTR	0.38	12.43	0.22	17.24	0.46	13.61	0.16	9.57	0.11	20.08

3.3 Feature Selection

Once the data has been cleaned, the next step is to compare four different feature scaling methods which includes: (i) Standard scaler, (ii) normalizer, (iii) min–max scaler, and (iv) robust scaler.

From both Tables 8 and 9, normalizer appears to have the highest consistency in improving the performance of the regression models out of the four scaling methods. It also yields the best performance (R^2 and RSME values) for the regression models compared to the other scalers, except for $PM_{2.5}$ dataset from Perlis. The highest R^2 values could be observed from Decision tree regression model + AdaBoost + Normalizer on Selangor PM_{10} dataset at 0.93. The other scaling methods appeared to perform similarly without much improvement compared to the prediction model without scaling. For $PM_{2.5}$ prediction with Penang and Perlis dataset, robust scaler scored a slight lower R^2 scores compared to the model without scaling. Thus, from the observation made, normalizer is selected as the most suitable scaler for the prediction dataset.

3.4 Feature Selection and Dimensionality Reduction

The final stage of the data pre-processing steps is data reductions, in which the performance between two features selection techniques and dimensionality reduction techniques were compared. In theory, features selection technique removes irrelevant features that does not have much impact on the prediction model, whereas dimensionality reduction technique works by searching for the best linear combination of features with less variables while maintaining as much of the information found in the original dataset.

By comparing the performance of the three data reduction techniques shown in Tables 10 and 11, model trained with principle component appeared to perform much better in predicting both PM_{10} and $PM_{2.5}$ concentrations. ETR and BTR prediction model have a very similar R^2 and RMSE scores. Although both models perform extremely well when compared with the other PCA treated models, BTR appeared to have an edge with consistently better performance across multiple datasets from different states. The best performance for PM_{10} predictions could be observed for both PCA treated ETR and BTR models on datasets from Kedah, Perak, and Selangor with R^2 values of 0.97. The same model configuration scored the best performance for $PM_{2.5}$ prediction as well on Perak dataset with R^2 scores of 0.92. This is a significant performance improvement when compared with the model that only utilizes normalized data. Figure 1 shows the scatter plot diagram of the best PM_{10} and $PM_{2.5}$ prediction performance for Perak dataset.

However, models with post features selection data did not perform as well with almost no improvement could be observed in term of prediction accuracy. In fact, certain models in some of the dataset performed worse with lower R^2 values when

Table 8 Performance for different PM_{10} regression models after feature scaling

PM_{10} prediction model	Penang		Kedah		Perak		Perlis		Selangor	
	R^2	RMSE	R^2	RMSE	R^2	RMSE	R^2	RMSE	R^2	RMSE
Standard scaler										
MLR	0.46	0.6591	0.68	0.3678	0.27	0.9339	0.44	0.6645	0.39	0.8273
FR	0.68	0.5063	0.84	0.3492	0.64	0.6529	0.62	0.5467	0.73	0.5534
ETR	0.74	0.4590	0.85	0.3614	0.75	0.5445	0.77	0.4257	0.79	0.4925
BTR	0.75	0.4394	0.84	0.5224	0.73	0.5671	0.76	0.4395	0.76	0.5246
Normalizer										
MLR	0.74	0.0066	0.68	0.0052	0.72	0.0085	0.44	0.0083	0.85	0.0083
FR	0.81	0.0058	0.81	0.0050	0.89	0.0052	0.74	0.0056	0.92	0.0059
ETR	0.80	0.0058	0.85	0.0052	0.92	0.0046	0.76	0.0054	0.92	0.0063
BTR	0.84	0.0049	0.84	0.0074	0.92	0.0046	0.77	0.0053	0.93	0.0056
Min–max scaler										
MLR	0.46	0.0475	0.50	0.0322	0.27	0.0714	0.44	0.0442	0.40	0.0769
FR	0.67	0.0371	0.77	0.0314	0.64	0.0505	0.69	0.0329	0.71	0.0533
ETR	0.74	0.0332	0.78	0.0326	0.75	0.0420	0.76	0.0289	0.79	0.0453
BTR	0.76	0.0315	0.77	0.0480	0.74	0.0437	0.74	0.0304	0.78	0.0467
Robust scaler										
MLR	0.46	0.8496	0.50	0.4394	0.27	1.2730	0.44	0.6255	0.40	1.2422
FR	0.67	0.6663	0.79	0.4495	0.65	0.8762	0.74	0.4250	0.72	0.8460
ETR	0.74	0.5850	0.78	0.4449	0.75	0.7414	0.77	0.4014	0.79	0.7344
BTR	0.78	0.5479	0.79	0.6815	0.73	0.7780	0.72	0.4434	0.79	0.7411

Table 9 Performance for different $PM_{2.5}$ regression models after feature scaling

$PM_{2.5}$ prediction model	Penang		Kedah		Perak		Perlis		Selangor	
	R^2	RMSE	R^2	RMSE	R^2	RMSE	R^2	RMSE	R^2	RMSE
Standard scaler										
MLR	0.32	0.8689	0.26	0.8306	0.38	0.7629	0.44	0.6488	0.27	0.2062
FR	0.45	0.7730	0.32	0.7993	0.26	0.8350	0.58	0.5586	0.28	0.2048
ETR	0.56	0.6977	0.40	0.7387	0.35	0.7808	0.64	0.5172	0.37	0.1907
BTR	0.53	0.7325	0.38	0.7634	0.38	0.7636	0.58	0.5619	0.27	0.2041
Normalizer										
MLR	0.23	0.0076	0.30	0.0067	0.71	0.0059	0.40	0.0050	0.66	0.0063
FR	0.54	0.0059	0.50	0.0056	0.63	0.0067	0.54	0.0044	0.62	0.0067
ETR	0.60	0.0055	0.49	0.0057	0.66	0.0064	0.62	0.0040	0.63	0.0066
BTR	0.65	0.0052	0.53	0.0055	0.73	0.0057	0.53	0.0044	0.68	0.0061
Min–max scaler										
MLR	0.32	0.0722	0.26	0.0676	0.38	0.0658	0.44	0.0489	0.25	0.0143
FR	0.42	−0.0666	0.33	0.0644	0.28	0.0708	0.59	0.0419	0.13	0.0154
ETR	0.58	0.0567	0.42	0.0600	0.35	0.0675	0.63	0.0394	0.20	0.0147
BTR	0.53	0.0598	0.37	0.0625	0.38	0.0658	0.56	0.0431	0.26	0.0142
Robust scaler										
MLR	0.32	0.8408	0.26	0.7920	0.38	0.8694	0.44	0.8070	0.26	0.3859
FR	0.39	0.7966	0.32	0.7595	0.30	0.9159	0.58	0.6996	0.25	0.3901
ETR	0.55	0.6803	0.42	0.7036	0.33	0.9034	0.63	0.6555	0.32	0.3712
BTR	0.53	0.6957	0.35	0.7458	0.36	0.8832	0.57	0.7075	0.29	0.3782

Table 10 Performance for different PM_{10} regression models after features selection and PCA

PM_{10} prediction model	Penang		Kedah		Perak		Perlis		Selangor	
	R^2	RMSE	R^2	RMSE	R^2	RMSE	R^2	RMSE	R^2	RMSE
Univariate linear regression										
MLR	0.74	0.0071	0.68	0.0075	0.70	0.0087	0.45	0.0082	0.45	0.0158
FR	0.80	0.0063	0.81	0.0057	0.89	0.0055	0.67	0.0064	0.82	0.0092
ETR	0.77	0.0062	0.83	0.0053	0.91	0.0048	0.65	0.0066	0.84	0.0086
BTR	0.84	0.0055	0.84	0.0053	0.92	0.0047	0.69	0.0062	0.83	0.0089
Univariate mutual info regression										
MLR	0.46	0.0095	0.64	0.0079	0.70	0.0087	0.44	0.0083	0.84	0.0087
FR	0.75	0.0064	0.83	0.0056	0.91	0.0049	0.63	0.0067	0.93	0.0058
ETR	0.80	0.0059	0.85	0.0055	0.91	0.0048	0.63	0.0067	0.92	0.0063
BTR	0.82	0.0055	0.82	0.0052	0.92	0.0047	0.64	0.0066	0.95	0.0051
Principle component analysis (PCA)										
MLR	0.59	0.0083	0.49	0.0092	0.63	0.0097	0.18	0.0100	0.82	0.0090
FR	0.93	0.0034	0.95	0.0029	0.96	0.0031	0.71	0.0060	0.97	0.0031
ETR	0.93	0.0033	0.97	0.0021	0.97	0.0020	0.83	0.0046	0.97	0.0027
BTR	0.94	0.0030	0.97	0.0023	0.97	0.0022	0.83	0.0046	0.97	0.0028

Table 11 Performance for different $PM_{2.5}$ regression models after features selection and PCA

$PM_{2.5}$ prediction model	Penang		Kedah		Perak		Perlis		Selangor	
	R^2	RMSE	R^2	RMSE	R^2	RMSE	R^2	RMSE	R^2	RMSE
Univariate linear regression										
MLR	0.27	0.0074	0.25	0.0069	0.38	0.0087	0.42	0.0049	0.48	0.0062
FR	0.56	0.0058	0.37	0.0063	0.49	0.0079	0.45	0.0048	0.38	0.0068
ETR	0.51	0.0061	0.41	0.0061	0.57	0.0072	0.49	0.0046	0.47	0.0063
BTR	0.55	0.0058	0.39	0.0062	0.55	0.0074	0.47	0.0047	0.52	0.0059
Univariate mutual info regression										
MLR	0.31	0.0072	0.26	0.0069	0.37	0.0087	0.43	0.0049	0.23	0.0075
FR	0.42	0.0067	0.33	0.0065	0.54	0.0075	0.45	0.0048	0.23	0.0075
ETR	0.54	0.0059	0.40	0.0062	0.58	0.0072	0.54	0.0044	0.33	0.0070
BTR	0.57	0.0057	0.33	0.0065	0.56	0.0073	0.50	0.0046	0.35	0.0069
Principle component analysis (PCA)										
MLR	0.11	0.0700	0.16	0.0073	0.63	0.0067	0.40	0.0050	0.46	0.0069
FR	0.69	0.0166	0.50	0.0057	0.91	0.0034	0.39	0.0050	0.81	0.0035
ETR	0.83	0.0137	0.65	0.0047	0.92	0.0030	0.52	0.0045	0.84	0.0032
BTR	0.85	0.0147	0.64	0.0048	0.92	0.0031	0.46	0.0048	0.85	0.0031

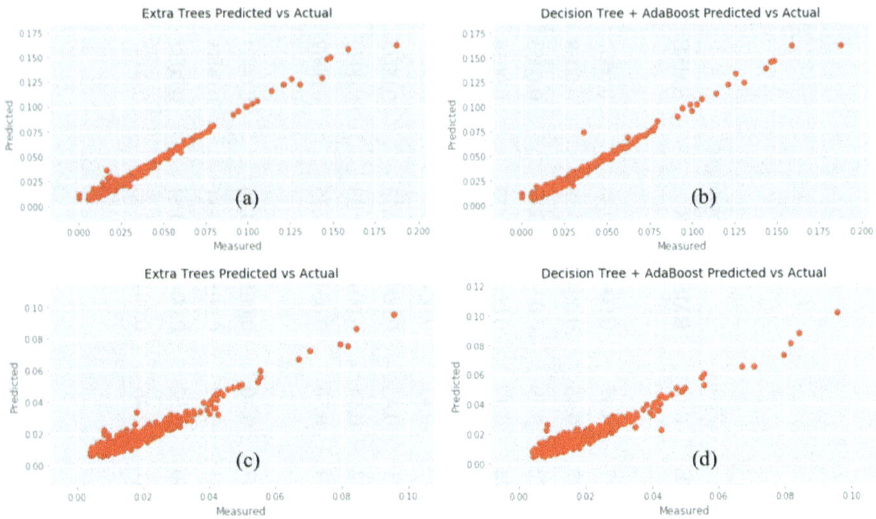

Fig. 1 Scatter plot diagram of the best PM_{10} and $PM_{2.5}$ prediction performance for Perak dataset. **a** ETR-PM_{10}, **b** BTR-PM_{10}, **c** ETR-$PM_{2.5}$, and **d** BTR-$PM_{2.5}$

compared with model that did not go through any type of data reduction (i.e. Selangor $PM_{2.5}$ dataset). This observation shows that significant amount of information may have been lost in the dataset when features selections were performed, which lead to the lower prediction performance. The performance is also much worse with $PM_{2.5}$ prediction, this could be due to the much lower data counts in $PM_{2.5}$ dataset (shown in Table 3) which cause the effect of information lost to be much more significant.

4 Conclusion

In this study we compared the performance of four machine learning algorithms to predict the concentration of PM_{10} and $PM_{2.5}$. Additionally, different techniques for pre-processing steps were compared and analyzed. Initially the regression models were ran without any data pre-processing, we observed that the regression models developed for PM_{10} predictions performed considerably well with R^2 values up to 0.79 for ETR and BTR models. Median value substitution was selected to impute the missing data, the results showed acceptable improvement in prediction accuracy especially for MLR and BTR models. Four different scaling method were then compared for the data transformation stage: Standard scaler, normalizer, min–max scaler, and robust scaler. Normalizer displayed the highest consistency in improving the prediction performance with an R^2 values up to 0.93. Lastly when three data reduction techniques were compared, models trained with PCA performed showed the most significant improvement in prediction accuracy with R^2 values up to 0.97.

Even though not many studies have covered the pre-processing aspects of PM_{10} and $PM_{2.5}$ predictions, the result from our study indicates that selecting the right pre-processing technique is as important as selecting the right prediction models.

Acknowledgements This work was supported by Universiti Sains Malaysia (USM), special gratitude to Department of Environmental (DOE) Malaysia for providing the air quality data for this study and Kementerian Pendidikan Malaysia (KPM) through Fundamental Research Grant Scheme (FRGS) grant number PJKIMIA/6071414.

References

1. IQAir (2019) 2019 World air quality report: region and city PM2.5 ranking. Retrieved from: https://www.iqair.com/us/world-most-polluted-cities
2. World Health Organization (2016) Health and the environment: draft road map for an enhanced global response to the adverse health effects of air pollution: report by the Secretariat. World Health Assembly, 69
3. Health Effects Institute (2019) States of global air 2019: a special report on global exposure to air pollution and its disease burden. Special Report. Retrieved from: https://www.stateofglobalair.org/sites/default/files/soga_2019_report.pdf
4. World Health Organization. Occupational and Environmental Health Team (2006) WHO air quality guidelines for particulate matter, ozone, nitrogen dioxide and sulfur dioxide: global update 2005: summary of risk assessment
5. Gokhale S, Raokhande N (2008) Performance evaluation of air quality models for predicting PM_{10} and $PM_{2.5}$ concentrations at urban traffic intersection during winter period. Sci Total Environ 9–24
6. Azarmi F, Kumar P, Marsh, D, Fuller G (2016) Assessment of the long-term impacts of PM_{10} and $PM_{2.5}$ particles from construction works on surrounding areas. Environmental Science: Process Impacts 208–221
7. Catino S, Tutino M, Ruggieri S, Marinaccio C, Giua R, de Gennaro G, Corsi P, Assennato G, Ribatti D (2017) Angiogenic activity in vivo of the particulate matter (PM_{10}). Ecotoxicol Environ Safety 156–161
8. Zanobetti A, Schwartz J, Dockery DW (2000) Airborne particles are a risk factor for hospital admissions for heart and lung disease. Environ Health Perspect 1071–1077
9. Mutalib SNSA, Juahir H, Azid A, Sharif SM, Latif MT, Aris AZ, Zain SM, Dominick D (2013) Spatial and temporal air quality pattern recognition using environmetric techniques: a case study in Malaysia. Environ Sci Process Impacts 1717–1728
10. Yahaya N Ali A, Ishak F (2006) Air pollution index (API) and the effects on human health: case study in Terengganu City, Terengganu, Malaysia. In: International association for people environment studies (IAPS) conference, Alexandria, Egypt
11. Latif MT, Othman M, Idris N, Juneng L, Abdullah AM, Hamzah WP, Khan MF, Nik Sulaiman NM, Jewaratnam J, Aghamohammadi N, Sahani M, Xiang CJ, Ahamad F, Amil N, Darus M, Varkkey H, Tangang F, Jaafar AB (2018) Impact of regional haze towards air quality in Malaysia: a review. Atmos Environ 28–44
12. Science G (2015) A case study of $PM_{2.5}$ characterization in Bangi, Selangor, Malaysia during the Southwest Monsoon Season. Aerosol Air Qual Res 2685–2691
13. Zaki T, Yusof M, Fitri NF, Shith S (2016) Morphology analysis of fine particles in background station of Malaysia. Sustainability in Environment
14. Khan MF, Sulong NA, Latif MT, Nadzir MSM, Amil N, Hussain DFM, Lee V, Hosaini PN, Shaharom S, Yusoff NAYM, Hoque HMS, Chung JX, Sahani M, Mohd Tahir N, Juneng L,

Maulud KNA, Abdullah SMS, Fujii Y, Tohno S, Mizohata A (2016) Comprehensive assessment of $PM_{2.5}$ physicochemical properties during the Southeast Asia dry season (southwest monsoon). J Geophys Res Atmos 121(24):14,589–14,611
15. Shi L-Z, Deng Q-H, Lu C, Liu W-W (2012) Prediction of PM_{10} mass concentrations based on BP artificial neural network. Zhongnan Daxue Xuebao (Ziran Kexue Ban)/J Cent South Univ (Sci Technol) 1969–1974
16. Alam MS, McNabola A (2015) Exploring the modeling of spatiotemporal variations in ambient air pollution within the land use regression framework: estimation of PM_{10} concentrations on a daily basis. J Air Waste Manag Assoc 628–640
17. Liu C, Henderson BH, Wang D, Yang X, Peng Z (2016) A land use regression application into assessing spatial variation of intra-urban fine particulate matter ($PM_{2.5}$) and nitrogen dioxide (NO_2) concentrations in City of Shanghai, China. Sci Total Environ 607–615
18. Ul-Saufie AZ, Yahaya AS, Ramli NA, Rosaida N, Hamid HA (2013) Future daily PM_{10} concentrations prediction by combining regression models and feedforward backpropagation models with principle component analysis (PCA). Atmos Environ 621–630
19. Chakrabarti S et al (2009) Data mining: know it all. M. Kaufmann, Burlington Mass.
20. Zamani Joharestani M, Cao C, Ni X, Bashir B, Talebiesfandarani S (2019) $PM_{2.5}$ prediction based on random forest, XGBoost, and deep learning using multisource remote sensing data. Atmosphere 373
21. Qu Y, Cai H, Ren K, Zhang, W, Yu Y, Wen Y, Wang J (2016) Product-based neural networks for user response prediction
22. Wang H, Wang S (2009) Mining incomplete survey data through classification. Knowl Inform Syst 221–233
23. Pyle D (1999) Data preparation for data mining. Morgan Kaufmann Publishers Inc., California
24. Little RJA, Rubin DB (2020) Statistical analysis with missing data. Wiley, Hoboken, NJ
25. García S, Luengo J, Herrera F (2015) Data preprocessing in data mining.
26. Thara DK, Prema Sudha BG, Xiong F (2019) Auto-detection of epileptic seizure events using deep neural network with different feature scaling techniques. Pattern Recogn Lett 544–550
27. Blum AL, Langley P (1997) Selection of relevant features and examples in machine learning. Artif Intell [online] 97(1–2):245–271
28. Saeys Y, Inza I, Larranaga P (2007) A review of feature selection techniques in bioinformatics. Bioinformatics 2507–2517
29. Al Sayyid A, Haque R, Taher Y, Makki S, Jaber A (2018) ProgMod: an analytical model for prognosis prediction of AML patients using survival regression and gene expression levels
30. Vlachogianni A, Kassomenos P, Karppinen A, Karakitsios S, Kukkonen J (2011) Evaluation of a multiple regression model for the forecasting of the concentrations of NOx and PM_{10} in Athens and Helsinki. Sci Total Environ 1559–1571
31. Stafoggia M, Bellander T, Bucci S, Davoli M, de Hoogh K, de' Donato F, Gariazzo C, Lyapustin A, Michelozzi P, Renzi M, Scortichini M, Shtein A, Viegi G, Kloog I, Schwartz J (2019) Estimation of daily PM_{10} and $PM_{2.5}$ concentrations in Italy, 2013–2015, using a spatiotemporal land-use random-forest model. Environ Int 170–179
32. Geurts P, Ernst D, Wehenkel L (2006) Extremely randomized trees. Mach Learn 3–42
33. Drucker H (1997) Improving regressors using boosting techniques. In: Proceedings of the 14th international conference on machine learning.

Characteristics and Empirical Modelling of Extract from *Hibiscus sabdariffa* Using Supercritical CO_2 Extraction with Ethanol-Water as Modifier

Zuhaili Idham, Nicky Rahmana Putra, Nor Faadila Mohd Idrus, Norlisa Mili, Muhammad Syafiq Hazwan Ruslan, Noor Azwani Mohd Rasidek, and Mohd Azizi Che Yunus

Abstract The main constituents of the extract from *Hibiscus sabdariffa* calyces are anthocyanins which gave intense natural red colorant and commonly used for centuries for therapeutic purposes. Supercritical carbon dioxide (SC-CO_2) technique could produce high-quality extracts from plant materials as it can exploit the non-toxic character and the gas–liquid properties of the SC-CO_2. This research aimed to characterise the colour and the pattern of *Hibiscus sabdariffa* SC-CO_2 extraction. The colour was evaluated by using a CIELAB method meanwhile the presence of anthocyanins in the red colour extract was confirmed by High-Performance Liquid Chromatograph (HPLC). Two types of empirical models (Modified Brunner's and Esquivel's models) were used to describe the extracts obtained at series of operational parameters (i.e. pressure and temperature) at 7.5 and 10% of modifier ratio, with a total solvent flow rate of 4 ml/min. Both models were analysed using a non-linear regression method. The modified Brunner's model shows the best agreement between experimental and model calculated data with the lowest *AARD* of 1.30% at 70 °C, 10 MPa, and 10% modifier ratio.

Z. Idham · N. R. Putra · N. F. M. Idrus · N. Mili · N. A. M. Rasidek · M. A. C. Yunus (✉)
Centre of Lipids Engineering and Applied Research, Ibnu Sina Institute for Scientific & Industrial Research, Universiti Teknologi Malaysia, Johor Bahru, Malaysia
e-mail: azizi@cheme.utm.my

Z. Idham
e-mail: zuhaili@cheme.utm.my

Z. Idham · N. F. M. Idrus · M. A. C. Yunus
Faculty of Engineering, School of Chemical and Energy Engineering, Universiti Teknologi Malaysia (UTM), 81310 Johor Bahru, Malaysia

N. Mili
Faculty of Engineering, Universiti Malaysia Sarawak (UNIMAS), 94300 Kota Samarahan, Sarawak, Malaysia

M. S. H. Ruslan
Department of Chemical Engineering, Universiti Teknologi PETRONAS, Bandar Sri Iskandar, Perak, Malaysia

M. A. A. Zaini et al. (eds.), *Proceedings of the 3rd International Conference on Separation Technology*, Lecture Notes in Mechanical Engineering,
https://doi.org/10.1007/978-981-16-0742-4_26

Keywords *Hibiscus sabdariffa* · Anthocyanins · Red colour · Supercritical carbon dioxide · Empirical model

1 Introduction

Hibiscus sabdariffa (H. sabdariffa) or Roselle is one of the highly preferable tropical plants for food and pharmaceutical industries. *H. sabdariffa* contains antioxidants that benefit humans by demonstrating positive effects on chronic diseases such as cancer or diabetes [1]. Since antioxidants are vastly found in plant foods, *H. sabdariffa* is commonly prepared in hot and cold beverages or herbal drinks [2]. Belong to a family of Malvaceae, *H. sabdariffa* is usually cultivated for its thick and fleshy green, red, or dark red calyxes [3]. The *H. sabdariffa* dried calyxes' extracts contain water-soluble antioxidants such as anthocyanins and polyphenols, as well as organic acids such as ascorbic and hibsicic acid [4]. Those antioxidants are widely studied as they are capable of scavenging reactive oxygen species (ROS), which can accept hydrogen atoms/electrons by reactive radical [5]. Also, anthocyanins could give a brilliant natural red colorant to the calyxes.

Several researchers employed supercritical carbon dioxide (SC-CO_2) for the selective extraction of colour or pigment compounds from plant materials [6–9]. Due to the higher initial capital investment, SC-CO_2 is less preferred compared to other extraction techniques. However, SC-CO_2 is a unique technology to extract high-quality bioactive compounds with slight pressure and temperature adjustment [10]. These parameters are crucial to avoid any degradation of thermolabile compounds due to high heat or contamination with solvent residues due to different solvent density. Furthermore, SC-CO_2 operates in the absence of oxygen, in contrast to steam distillation, maceration, or Soxhlet extraction in traditional techniques. This condition makes SC-CO_2 able to prevent the oxidation process that potentially degrades particular bioactive compounds. In this study, the use of SC-CO_2 with the addition of mixed ethanol and water as a modifier would help to increase the SC-CO_2 polarity and matrix swelling. The polar modifier molecules could accelerate the desorption process by competing with the analytes for binding sites and disrupting the plant matrix structure [11], which later enhance the contact between both solute and the SC-CO_2.

Although the application of SC-CO_2 as a replacement of bulk organic solvent in *H. sabdariffa* calyxes' extraction has been demonstrated by several researchers [12–14], the evaluation of extraction kinetics from the established model for the total extract is scarce. The mathematical modelling of SC-CO_2 extraction curves is essential to assess the kinetic behaviour of the experimental data. The model could establish studied process and operational parameters and generates sample data for scale-up purposes [15]. However, certain conditions applied, such as the types of raw material used for extraction and commercial feasibility for industrial implementation. Therefore, this study aimed to characterize the red colour extract and evaluate the SC-CO_2 extraction kinetics of *H. sabdariffa* with modified Brunner's and Esquivel's models, on top of determining the dominant extraction kinetics between the two models. The effects of pressure and temperature will also be discussed and evaluated.

2 Materials and Method

2.1 Plant Materials

H. sabdariffa dry calyxes (var. UMKL-1) was obtained from Ladang Setiu, Terengganu, Malaysia. The dry calyxes were pulverized (Waring Commercial Blender) and sieved to the average size of 200–355 μm.

2.2 Chemicals

Liquid carbon dioxide (CO_2) (99% purity) was purchased from Kras Instrument, Johor Bahru, Malaysia and applied in the supercritical extraction equipment. An analytical grade of ethanol (99.86% purity) was bought from Sigma Aldrich.

2.3 Supercritical CO_2 Extraction

The extraction setup was conducted at different pressures (10, 20, and 30 MPa), temperatures (40, 55, and 70 °C) and modifier ratio (7.5 and 10% from 4 ml/min of total solvent flow rate). The range of conditions was selected based on the limitation of equipment and bioactive compound. Chiller was set to 6 °C while the heater of the back pressure regulator was set to 50 °C. Approximately 1.5 ± 0.005 g of sample was placed in the 3L extraction vessel. 75:25 (v/v) of ethanol and water was used as a modifier. Next, by using a supercritical pump, the mixture of modifier and liquid CO_2 was pumped out from the gas cylinder into the system. The quantity of extract obtained at regular time intervals was weighed (balance precision ± 0.00001 g). The extraction was stopped at 120 min. Next, by using a vacuum concentrator (Mivac Concentrator) with a temperature of 40 °C, the ethanol was removed until the *H. sabdariffa* extract is dried.

2.4 Colour Characteristics

The extract characteristics were selected from the best condition of SC-CO_2 yield extraction based on the colour and presence of anthocyanins. The evaluation of colour was measured in terms of the CIE L*, a*, and b* values using a C-10 Colour Reader (Konica Minolta, Osaka, Japan). The L* lightness characteristics are white = 100, and black = 0. The value of a* are +a* = red and −a* = green, while the value of b* are +b* = yellow and −b* = blue.

Identification of anthocyanins was carried out using the Waters Breeze 2 High-Performance Liquid Chromatography (HPLC) equipped with 2475 multi λ fluorescence detector, 717 plus autosampler, 1525 Binary HPLC pump, and temperature control module. The column temperature was 30 °C and injection volume was 20 μL. Acetonitrile: water (65:35) was used as mobile phase solvent. The flow rate was 1.0 mL/min and the column were monitored by a UV detector set at wavelength 520 nm. Their relative concentrations were determined from respective peak areas of chromatograms, using cyanidin 3-glucoside (kuromanin chloride) as standard.

2.5 Mathematical Modelling

The extraction curves of *H. sabdariffa* extract were modified using two models, as previously mentioned in the literature. Experimental data were compared to the analysed model. The following Eq. (1) represents the empirical Esquivel's model [16]:

$$m_{ext} = x_0 F\left(\frac{t}{b+t}\right) \tag{1}$$

where m_{ext} is the mass of the extract, x_0 is the initial solute mass ratio in the solid phase (kg kg^{-1}), F is the mass of the solid material (kg), t is the extraction time (s), and b is an adjustable parameter (s). In this study, the Eq. (2) was improved with the addition of one adjustable parameter [16] as reported in our previous study [17];

$$Y_t = Y_1\left(\frac{t}{k_1+t}\right) \tag{2}$$

where Y_t is the total mass of extraction yield (g), Y_1 is the predicted total mass of extraction yield (g), k_1 is the adjustable parameter, and t is time (s). Meanwhile, the Eq. (3) for the empirical modelling of SC-CO_2 extraction using Brunner's model [18] was also modified by adding of one adjustable parameter [16, 17];

$$Y_t = Y_2\left(1 - e^{-k_2 t}\right) \tag{3}$$

where Y_t is the total mass of extraction yield (g), Y_2 (g) is the predicted total mass of extraction yield, k_2 is the adjustable parameter and t is time (s).

2.6 Statistical Analysis

To assess the correlation between the experimental data and calculated values using Brunner and Esquivel's fitted models, the coefficient of determination (R^2) was applied. An average absolute relative deviation ($AARD$) was computed to correlate between the experimentally determined extraction yield and the calculated value from different mathematical models. The $AARD$ equation is as follows:

$$AARD = \frac{1}{n}\sum_{i=1}^{n}\left|\frac{y_{exp} - y_{cal}}{y_{exp}}\right| \tag{4}$$

3 Result and Discussion

3.1 Supercritical CO_2 Yield and Colour Characteristics

H. sabdariffa extractions were performed at different sets of parameters, and total extraction values are shown in Table 1. The results show that the extraction yield was in a range of 0.466–1.182 g. The highest yield was obtained at 55 °C, and 30 MPa (Run 8) and the lowest was at 40 °C and 10 MPa (Run 3) from the same modifier ratios used. The yield of total extract was around 30–78% of the total sample. These findings are improved from our previous research, which used a low range of pressure near-critical value (i.e. 8–12 MPa) [14].

It is vital to measure colour quantitatively to evaluate the colour quality in a systematically. Colour denotes the product's visual appearance while pigments or colorants are the chemical compounds imparting the colour observed. In this study, the highest yield from the SC-CO_2 extraction condition was characterized for the colour and the presence of anthocyanins. Figure 1 shows the colour of *H. sabdariffa* extracted from SC-CO_2 for every fraction of time. Meanwhile, the measurement of colour intensity is expressed in terms of their L*a*b* values. The L* values indicate perceived lightness, white (100) or darkness, black (0). The value of a* indicates red (positive a*) and green (negative a*), while b* represents yellow (positive b*) and blue (negative b*). These are tabulated in Table 2. The figures visibly display the dark red colour, started from 30 min until 60 min, showing that the exhaustive extraction has occurred. This was proved by the lower L* value (Table 2). All the samples presented both positive a* and negative b* values with different intensities towards the time of extraction. The mixture of redness and blueness made the extract became purplish-red coloured. The presence of anthocyanins in the extract was confirmed by the identification of anthocyanins using cyanidin 3-glucoside at 1.333 retention time, as shown in Fig. 2.

Table 1 Experimental conditions used in this study

Run	Pressure (MPa)	Temperature (°C)	Modifier ratio (%)	Yield (g)
1	10	70	10	0.547
2	10	55	10	0.473
3	10	40	10	0.466
4	20	70	10	0.872
5	20	55	10	0.838
6	20	40	10	0.798
7	30	70	10	0.827
8	30	55	10	1.182
9	30	40	10	0.837
10	10	70	7.5	0.659
11	10	55	7.5	0.610
12	10	40	7.5	0.562
13	20	70	7.5	0.810
14	20	55	7.5	0.774
15	20	40	7.5	0.712
16	30	70	7.5	0.815
17	30	55	7.5	0.816
18	30	40	7.5	0.698

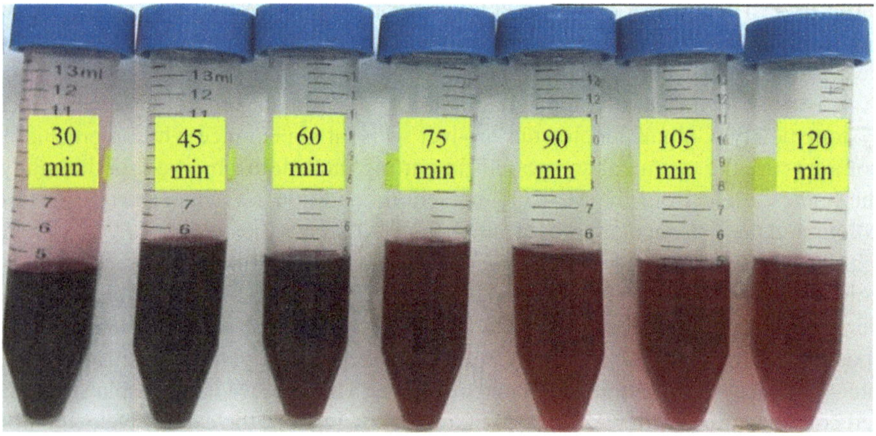

Fig. 1 Red colour extract from the highest SC-CO_2 extraction yield at condition 55 °C, 30 MPa and 10% of modifier ratio

Table 2 Colour intensity of the *H. sabdariffa* SC-CO_2 extract of different time fractions at 55 °C, 30 MPa, and 10% modifier ratio

Sample	L^*	a^*	b^*
30 min	21.0	10.3	−6.4
45 min	23.8	11.1	−5.6
60 min	22.0	10.9	−4.8
75 min	24.5	8.5	−4.5
90 min	26.7	8.6	−3.9
105 min	27.8	7.9	−2.7
120 min	29.3	7.8	−3.1

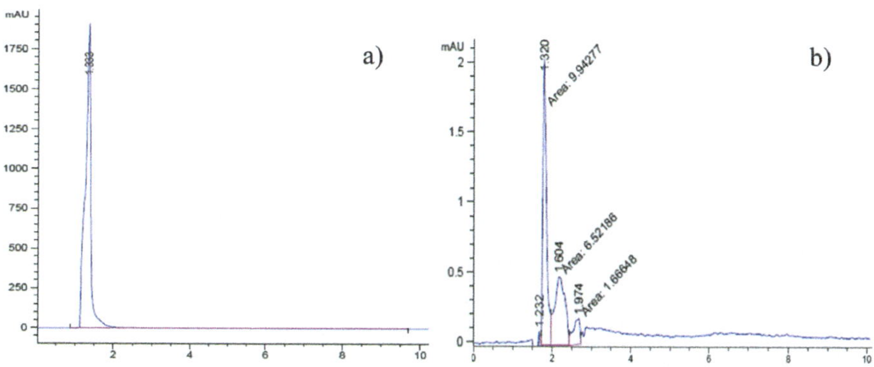

Fig. 2 Identification of anthocyanins in the red colour extract **a** Standard of cyanidin 3-glucoside, **b** red colour SC-CO_2 extract

3.2 *Mathematical Model*

The mathematical models used to describe the SC-CO_2 process usually has at least one parameter that needs to be calculated or adapted to experimental data. An empirical model can evaluate the extraction curves, according to Esquivel et al. [19]. Esquivel's model defines the overall extraction curve as a typical hyperbole of systems by which solvent is easily infiltrated into the solution [15]. Brunner's model assumed that the extraction process occurred in one step when only internal heat transfer controlled the diffusion [18]. In short, the supercritical extraction process for soluble compounds from solid plant material takes place in several consecutive steps, according to Brunner [20]. The plant matrix must absorb the supercritical solvent and other fluids, which are applied to the extraction process deliberately. The cell structure swells, thus widening the membranes of the cells and the intercellular channels—such steps raising the resistance to mass transport. At the same time, the solvent dissolves the extract compounds. Prior to solvation, a chemical reaction may occur. The dissolved compounds are then transported to the outer surface of the solid, whereby the essential transport mechanism being diffusion. This process may lead to a phase change. The compounds are then transported from the surface layer to the bulk of the supercritical solvent and separated from the solid material with the solvent.

The calculate responses in Esquivel and Brunner's equations were obtained by having the lowest $AARD$ value of the yield calculated in both model and experimental data. Approximation of $AARD$ value of up to 5% is acceptable and approved in Applied Mathematical Models. However, >10% indicates sparse estimation of experimental and model-predicted data values [21]. Based on Table 3, at a 10% modifier ratio, results show that the range of $AARD$ is 1.81–15.85% for the modified Esquivel's model and 1.30–13.16% for modified Brunner's. While at 7.5% modifier ratio (Table 4), the range of the $AARD$ value is 2.79–30.19% for modified Esquivel's and 1.61–29.31% for modified Brunner's model. Therefore, these findings proved that the modified Brunner's model shows slightly better agreement with the experimental data with 0.999, even though there is no significant difference of $AARD$ value. The poor $AARD$ values occurred in both models at different operating conditions were due to the limited adjustable parameters applied (i.e. only two). These limited adjustable parameters could not identify the mass transfer mechanisms, such

Table 3 Calculated responses using applied mathematical models (modified Brunner and Esquivel's models) at a 10% modifier ratio

P (MPa)	T (°C)	Modified Esquivel's model				
		Y_1 (g)	k_1 (sec)	Y_1/k_1 (g/sec)	% AARD	R^2
10	70	0.733	2455	0.00030	1.81	0.998
10	55	1.448	13520	0.00010	3.37	0.997
10	40	0.534	920	0.00058	3.94	0.988
20	70	1.442	4274	0.00034	5.60	0.989
20	55	1.650	6266	0.00026	11.49	0.976
20	40	4.318	28632	0.00015	8.94	0.976
30	70	1.853	10502	0.00018	15.85	0.957
30	55	1.537	1831	0.00084	3.32	0.994
30	40	1.873	7815	0.00024	9.04	0.986
P (MPa)	T (°C)	Modified Brunner's model				
		Y_2 (g)	$1/k_2$ (1/sec)	Y_2/k_2 (g/sec)	% AARD	R^2
10	70	0.585	0.00038	0.00022	1.30	0.999
10	55	1.144	0.00012	0.00010	3.23	0.997
10	40	0.471	0.00066	0.00031	2.61	0.995
20	70	1.013	0.00030	0.00030	4.85	0.992
20	55	1.219	0.00018	0.00022	11.25	0.976
20	40	1.417	0.00013	0.00018	10.28	0.981
30	70	2.326	0.00006	0.00014	13.16	0.971
30	55	1.215	0.00051	0.00062	2.19	0.997
30	40	1.075	0.00022	0.00024	8.47	0.987

Table 4 Calculated responses using applied mathematical models (modified Brunner and Esquivel's models) at 7.5% modifier ratio

P (MPa)	T (°C)	Modified Esquivel's model				
		Y_1 (g)	k_1 (s)	Y_1/k_1 (g/s)	% AARD	R^2
10	70	2.562	18998	0.00013	3.12	0.996
10	55	2.652	32928	0.00008	30.19	0.961
10	40	3.941	65775	0.00006	29.20	0.976
20	70	2.219	12143	0.00018	6.85	0.992
20	55	0.942	1410	0.00067	2.63	0.995
20	40	2.032	12559	0.00016	9.15	0.989
30	70	2.26	11479	0.00019	6.32	0.990
30	55	1.36	4137	0.00032	5.09	0.991
30	40	6.55	60427	0.00011	18.34	0.973
P (MPa)	T (°C)	Modified Brunner's model				
		Y_2 (g)	$1/k_2$ (1/s)	Y_2/k_2 (g/s)	% AARD	R^2
10	70	1.452	0.00009	0.00013	2.80	0.997
10	55	4.324	0.00002	0.00007	27.67	0.965
10	40	4.026	0.00001	0.00006	28.20	0.979
20	70	1.290	0.00014	0.00018	6.67	0.993
20	55	0.796	0.00052	0.00041	1.61	0.999
20	40	1.227	0.00013	0.00016	8.93	0.990
30	70	1.926	0.00009	0.00018	6.56	0.987
30	55	0.948	0.00031	0.00029	4.29	0.994
30	40	5.056	0.00002	0.00010	17.89	0.972

as the possibility of the solute to permeate into the solid particles, as the model does not represent this.

From the calculated adjustable parameter, both the highest and lowest extraction rates were simply predicted by the modified Esquivel's model. The highest extraction rate, 0.00084 g s^{-1}, was obtained at a 10% modifier ratio during 30 MPa and 55 °C while the lowest extraction rate, 0.00005 g s^{-1} was at 7.5% modifier ratio, 10 MPa, and 40 °C. Modified Brunner's model predicted under the same operating conditions but with a lower value of extraction rate, slightly lower $AARD$ with experimental data and higher correlation value ($R^2 = 0.997$). Therefore, the modified Brunner's model is preferred in this study. The kinetic extraction curve (Figs. 3, 4, 5 and 6) illustrates the role of extraction rate to achieve the extraction yield at the initial time of 30 min.

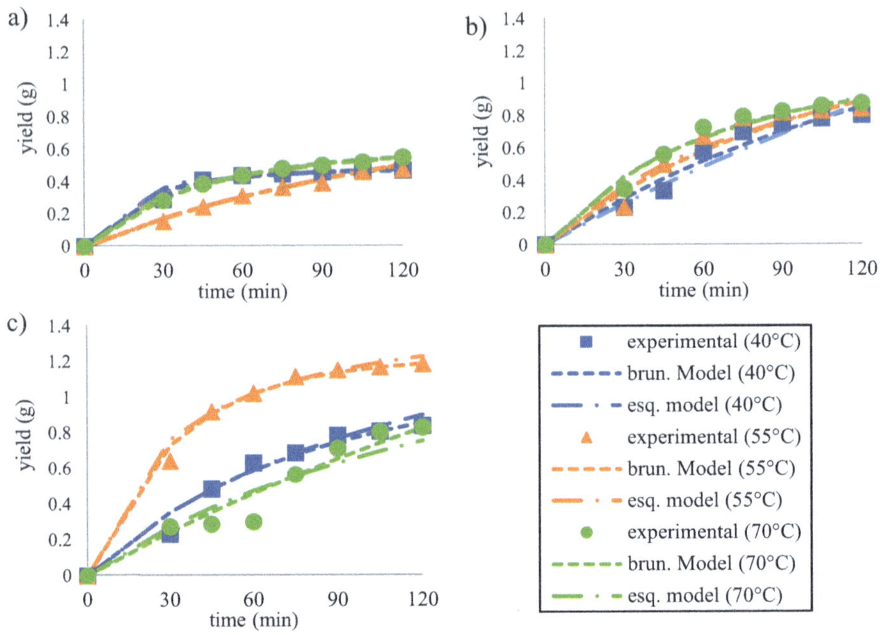

Fig. 3 Effect of temperature on overall extraction curve at **a** 10 MPa, **b** 20 MPa, and **c** 30 MPa of 10% modifier ratio

3.3 Effect of Pressure and Temperature on the Kinetic Extraction Curve

The temperature and pressure effects on the extraction of kinetic curves at each experimental condition are presented in Figs. 3, 4, 5 and 6, as a function of accumulative yield (g) versus the extraction time (min). Figures 3 and 4 observe the effect of temperature, while Figs. 5 and 6 clearly show the impact of pressure at a constant temperature. The extraction kinetics were monitored every 15 min. The overall extraction curve shows that during the first 60 min, there was a rapid extraction process, before slowly decrease and resulting in a lower at the end of the process. Since the stationary phase is started after 30 min of the extraction process in almost all operating conditions, therefore, 120 min of extraction time was sufficient to describe the extraction pattern since the maximum total yield achieved was nearly 78% of the dry weight of the total sample.

The effect of temperatures on extraction kinetic with comparison to both models at 10 and 7.5% modifier ratio are shown in Figs. 3 and 4. Higher temperatures led to increased extraction rates at constant pressures of 10 and 20 MPa. In contrast, at a pressure of 30 MPa, the operating temperature had a negligible effect on the extraction output. Similar patterns were observed by Maran et al. [22] at supercritical of phenolic extract from Jamun fruit and Fragoso-Jiménez et al. [23] from *Polianthes tuberosa*

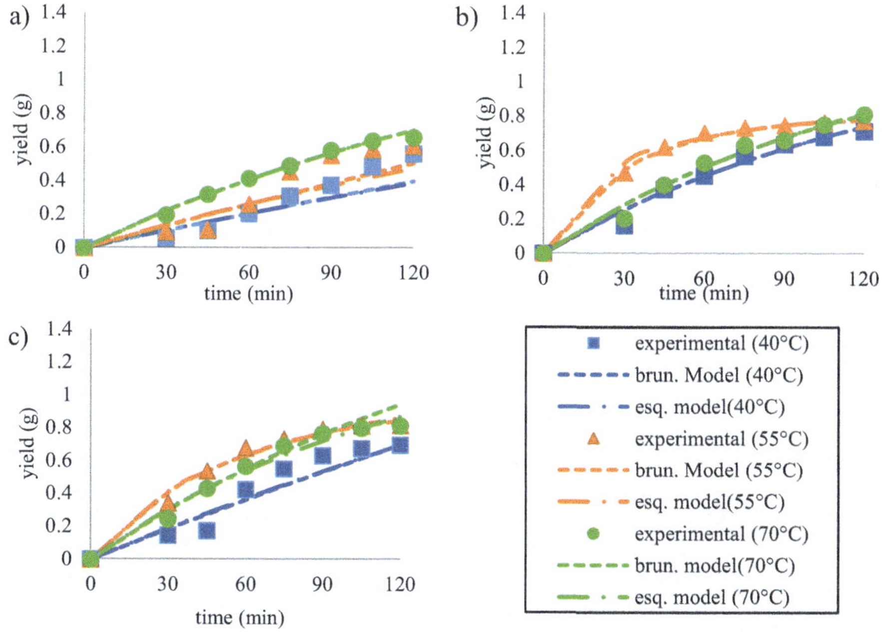

Fig. 4 Effect of temperature on extraction yield profile at **a** 10 MPa, **b** 20 MPa, and **c** 30 MPa of 7.5% modifier ratio

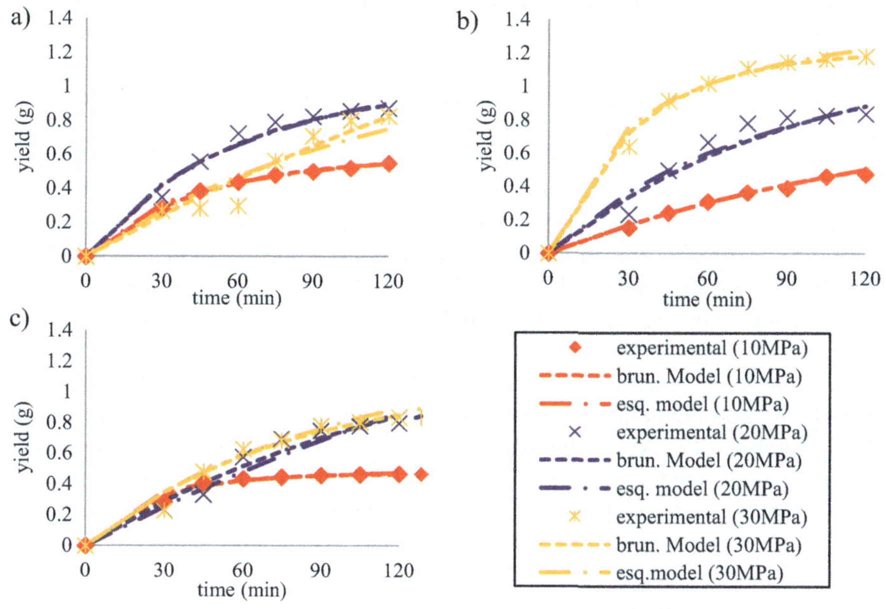

Fig. 5 Effect of pressure on extraction yield profile at **a** 70 °C, **b** 55 °C, and **c** 40 °C of 10% modifier ratio

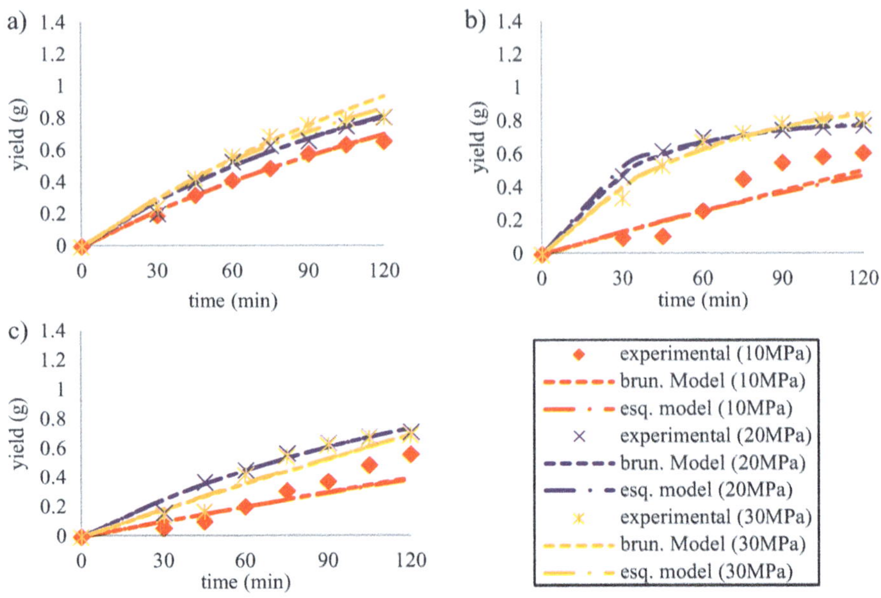

Fig. 6 Effect of pressure on extraction yield profile at **a** 70 °C, **b** 55 °C, and **c** 40 °C of 7.5% modifier ratio

flower extract. At 30 MPa, 55 °C was the ideal kinetic extraction temperature. The increase in yield with an increase in temperature is due to the increased of solute vapour pressure that overcomes the reduction in solvent density.

Results also showed that the overall increase in pressure is directly proportional to the rise in the obtained extract. A substantial increase in pressure leads to an increase in extraction yield at a constant temperature of 55 °C (Fig. 5b). The increase in pressure increases the density of the supercritical fluid, hence increases the strength of the solvent (i.e. SC-CO_2). Meanwhile, 10 MPa shows the lowest profile of the extraction curve. The decrease in solvent power at low pressure is due to a reduction in solvent density [15]. We have detected an increase in yield at a constant temperature from the behaviour of the extraction curve at all stages of the process. This is due to the increase in solvent density and solvent power, as explained earlier. The solvation power of SC-CO_2 increases due to the higher solvent density at a constant temperature. The highest extraction rates were therefore expected at 30 MPa. Our findings are comparable with other studies reported here [24–26]. A good correlation between the experimental and the calculated model has been observed by the extraction curve plot function, except under the low operational conditions, i.e. 10 MPa, 55 °C, and 7.5% modifier ratio (Fig. 4b), and 10 MPa, 40 °C, 7.5% modifier ratio (Fig. 6c). This was due to the low extraction yield from the initial extract (first 30 min), and rapid extraction, which started after 60 min.

4 Conclusions

In this study, the colour characteristics of *H. sabdariffa* calyxes obtained by SC-CO_2 extraction with ethanol-water as a modifier was measured as purplish-red colour and confirmed by the presence of anthocyanins. The range of parameters of applied models was obtained based on mathematical modelling by minimizing the value of the deviation of calculated output. The extraction yield calculated using the modified Brunner's model has shown the highest accuracy with the collected experimental data. The lowest $AARD$ between experimental and model data was 1.61% at 55 °C and 20 MPa (7.5% modifier ratio). This study also reveals that pressure and temperature have significant effects on the *H. sabdariffa* extraction yield. A better extraction yield during the extraction time was achieved with high operating pressure (20–30 MPa) and temperature (55 and 70 °C). On the other hand, low operating conditions were poorly agreed with both models due to the lower yield during initial extraction. This modelling study could indicate a better estimate of the bioactive compounds of *H. sabdariffa* calyxes for the food and bioprocess industries.

Acknowledgements All authors would like to acknowledge the Fundamental Research Grant Scheme, vote number: 4F997 (FRGS/1/2018/STG07/UTM/02/13) from Minister of Higher Education and Universiti Teknologi Malaysia for the UTMLead scholarship and UTM Industry International Incentive Grant (01M76) for financial support. We were also grateful to the Centre of Lipids Engineering and Applied Research (CLEAR), UTM for the facilities provided.

References

1. Riaz G, Chopra R (2018) A review on phytochemistry and therapeutic uses of Hibiscus sabdariffa L. Biomed Pharmacother 102:575–586
2. Kao ES, Yang MY, Hung CH, Huang CN, Wang CJ (2016) Polyphenolic extract from *Hibiscus sabdariffa* reduces body fat by inhibiting hepatic lipogenesis and preadipocyte adipogenesis. Food Funct 7(1):171–182
3. Andzi Barhé T, Feuya Tchouya GR (2016) Comparative study of the anti-oxidant activity of the total polyphenols extracted from Hibiscus Sabdariffa L., Glycine max L. Merr., yellow tea and red wine through reaction with DPPH free radicals. Arab J Chem 9(1):1–8
4. Da-Costa-Rocha I, Bonnlaender B, Sievers H, Pischel I, Heinrich M (2014) Hibiscus sabdariffa L.—a phytochemical and pharmacological review. Food Chem 165:424–443
5. Losada-Barreiro S, Bravo-Díaz C (2017) Free radicals and polyphenols: the redox chemistry of neurodegenerative diseases. Eur J Med Chem 133:379–402
6. Chatterjee D, Jadhav NT, Bhattacharjee P (2013) Solvent and supercritical carbon dioxide extraction of color from eggplants: characterization and food applications. LWT—Food Sci Technol 51(1):319–324
7. del Pilar Garcia-Mendoza M, Espinosa-Pardo FA, Baseggio AM, Barbero GF, Maróstica Junior MR, Rostagno MA, Martinez J (2017) Extraction of phenolic compounds and anthocyanins from juçara (Euterpe edulis Mart.) residues using pressurized liquids and supercritical fluids. J Supercrit Fluids 119:9–16
8. Nobre BP, Mendes RL, Queiroz EM, Pessoa FLP, Coelho JP, Palavra AF (2006) Supercritical carbon dioxide extraction of pigments from bixa orellana seeds (experiments and modeling). Braz J Chem Eng 23(2):251–258

9. Talib N, Ahmad MR, Ismail K, Ab Kadir MI (2016) Comparison of supercritical fluid extraction and ultrasound-assisted extraction of natural dyes from a brown seaweed (Sargassum spinosum). Int J Text Sci 5(6):141–144
10. Gadkari PV, Balaraman M (2015) Extraction of catechins from decaffeinated green tea for development of nanoemulsion using palm oil and sunflower oil based lipid carrier systems. J Food Eng 147:14–23
11. Kumar S (2015) Analytical techniques in natural product research. In: Analytical techniques for natural product research. CABI, Wallingford, United Kingdom, pp 1–10
12. Lukmanto S, Roesdiyono N, Ju YH, Indraswati N, Soetaredjo FE, Ismadji S (2013) Supercritical CO_2 extraction of phenolic compounds in Roselle (Hibiscus Sabdariffa L.). Chem Eng Commun 200(9):1187–1196
13. Pimentel-Moral S, Borrás-Linares I, Lozano-Sánchez J, Arráez-Román D, Martínez-Férez A, Segura-Carretero A (2019) Supercritical CO_2 extraction of bioactive compounds from Hibiscus sabdariffa. J Supercrit Fluids 147:213–221
14. Idham Z, Nasir HMM, Yunus MAC, Yian LN, Peng WL, Hassan H, Mohd Setapar SH (2017) Optimisation of supercritical CO_2 extraction of red colour from roselle (Hibiscus Sabdariffa Linn.) calyces. Chem Eng Trans 56:871–876
15. Comim SRR, Madella K, Oliveira JV, Ferreira SRS (2010) Supercritical fluid extraction from dried banana peel (Musa spp., genomic group AAB): extraction yield, mathematical modeling, economical analysis and phase equilibria. J Supercrit Fluids 54(1):30–37
16. Zeković Z, Filip S, Vidović S, Jokić S, Svilović S (2014) Mathematical modeling of ocimum basilicum L. supercritical CO_2 extraction. Chem Eng Technol 37(12):2123–2128
17. Putra NR, Idham Z, Machmudah S, Ruslan MSH, Che Yunus MA (2018) Extraction of peanut skin oil by modified supercritical carbon dioxide: empirical modelling and optimization. Sep Sci Technol 53(17):2695–2703
18. Brunner G (1984) Mass transfer from solid material in gas extraction. Ber der Bunsenges/Phys Chem Chem Phys 88(9):887–891
19. Esquível MM, Bernardo-Gil MG, King MB (1999) Mathematical models for supercritical extraction of olive husk oil. J Supercrit Fluids 16:43–58
20. Brunner G (1996) Gas extraction—an introduction to fundamentals of supercritical fluids and the application to separation processes. Topics in physical chemistry. In: Baumgärtel H, Franck EU, Grunbein W (eds) Berichte der Bunsengesellschaft für physikalische Chemie, vol 4. Springer, New York LLC
21. Jokic S, Svilovic S, Vidovic S (2015) Modelling the supercritical CO_2 extraction kinetics of soybean oil. Croat J Food Sci Technol 7(2):52–57
22. Maran JP, Priya B, Manikandan S (2014) Modeling and optimization of supercritical fluid extraction of anthocyanin and phenolic compounds from Syzygium cumini fruit pulp. J Food Sci Technol 51(9):1938–1946
23. Fragoso-Jiménez J, Tapia-Campos E, Estarron-Espinosa M, Barba-Gonzalez R, Castañeda-Saucedo M, Castillo-Herrera G (2019) Effect of supercritical fluid extraction process on chemical composition of Polianthes tuberosa flower extracts. Processes 7(2):60
24. Hasmida MN, Liza MS, Nur Syukriah AR, Harisun Y, Mohd Azizi CY, Fadzilah Adibah AM (2015) Total phenolic content and antioxidant activity of Quercus infectoria galls using supercritical CO_2 extraction technique and its comparison with Soxhlet extraction. Pertanika J Sci Technol 23(2):287–295
25. Ruslan MSH, Idham Z, Nian Yian L, Ahmad Zaini MA, Che Yunus MA (2018) Effect of operating conditions on catechin extraction from betel nuts using supercritical CO_2-methanol extraction. Sep Sci Technol 53(4):662–670
26. Vallecilla-Yepez L, Ciftci ON (2018) Increasing cis-lycopene content of the oleoresin from tomato processing byproducts using supercritical carbon dioxide. LWT—Food Sci Technol 95:354–360

Effect of Inulin on the Formation Kinetics of Methane Hydrate

Sana Yaqub, Bhajan Lal, Abdullah Al-Mubarak B Md Jalil, and Arul Bharti

Abstract In the petroleum industry, clathrate formation during natural gas transportation is the dominant flow assurance problem. The use of kinetic hydrate inhibitor (KHI) is one of the optimum approaches to inhibit hydrate formation and provide flow assurance in offshore gas pipelines. Therefore, the performance of inulin on the formation kinetics of methane (CH_4) clathrate is examined. At 274 K and 7.5 MPa a sapphire hydrate reactor is used to perform kinetic experiments. Gas hydrate kinetics dealt with fundamental knowledge about hydrate onset time, the initial formation rate, and the amount of gas consumed at numerous concentrations (0.12, 0.5, and 1wt %) of biopolymer. Results reveal that the addition of inulin forms substantial hydrogen bonding with water (H_2O) molecules and efficiently inhibits the CH_4 hydrate formation for 37 min. The increased biopolymer concentration to 1wt% showed increasing KHI performance. In addition, the hydrate formation rate is reduced by 51% better than H_2O. While by adding small amounts of inulin, gas consumption is also significantly (65%) decreased. It is concluded that inulin can be an imminent choice for inhibiting CH_4 hydrate formation in offshore gas pipelines.

Keywords Gas hydrates · Inulin · Kinetic hydrate inhibitor

S. Yaqub · B. Lal (✉) · A. Bharti
Chemical Engineering Department, Universiti Teknologi PETRONAS, 32610 Bandar Seri Iskandar, Perak Darul Ridzuan, Malaysia
e-mail: bhajan.lal@utp.edu.my

S. Yaqub · B. Lal
Research Centre for CO2 Capture (RCCO2C), Universiti Teknologi PETRONAS, 32610 Bandar Seri Iskandar, Perak, Malaysia

A. A.-M. B. Md Jalil
R&D Gas Sustainability Technology, PETRONAS Research Sdn. Bhd. Lot, 3288 & 3289, Off Jalan Ayer Itam Kawasan Institusi Bangi, 43000 Kajang, Selangor, Malaysia

M. A. A. Zaini et al. (eds.), *Proceedings of the 3rd International Conference on Separation Technology*, Lecture Notes in Mechanical Engineering,
https://doi.org/10.1007/978-981-16-0742-4_27

1 Introduction

In natural gas pipelines along with hydrocarbons and non-hydrocarbons, a significant amount of H_2O is also present, which under suitable temperature, pressure conditions are known to form hydrates [1]. Gas hydrates, also known as clathrates are made at lesser temperatures and greater pressure by the combination of H_2O and gas guest molecules [2]. The guest molecules are confined inside H_2O cages and according to the size of a guest molecule; the clathrates are crystallized into three structures known as structure I (sI), structure II (sII) and structure H (sH) [3]. Clathrates are naturally existed in marine sediments and distributed globally in permafrost zones [4]. These natural gas hydrates (NGHs) are the cause of massive CH_4 release, and can be a better energy substitute [5]. Other applications of gas hydrates include sea H_2O desalination [6], hydrogen storage [7], CO_2 sequestration [8], gas separation [9] and solidified natural gas storage/transportation [8].

However, the development of clathrates in natural gas pipelines is a nuisance for the flow assurance engineering [10]. Their formation causes plugging and blockage of pipelines, valves and other safety equipment resulting in shutdowns, stops production, sudden explosions and economic losses [11]. Two chemical methods, namely thermodynamic and kinetics, are adapted to restrict the clathrate formation in NG pipelines. In the thermodynamic method, chemicals known as thermodynamic hydrate inhibitors (THIs) are injected in the pipeline [12]. Where, they change the equilibrium conditions and reduce the hydrate stability zone. These chemicals are derivatives of alcohol and are volatile. Because of their volatile nature, they are injected in huge quantities (10-60wt %) [13]–[15]. Injection of such a large amount is hazardous as well as requires large storage space on offshore platforms. In the kinetic method, this problem is solved by injecting H_2O-soluble polymers known as kinetic hydrate inhibitors (KHIs). These chemicals are useful in small quantities (>1wt %) and postpone the hydrate formation for an extended time beyond the fluid residence time in a pipeline [12, 16]. Conventional KHIs are mostly polyvinylpyrrolidone (PVP) and polyvinyl caprolactam (PVCap). The pyrrolidone $(CH_2)_4NH$ group in PVP and caprolactam group $(CH_2)_5(CNH)$ in PVCap are the active functional groups [17]. These active functional groups form hydrogen bonding with H_2O molecules giving them good KHI properties. However, conventional KHIs are non-biodegradable and damage aquatic life [18]. Despite their good kinetic inhibition performance, degradable additives such as amino acids, starch, cellulose, pectin and gums etc., have gained attention [19]–[21]. Gas hydrate kinetics dealt with fundamental knowledge about hydrate onset time, initial hydrate formation rate and gas consumption during clathrate formation. Hydrate onset time is a time upon which a visible hydrate nuclei is made [22]. A good KHI either increases the onset time or decrease the clathrate formation rate and gas consumption. Moreover, a KHI can also possess all three properties.

Xu et al. [23] investigated that by the addition of 0.5wt% chitosan at 274 K and 4.5 MPa, the CH_4 hydrate formation was delayed for 126.7 min. They also observed that by increasing chitosan concentration (0.2–0.8wt %), the onset time was linearly increased. Furthermore, Lee et al. [19] examined the kinetic performance of numerous types of starch, including; Raisamyl, Raifix, Raisabond and tapioca starch. Where, at 274 K and 3.5 MPa, 0.5wt% Raisamyl starch delayed CH_4 hydrate nucleation for 94.7 min, followed by 1 wt% Raifix and Raisabond with an onset time of 15 min and 18 min respectively. In addition, 0.5wt% tapioca starch delayed the CH_4/propane hydrate formation for 192 min at 279 K and 3.0 MPa. Similar to PVP and PVCap, the inhibition mechanism of these biopolymers were related to their active functional groups.

Daraboina et al. [24] proposed that the addition of antifreeze protein (AFP) inhibits the NGH for 19.6 min. However, Xu et al. [21] observed that in the presence of 0.5wt% pectin, extracted from pomelo peel, the CH_4 hydrate formation is delayed for $\approx$180 min. While comparing the pectin performance with PVCap, they found a better performance of later with 360 min. In another study, Yaqub et al. [25] found the reduced performance of commercial pectin at similar experimental conditions. They found that pectin can delay CH_4 hydrate nucleation for 27 min. The reduced onset time might be due to the difference in the purity of the additive. Mazlin et al. [26] performed preliminary investigation for preventing CH_4 hydrate using banana peel cellulose (BPC), stevia and table sugar at 5.0 and 7.0 MPa. When compared, the performance of H_2O and PVP results revealed the promotional behavior of BPC at both pressures. However, at low pressure by the addition of table sugar, the onset time was enhanced from 75 to 82 min. It was justified that the potential of table sugar to act as KHI was attributed to the hydroxyl functional group present on its backbone. According to the concept of hydrogen bonding between the hydroxyl group and H_2O molecules, new additives with potential hydroxyl groups are required to explore at different experimental conditions.

In this work, to prevent the CH_4 hydrate formation, the effect of a novel biopolymer, namely inulin is investigated at 7.5 MPa and 274 K. To our knowledge, to date, no experimental data has been stated on the kinetics of inulin for CH_4 hydrate. Three kinetic parameters, i.e., hydrate onset time, initial formation rate and total moles of gas consumed, have been measured and compared with H_2O.

2 Experimental

2.1 Materials

Table 1 enlisted the materials that were used for this research. All the material were used in their pure form. However, deionized H_2O was used to make the aqueous solution of powdered inulin. The weight of powdered inulin was measured using the analytical balance of precision $\pm$ 0.01 mg. Figure 1 reflects the chemical structure of inulin. Inulin is included in a family of carbohydrates known as fructans. In general, fructan is comprised of fructose-fructose linkages through glycosidic bonds [27].

Table 1 List of materials used in gas hydrate study

Materials	Supplier	Chemical formula	Purity (%)
Deionized H_2O	Own	H_2O	18MΩ[1]
Inulin (chicory)	Sigma Aldrich	C_{6n} $H10n_{+2}O_{5n+1}$	≥ 99.0
CH_4	Linde Malaysia	CH_4	≥ 99.9%

[1]Resistivity

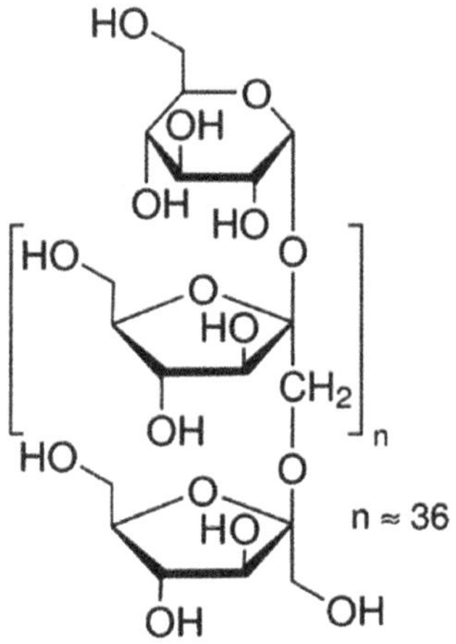

Fig. 1 Chemical structure of inulin, representing repeating unit C_{6n} $H10n_{+2}O_{5n+1}$, where n = 36

2.2 *Equipment*

The equipment used for gas hydrate study is a sapphire hydrate reactor. The details about the equipment are explained and represented in [25, 28, 29].

2.3 *Methods*

The kinetic hydrate experiments are performed using a constant cooling method. In this method the system temperature (285 K) is decreased to experimental temperature (274 K and 277 K). After vacuuming all the contaminants, the CH_4 gas is injected in a sapphire reactor at experimental pressure of 9.5 MPa. Further details about the method are explained in [25, 28, 29]. However, three kinetic performance indictors namely onset time, initial formation rate and total mole consumption are calculated using the following equation

$$t_o = t_s - t_h \tag{1}$$

where, t_o = onset time.

t_s = start time.
t_h = time of hydrate formation

$$r(t) = -\left[\left(n_i^{i-1-}n_i^{i+1}\right)/(t_{i-1} - t_{i+1})\right] \quad (2)$$

where, r(t) = hydrate formation rate.
n_i^{i-1} = initial gas consumption.
n_i^{i+1} = final gas consumption

$$\Delta n_H = V_g/R\left[\left(P_g/zT_g\right)_i - \left(P_g/zT_g\right)_f\right] \quad (3)$$

where, V_g = volume of gas phase.
P_g = pressure of gas phase.
T_g = temperature of gas phase.
R = real gas constant.
z = compressibility factor [12].

3 Results and Discussion

3.1 Influence of Inulin on the Onset Time of CH_4 Hydrate

Figure 2 represents the impact of numerous concentrations of inulin on the onset time of CH_4 hydrate at 7.5 MPa and 274 K (sub-cooling temperature (Teq.-Texp.) is 9.6 K). In a pure H_2O system, the clathrates are made in 21.5 min, while the addition of 1wt% inulin increased the onset time to 36.5 min. The effective performance of inulin at such a higher driving force is associated with its active functional groups, i.e., a hydroxyl group. Which forms strong hydrogen bonding with H_2O molecules

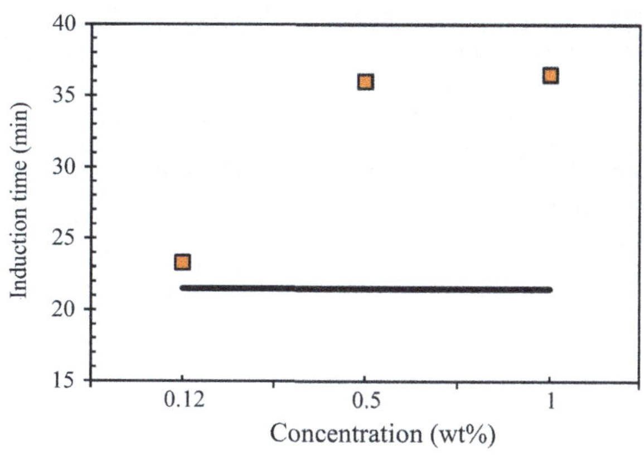

Fig. 2 Influence of different concentrations of Inulin on the onset time of CH_4 hydrate at 7.5 MPa and 274 K. The black line shows H_2O data for comparison

and mitigates clathrate formation. A similar performance was achieved by Mazlin et al. [26] by the addition of table sugar, having an enormous amount of hydroxyl functional group on the parent chain.

In addition, the KHI performance is raised with ascending concentration. Compared to 0.12wt% and 1wt% inulin, 0.5 wt% inulin shows a better onset time. The improved KHI performance with increasing concentration depends upon the increased number of active functional groups. With increasing the KHI concentration, the number of hydroxyl groups comes in contact with H_2O molecules is increased and makes a barrier for the upcoming gas to trapped inside H_2O cages. Xu et al. [23] noticed a comparable behavior in the presence of increasing concentration (0.2wt%-0.8wt%) of chitosan for preventing CH_4 hydrates. However, considering the cost of inulin 0.5wt% is considered as the optimum concentration of inulin for practical usage.

3.2 *Influence of Inulin on the Initial Rate of CH_4 Hydrate*

Figure 3 shows the impact of different concentrations of inulin on the initial rate of CH_4 hydrate at 7.5 MPa and 274 K. In a non-KHI system rate of the clathrate formation is two times more than in a KHI system. The addition of 1wt% inulin diminished the hydrate formation rate two times than H_2O. However, in Fig. 3, the formation rate follows a descending trend with decreased concentration. Where 0.5 wt% inulin reduced the formation rate 50% better than 1wt% inulin. Likely, 0.12wt% performed better than 0.5wt% inulin.

The limiting parameter for gas hydrate formation is a mass transfer from the gas phase to the bulk of H_2O [30]. Increasing the concentration is reducing this barrier and consequently increasing the hydrate formation rate. The improved KHI performance at lower concentration depends upon the decreased surface tension at 0.12wt% than at increased concentrations. Literature supported that as the biopolymer concentration

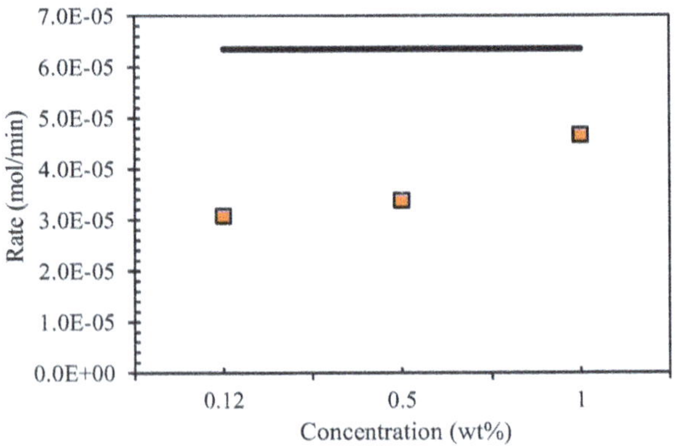

Fig. 3 Influence of different concentrations of Inulin on the Initial CH_4 hydrate formation rate at 7.5 MPa and 274 K. The black line shows H_2O data for comparison

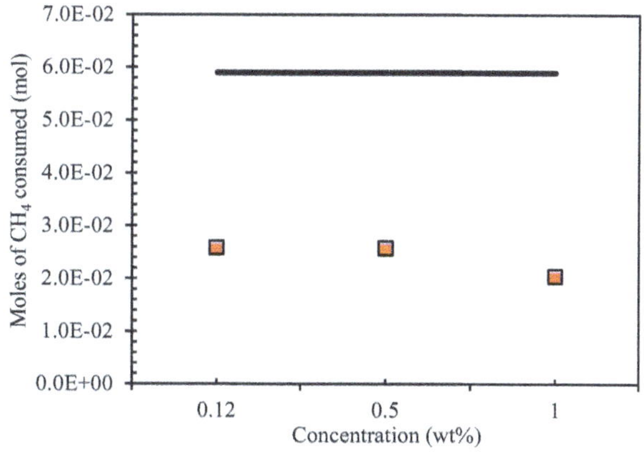

Fig. 4 Impact of numerous concentrations of Inulin on the total moles of CH_4 consumed, at 7.5 MPa and 274 K. The black line shows H_2O data for comparison

is increased, its surface tension is decreased [31], which increases the mass transfer and hydrate formation rate.

3.3 Influence of Inulin on the Mole Consumption of CH_4

Figure 4 depicts the CH_4 consumption in the presence of various concentrations of inulin. Gas consumption is an indication of H_2O to hydrate conversion, where more gas consumption indicates the poor inhibition performance of a KHI. Compared to a non-KHI system, the system contains KHI significantly consumed less gas.

The descending trend in gas consumption is observed with the ascending concentration of inulin. Where 1w% inulin effectively reduced the hydrate growth. Because an increased amount of inulin provides more hydrophobic groups, which adsorbed at the surface of initially formed hydrate and prohibited further entrapping of gas [32, 33]. However, at a lower concentration, i.e., 0.12wt% the available amount of hydrophobic groups are less; therefore, the hydrate growth inhibition is less than 1wt%.

4 Conclusions

In this work, the effect of inulin for preventing CH_4 hydrates is investigated on a sapphire hydrate reactor at 274 K and 7.5 MPa. Results reveal that the addition of inulin forms strong hydrogen bonding with H_2O molecules and efficiently mitigates the CH_4 clathrates for 37 min. Nevertheless, the clathrate formation rate is reduced by about 51% than H_2O. The onset time and hydrate formation rate follows an ascending trend with concentration. While by adding small amounts of inulin, gas consumption is also significantly (65%) reduced. Therefore, it is concluded that inulin can be a potential KHI for CH_4 hydrate mitigation in flow assurance strategies.

Acknowledgements Thanks to Universiti Teknologi PETRONAS and Research Centre for CO_2 Capture (RCCO2C) for all the facilities.

References

1. Wilcox WI, Carson DB, Katz DL (1941) Natural gas hydrates. Ind Eng Chem 33(5):662–665
2. Kelland MA, Svartaas TM, Dybvik L (1995) A new generation of gas hydrate inhibitors. In SPE Annual Technical Conference and Exhibition. Society of Petroleum Engineers. https://doi.org/10.2118/30695-MS
3. Carroll J (2003) Natural gas hydrates
4. Kvenvolden KA, McMenamin MA (1980) Hydrates of natural gas: a review of their geologic occurrence
5. Makogon YF, Holditch SA, Makogon TY (2007) Natural gas-hydrates—a potential energy source for the 21st century. J Pet Sci Eng 56(1–3):14–31
6. Kang KC, Linga P, Park K, Choi S-J, Lee JD (2014) Seawater desalination by gas hydrate process and removal characteristics of dissolved ions (Na+, K+, Mg2+, Ca2+, B3+, Cl−, SO42−). Desalination 353:84–90
7. Kim D-Y, Park Y, Lee H (2007) Tuning clathrate hydrates: application to hydrogen storage. Catal Today 120(3–4):257–261
8. Masoudi R, Tohidi B (2005) Gas hydrate production technology for natural gas storage and transportation and CO_2 sequestration
9. Eslamimanesh A, Mohammadi AH, Richon D, Naidoo P, Ramjugernath D (2012) Application of gas hydrate formation in separation processes: a review of experimental studies. J Chem Thermodyn 46:62–71
10. Sloan ED (2010) Natural gas hydrates in flow assurance. Gulf Professional Publishing
11. Sloan ED (2003) Fundamental principles and applications of natural gas hydrates. Nature 426(6964):353–363
12. Lal B, Nashed O (2019) Chemical additives for gas hydrates. Springer Nature
13. Sloan ED (2005) A changing hydrate paradigm—from apprehension to avoidance to risk management. Fluid Phase Equilib. 228:67–74. https://doi.org/10.1016/j.fluid.2004.08.009
14. Valberg T (2006) Efficiency of thermodynamic inhibitors for melting gas hydrates. Master's Thesis, Nor Univ Sci Technol Trondheim, Norw
15. Peng X, Hu Y, Liu Y, Jin C, Lin H (2010) Separation of ionic liquids from dilute aqueous solutions using the method based on CO2 hydrates. J Nat Gas Chem 19(1):81–85. https://doi.org/10.1016/S1003-9953(09)60027-X
16. Kamal MS, Hussein IA, Sultan AS, von Solms N (2016) Application of various water soluble polymers in gas hydrate inhibition. Renew Sustain Energy Rev 60:206–225
17. Koczkur KM, Mourdikoudis S, Polavarapu L, Skrabalak SE (2015) Polyvinylpyrrolidone (PVP) in nanoparticle synthesis. Dalt Trans 44(41):17883–17905
18. Wan L, Zhang N, Liang D-Q (2019) Inhibition effects of polysaccharides for gas hydrate formation in methane–water system. J Mol Liq, 111435
19. Lee JD, Wu HJ, Englezos P (2007) Cationic starches as gas hydrate kinetic inhibitors. Chem Eng Sci 62(23):6548–6555. https://doi.org/10.1016/j.ces.2007.07.041
20. Srungavarapu M, Patidar KK, Pathak AK, Mandal A (2018) Performance studies of water-based drilling fluid for drilling through hydrate bearing sediments. Appl Clay Sci 152:211–220. https://doi.org/10.1016/j.clay.2017.11.014

21. Xu SR, Fan SS, Fang ST, Lang XM, Wang YH, Chen J (2016) Pectin as an extraordinary natural kinetic hydrate inhibitor. Sci Rep 6:23220. https://doi.org/10.1038/Srep23220
22. Khan MS, Lal B, Keong LK, Ahmed I (2019) Tetramethyl ammonium chloride as dual functional inhibitor for methane and carbon dioxide hydrates. Fuel 236:251–263
23. Xu YJ, Yang ML, Yang XX (2010) Chitosan as green kinetic inhibitors for gas hydrate formation. J Nat Gas Chem 19(4):431–435. https://doi.org/10.1016/S1003-9953(09)60083-9
24. Daraboina N, Ripmeester J, Walker VK, Englezos P (2011) Natural gas hydrate formation and decomposition in the presence of kinetic inhibitors. 3. Structural and compositional changes. Energy Fuels 25(10):4398–4404
25. Yaqub S, Bin Mohd Shariff A, Mellon NB (2019) Unraveling the effect of sub-cooling temperatures on the kinetic performance of biopolymers for methane hydrate. J Nat Gas Sci Eng
26. Idress M, Jasamai M, Yuhaznel FN, Peng W, Karimi N (2018) Preliminary study of natural polymer as kinetic hydrate inhibitor. Mater Today Proc 5(10):21667–21671
27. Kaur N, Gupta AK (2002) Applications of inulin and oligofructose in health and nutrition. J Biosci 27(7):703–714
28. Yaqub S, Keong LK (2019) Thermodynamic and Kinetic effect of Biodegradable Polymers on Carbondioxide Hydrates. J Ind Eng Chem
29. Bavoh CB, Partoon B, Lal B, Keong LK (2017) Methane hydrate-liquid-vapour-equilibrium phase condition measurements in the presence of natural amino acids. J Nat Gas Sci Eng 37:425–434. https://doi.org/10.1016/j.jngse.2016.11.061
30. Englezos P, Kalogerakis N, Dholabhai PD, Bishnoi PR (1987) Kinetics of formation of methane and ethane gas hydrates. Chem Eng Sci 42(11):2647–2658
31. Lee B-B, Chan E-S, Ravindra P, Khan TA (2012) Surface tension of viscous biopolymer solutions measured using the du Nouy ring method and the drop weight methods. Polym Bull 69(4):471–489
32. O'Reilly R, Ieong NS, Chua PC, Kelland MA (2011) Crystal growth inhibition of tetrahydrofuran hydrate with poly (N-vinyl piperidone) and other poly (N-vinyl lactam) homopolymers. Chem Eng Sci 66(24):6555–6560
33. Yagasaki T, Matsumoto M, Tanaka H (2015) Adsorption mechanism of inhibitor and guest molecules on the surface of gas hydrates. J Am Chem Soc 137(37):12079–12085

Improving the Vitamin-E Concentration in Crude Palm Oil Using Hot Compressed Water Technology

Mohd Sharizan Md Sarip, Nor Azian Morad, Mohd Al Hafiz Mohd Nawi, Mohd Rizuan Mansor, and Lokman Hakim Ibrahim

Abstract Vitamin E are known as one of nutritional benefit to human being. Palm oil fruitlet contain a significant amount vitamin E which normally being trapped in palm oil fibre after the extraction process. This due the low vitamin E extractability in the palm oil mill caused by conventional screw press system. The hot compressed water extraction (HCWE) process is introduced as the green method to enrich the concentration of vitamin-E in palm oil. The main objective was to study the effect of HCWE operating condition that is temperature and pressure to the vitamin-E extraction that are α-tocopherol, α-tocotrienol and γ-tocotrienol. HCWE temperature from 120 until 180 °C with 10 °C internal and HCWE pressure from 30 to 50 bar with 10 bar interval was studied. It is observed that the HCWE process significantly improved the concentration of α-tocopherol compared with the commercial CPO and CPO extracted using solvent (hexane), aqueous oil extraction (AOE) and aqueous enzymatic extraction (AEE). However, this process shows poor extractive capability for γ-tocotrienol. The highest total valued of vitamin E (α-tocopherol, α-tocotrienol and γ-tocotrienol) was obtained at the temperature of 180 °C and pressure of 40 bar with valued of 2204 ppm. It is comparable to solvent (hexane) extraction which extracted the total vitamin E of 2296 ppm.

Keywords Crude Palm Oil · Hot Compressed Water Extraction · Vitamin E

M. S. Md Sarip (✉) · M. R. Mansor · L. H. Ibrahim
Faculty of Chemical Engineering Technology, Universiti Malaysia Perli (UniMAP), 02100 Sungai Chuchuh, Padang Besar, Perlis, Malaysia
e-mail: sharizan@unimap.edu.my

N. A. Morad
Universiti Teknologi Malaysia, Jalan Sultan Yahya Petra, 54100 Kuala, Lumpur, Malaysia

M. A. H. Mohd Nawi
Faculty of Chemical Engineering Technology, Universiti Malaysia Perlis (UniMAP), 02100 Sungai Chuchuh, Padang Besar, Perlis, Malaysia

M. A. A. Zaini et al. (eds.), *Proceedings of the 3rd International Conference on Separation Technology*, Lecture Notes in Mechanical Engineering,
https://doi.org/10.1007/978-981-16-0742-4_28

1 Introduction

Vitamin E was a family of eight neutral plant lipid compounds from tocopherol and tocotrienol which designated by alpha, beta, delta and gamma. It was differentiate by the methylation pattern on the chromanol ring [1]. There are known that tocochromanol ring in vitamin E structure was capable to quenching the carbon radicals by the electrophilic hydroxyl group on this chroman ring [1]. This capability is rendering the properties of vitamin E as an antioxidant. Furthermore, vitamin E was also proven to play an important role in neurological health, cystic fibrosis, friedrich ataxia, nieman-pick disease type C and etc. as review by Ulatowski and Manor (2015) [1]. Vitamin-E was naturally can be found in various vegetable oil, nut, seed and green leafy vegetable such as broccoli and spinach. Palm oil is one of the vitamin-E sources and known as the main sources of tocotrienol compare to the other vegetable oil [2, 3].There are about 600 to 1000 ppm vitamin E exist in commercial crude palm oil (CPO) with different type and composition.

Meanwhile, it was approximately 2000 to 4000 ppm vitamin-E is found in the palm oil pressed fiber (PPF) [4]. PPF are the byproduct of the screw press system in the palm oil mill. This indicated the current screw press system in the mill are ineffective to extract vitamin-E from oil palm and leave it un-extracted in PPF. The higher concentration of vitamin-E would create the added value in palm oil product [5].

In this study, the hot compressed water extraction (HCWE) process was introduced as the green method to enrich the concentration of vitamin-E in palm oil. HCWE process is used of water as a solvent from 100 to 372 °C in compressed condition. It was receiving the great attention from various researcher for extraction of environmental sample [6], soil remediation [7, 8], plant material [9] and etc. The main parameter involved in HCWE process is temperature and pressure. The main objective was to study the effect of HCWE operating condition that is temperature and pressure to the vitamin-E extraction that are α-tocopherol, α-tocotrienol and γ-tocotrienol.

2 Methodology

2.1 Material

Raw material palm oil fruit was collected from Bentong, Pahang, Malaysia. The fruit undergo pre-treatment process though sterilization at 130 °C and 40 min using autoclave. The sterilized fruitlet was shredded into 3.5 to 4.0 cm of pieces palm oil mesocarp. This material is sealed in plastic containers and stored at –5 °C prior to the experimental procedure.

Vitamin E standard were obtained from Sigma Aldrich, USA and Chromodex, USA. HCWE system were utilizing the distilled water for extraction process.

2.2 Hot Compressed Water Extraction (HCWE)

Fabricated HCWE equipment located at Shizen Laboratory, Universiti Teknologi Malaysia (UTM) was utilized in this study. The detail diagram for the equipment was in Sarip et al. [10]. The procedure are explained on that manuscript where temperature were varied from 120 to 180 °C and pressure from 30 to 50 bar [10]. The extraction time of 60 min was used with semi-batch process [10]. The extract was going through the post treatment process before the analysis.

2.3 Post Treatment of Crude Palm Oil Collection

The post treatment process of the extract was utilizing the clarifying and centrifuging process as explain earlier in Sarip et al. [10]. The clarifying process was done using separating funnel for 60 min at a constant temperature of 60 °C. Then, the solution is form three layers of oil, oil–water emulsion, and water. The bottom layer was drained out from separating funnel leaving the top two layers of oil and oil–water emulsion. These two layers were centrifuged for 10 min at 7000 rpm in order to obtain crude palm oil [10].

2.4 Analysis of α-Tocopherol, α-Tocotrienol and γ-Tocotrienol

High performance Liquid Chromatography (HPLC) (Agilent, USA) was used to analyzed the concentration of vitamin E namely α-tocopherol, α-tocotrienol and γ-tocotrienol. The UV fluorescence detector with Zorbax RX-SIL normal phase silica column (5 μm, 250 mm × 4.6 mm i.d.) (Agilent, USA), mobile phase of n-hexane/THF/2-propanol (1000:60:4, by volume) (HPLC grade) and flow rate of 1.0 $mLmin^{-1}$ are used [11]. An emission wavelength of 326 nm and excitation wavelength of 292 nm was used to detected the compound. The detail description are explain in the Sarip et al. [10]. The α-tocopherol, α-tocotrienol and γ-tocotrienol compound was identifying by referring with the retention time of the standard. In this study, the resolution obtained for α-tocopherol and α-tocotrienol is 3.7060 meanwhile resolution for α-tocotrienol and γ-tocotrienol is 12.7908. This indicated all the components are well separated in this HPLC condition and column. Meanwhile, the calibration curve for all compounds showing the good linearity with the r^2 value of 0.9991, 0.9994 and 0.9971 for α-tocopherol, α-tocotrienol and γ-tocotrienol, respectively.

3 Result and discussion

3.1 Effect of Pressure and Temperature in HCWE on α-Tocopherol

Three main vitamin E component were evaluated namely α-tocopherol, α-tocotrienol and γ-tocotrienol. Figure 1 shows the effect of pressure on α-tocopherol concentration at constant temperature of 120, 140, 160 and 180 °C. The concentrations of α-tocopherol in commercial CPO, solvent (hexane), aqueous oil and aqueous enzymatic extractions are also shown in the Fig. 1 for comparison purposes.

As can be seen in Fig. 1, a different effect of pressure can be seen at a different temperature. At 120 °C, the concentration of α-tocopherol was increased from 1135.6 to 1346.2 ± 0.1 ppm at 30 to 40 bar. The concentration was slightly decreased to 1025.1 ± 0.1 ppm from 40 to 50 bar. At 140 °C, the concentration of α-tocopherol was slightly decreased from 1222.9 to 1024.9 ± 0.1 ppm as the pressure was increased at 30 to 40 bar. The concentration was increased to 1267.1 ± 0.1 ppm at 50 bar. Similar trend can be observed at a temperature of 160 °C at the maximum concentration of 1388.2 ± 0.1 ppm at 50 bar. At 180 °C, the concentration of this compound was decreased from 1222.1 to 987.2 ± 0.1 ppm as pressure was increased from 30 to 50 bar. However, the effect of pressure and temperature is insignificant as determined by f value (2,16) = 2.96, $p < 0.08$ and f value (3,16) = 2.05, $p < 0.15$ respectively.

In general, a much lower concentrations of α-tocopherol by about 6 folds are detected in commercial CPO at 212 ppm [11] and CPO extracted using aqueous oil

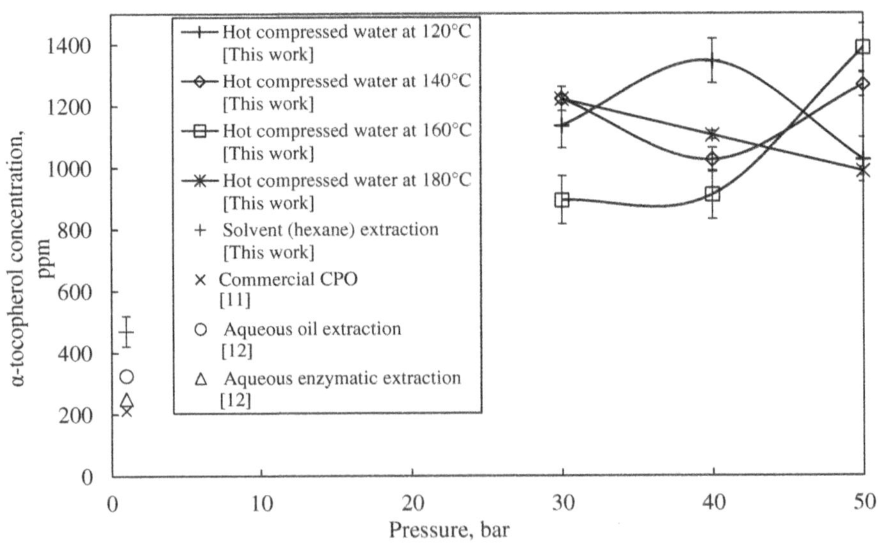

Fig. 1 The effect of pressure on the α-tocopherol concentration at constant temperatures of 120, 140, 160 and 180 °C

(AOE) and aqueous enzymatic extractions (AEE) at 325 and 251 ppm respectively [12] as shown in Fig. 1. However, the extraction of PPF had obtained a higher concentration of α-tocopherol by 67 to 70% of the total vitamin E or 1530 to 2144 ppm [13, 14]. This indicated that α-tocopherol is the main vitamin E component which is difficult to be extracted since the commercial CPO has only 21% of this component [11]. Meanwhile, only 10 to 12% of this component is being extracted using AOE and AEE respectively [12].

Unlike carotenoid, vitamin E is more heat stable with a loss of 13.3% during the extrusion process from the temperature of 95 to 145 °C compared with carotenoid loss of 25.2% for the same conditions [15]. A further increased losses of up to 20% was observed when heating continue at 255 °C for 15 min [16]. The loss rates between vitamin E isomers is differed with the α-tocopherol is most heat stable component followed by α-tocotrienol and γ-tocotrienol [17]. Therefore, controlled temperature as in this study is not expected to cause severe loss of the component. It is also believed that the component is located deep inside the cell walls. Therefore, increasing pressure significantly improves the mass transfer of α-tocopherol by improving the HCW penetration which helps the entrainment of this component.

3.2 *Effect of Pressure and Temperature in HCWE on Tocotrienol*

Another important component of vitamin E in CPO is tocotrienol. Tocotrienol is unpopular component of vitamin E due to its low concentration in food sources. It also had very limited studies on its bioavailability and nutritional effect to human body as compare to tocopherol [18]. Tocotrienol was proven to exhibit biological effect such as neuroprotection, radio-protection, lipid lowering properties, anti-flammatory and anti-cancer [18]. There are two main prominent species for tocotrienol in palm oil namely α-tocotrienol and γ-tocotrienol. The effect of the temperature and pressure on these components were observed in this study. Figure 2 shows the effect of pressure on α-tocotrienol concentration at constant temperature of 120, 140, 160 and 180 °C. The concentration of α-tocotrienol in commercial CPO and CPO extracted using solvent(hexane) is also shown in the Fig. 2 for comparison purposes.

At 120 °C, the concentration of α-tocotrienol was increased as pressure increases from 69.6 to 342.6 ± 0.1 ppm at 30 to 40 bar. At 50 bar, its concentration was decreased to 121.4 ± 0.1 ppm. Similar observation can be seen for the temperature of 140, 160 and 180 °C with higher concentrations obtained at the temperatures of 160 and 180 °C with the maximum concentration of 555.4 ± 0.1 ppm. The significant effect of pressure was observed for α-tocotrienol compared with temperature as determined by f value $(2,18) = 41.26$, $p < 1.89 \times 10^{-7}$. Therefore, 40 bar is identified as best optimum pressure for this component. Meanwhile, the insignificant effect of temperature was observed as determined by f value $(3,18) = 3.05$, $p < 0.06$. The CPO extracted using HCWE process shows a slightly higher concentration of this

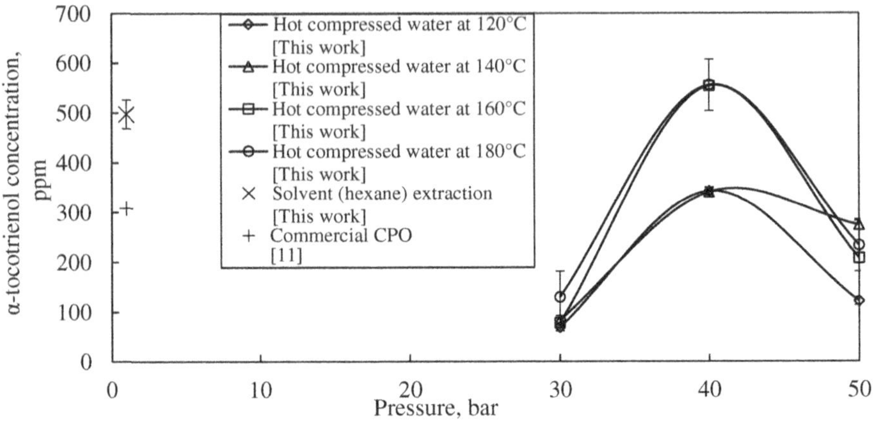

Fig. 2 The effect of pressure on the α-tocotrienol concentration at constant temperatures of 120, 140, 160 and 180 °C

component compared to solvent (hexane) extraction and commercial CPO as shown in Fig. 2.

Figure 3 shows the effect of temperature on the γ-tocotrienol concentration at constant pressure of 30, 40 and 50 bar. A different observation can be seen for γ-tocotrienol concentration where temperature and pressure have significant effects as determined by f value $(3,18) = 23.21$, $p < 2.07 \times 10^{-6}$ and f value $(2,18) =$

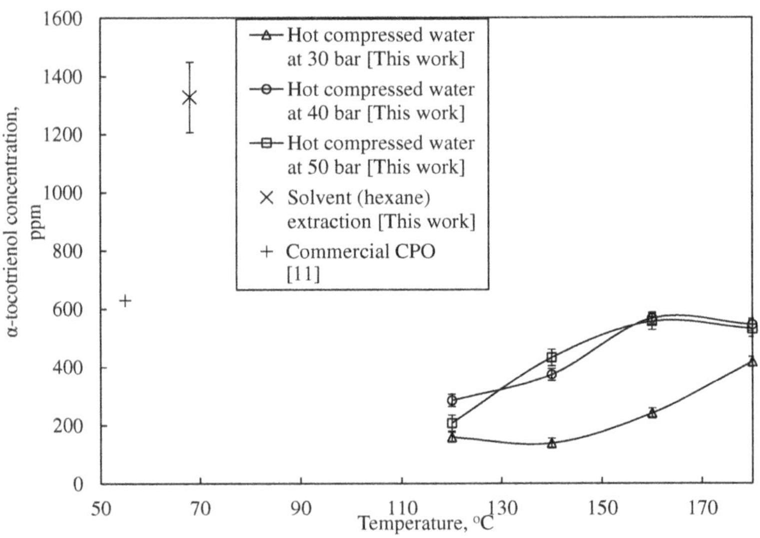

Fig. 3 The effect of temperature on the α-tocotrienol concentration at constant pressure of 30, 40 and 50 bar

27.01, $p < 3.80 \times 10^{-6}$ respectively. Pressure was more significant compared to temperature based on the higher f value. As can be seen in the Fig. 3, the concentration of γ-tocotrienol was increased from 159.1 to 417.9 ± 0.1 ppm at 120 to 180 °C at pressure of 30 bar. At constant pressure of 40 bar, the concentration of this component was increased from 285.2 to 566.9 ± 0.1 ppm at 120 to 160 °C before it slightly dropped to 544.8 ± 0.1 ppm at 180 °C. A similar observation can be seen for the pressure of 50 bar. The comparison between HCWE and solvent(hexane) extraction indicated that the concentration of γ-tocotrienol in CPO extracted using solvent (hexane) is appoximately 3 folds higher compared to HCWE. The concentration of this component through HCWE extracted CPO was slightly lower compared with commercial CPO as referred from Puah et al. (2007) [11].

4 Conclusion

In comparing different methods employed in the experiment, it is observed that the HCWE process significantly improved the concentration of α-tocopherol compared with the commercial CPO and CPO extracted using solvent (hexane), aqueous oil extraction (AOE) and aqueous enzymatic extraction (AEE). However, this process shows poor extractive capability for γ-tocotrienol. The highest total valued of vitamin E (α-tocopherol, α-tocotrienol and γ-tocotrienol) was obtained at the temperature of 180 °C and pressure of 40 bar with valued of 2204 ppm. It is comparable to solvent (hexane) extraction which extracted the total vitamin E of 2296 ppm. This process also shows significant increase of vitamin E by about 2 folds in comparison with commercial CPO [11, 19] and SFE CO_2 [4]. However, the AOE and AEE show higher concentrations of vitamin E at 2686 and 2381 ppm respectively [12]. It is interesting to note that, the HCWE process is believed to share a similar mechanism with AOE and AEE which effectively extract vitamin E.

Acknowledgements The research is funded by Universiti Teknologi Malaysia (UTM), Universiti Malaysia Perlis (UniMAP) and Malaysian Ministry of Higher Education (MoHE). Palm oil sample was collected with the assistance from En. Nasir, Felda Mempaga Satu, Bentong, Pahang.

References

1. Ulatowski LM, Manor D (2015) Vitamin E and neurodegeneration. Neurobiol Dis 84:78–83. https://doi.org/10.1016/j.nbd.2015.04.002
2. Qureshi AA, Mo H, Packer L, Peterson DM (2000) Isolation and Identification of Novel Tocotrienols from Rice Bran with Hypocholesterolemic, Antioxidant, and Antitumor Properties. J Agric Food Chem 48:3130–3140. https://doi.org/10.1021/jf000099t
3. Ng M, Choo Y, Ma A, Chuah C, Hashim M (2004) Separation of vitamin E (tocopherol, tocotrienol, and tocomonoenol) in palm oil. Lipids 39:1031–1035. https://doi.org/10.1007/s11745-004-1327-y

4. Lau H, Choo Y, Ma A, Chuah C (2006) Quality of residual oil from palm-pressed mesocarp fiber (Elaeis guineensis) using supercritical CO_2 with and without ethanol. J Amer Oil Chem Soc 83:893–898. https://doi.org/10.1007/s11746-006-5043-9
5. Ooi CK, Choo YM, Yap SC, Basiron Y, Ong ASH (1994) Recovery of carotenoids from palm oil. J Am Oil Chem Soc 71:423–426. https://doi.org/10.1007/BF02540524
6. Lagadec AJM, Miller DJ, Lilke AV, Hawthorne SB (2000) Pilot-scale subcritical water remediation of polycyclic aromatic hydrocarbon and pesticide contaminated Soil. Environ Sci Technol 34:1542–1548. https://doi.org/10.1021/es990722u
7. Amashukeli X, Pelletier CC, Kirby JP, Grunthaner FJ (2007) Subcritical water extraction of amino acids from Atacama desert soils. JGR: Biogeosciences 112:1–10. doi: https://doi.org/https://doi.org/10.1029/2006JG000308
8. Soltanali S, Hagani ZS, Rouzbahani V (2009) Investigation of operating conditions for soil remediation by subcritical water. Chem Ind Chem Eng Q 15:89–94. https://doi.org/10.2298/CICEQ0902089S
9. Nastić N, Švarc-Gajić J, Delerue-Matos C, Barroso MF, Soares C, Moreira MM, Morais S, Mašković P, Srček VG, Slivac I, Radošević K, Radojković M (2018) Subcritical water extraction as an environmentally-friendly technique to recover bioactive compounds from traditional Serbian medicinal plants. Ind Crops Prod 111:579–589. https://doi.org/10.1016/j.indcrop.2017.11.015
10. Sarip MSM, Morad NA, Yamashita Y, Tsuji T, Mohd Azizi CY, Abd Aziz MK, Lam HL (2016) Crude palm oil (CPO) extraction using hot compressed water (HCW). Sep Purif Technol 169:103–112. https://doi.org/10.1016/j.seppur.2016.06.001
11. Puah CW, Choo YM, Ma AN, Chuah CH (2007) The Effect of Physical Refining on Palm Vitamin E (Tocopherol, Tocotrienol and Tocomonoenol). Am J Appl Sci 6:374–377. https://doi.org/10.3844/ajassp.2007.374.377
12. Teixeira CB, Macedo GA, Macedo, da Silva LHM, Rodrigues AMD, (2013) Simultaneous extraction of oil and antioxidant compounds from oil palm fruit (Elaeis Guineensis) by an aqueous enzymatic process. Bioresour Technol 129:575–581. https://doi.org/doi:10.1016/j.biortech.2012.11.057
13. Choo YM, Yap SC, Ooi CK, Ma AN, Goh SH, Ong ASH (1996) Recovered oil from palm-pressed fiber: a good source of natural carotenoids, vitamin E, and sterols. J Am Oil Chem Soc 73(5):599–602. https://doi.org/10.1007/BF02518114
14. Lau HLN, Choo Y, Ma AN, Chuah C (2006) Quality of residual oil from palm-pressed mesocarp fiber (Elaeis Guineensis) using supercritical CO_2 with and without ethanol C. J Am Oil Chem Soc 83(10):893–898. https://doi.org/doi:10.1007/s11746-006-5043-9
15. Riaz MN, Asif M, Ali R (2009) Stability of vitamins during extrusion. Crit Rev Food Sci Nutr 49(4):361–368. https://doi.org/10.1080/10408390802067290
16. Kuppithayanant N, Hosap P, Chinnawong N (2014) The effect of heating on vitamin e decomposition in edible palm oil. Int J Environ Rural Dev 5(2):121–125
17. Young-Sang L, Soon-Ryang P (2008) Comparative Stability of vit E isomer Extracted from Unsaponifiable Fractions of Rice Bran Oil under Various Temperature and Oxygen Conditions. Korean J Plant Res 21(6):435–439. http://www.koreascience.or.kr/article/JAKO200805441027412.pub
18. Fu JY, Che HL, Tan DMY, Teng KT (2014) Bioavailability of tocotrienols: evidence in human studies. Nutr Metab Nutrition 11(1):1–10. https://doi.org/10.1186/1743-7075-11-5
19. Nagendran B, Unnithan UR, Choo YM, Sundram K (2000) Characteristics of red palm oil, a carotene- and vitamin E–rich Refined oil for food uses. Food Nutr Bull 21(2):189–194. https://doi.org/10.1177/156482650002100213

Removal of Chromium (VI) Using Synergistic Liquid–Liquid Extraction Through LIX63/Cyanex 302 System

Raja Norimie Raja Sulaiman and Norasikin Othman

Abstract The use of single carrier system in liquid–liquid extraction (LLE) offers a slow kinetic extraction and low capacity of targeted solute in the organic phase which adversely affects the extraction efficiency. Hence, throughout this work, a mixed carrier system containing of 8-diethyl-7-hydroxyldodecane-6-oxime (LIX63) and Di-2,4,4-trimethylpentyl mono-thio-phosphinic acid (Cyanex 302) has been investigated for the synergistic LLE of chromium (VI) from electroplating wastewater. An organic phase consists of a mixture of known concentration of LIX63 and Cyanex 302 in kerosene. Meanwhile, the aqueous phase represents an electroplating wastewater containing of 40 ppm of chromium (VI). Results demonstrated that about 98% of chromium (VI) have been synergistically extracted using a binary mixed carrier system consisting of 0.05 M Cyanex 302 + 0.05 M LIX63 with maximum synergistic factor, R_{max} of 27.6. Both LIX63 and Cyanex 302 behaved as a carrier and synergist carrier, respectively. Besides, the stripping performance of chromium from loaded organic phase containing LIX63/Cyanex 302 system showed that Cr could be stripped using 7.0 M of sulfuric acid solutions. It can be concluded that that the synergism behavior of LIX63/Cyanex 302 system in LLE appears as a new formulation which significantly improved the removal of toxic Cr(VI) from electroplating wastewater as well as reducing the use of the expensive and toxic organophosphorus extractants.

Keywords Removal · Chromium (VI) · Synergistic · Liquid–Liquid Extraction

---

R. N. R. Sulaiman · N. Othman (✉)
School of Chemical and Energy Engineering, Faculty of Engineering, Universiti Teknologi Malaysia, 81310 Skudai, Johor, Malaysia
e-mail: norasikin@cheme.utm.my

Centre of Lipids Engineering and Applied Research (CLEAR), Ibnu Sina Institute for Scientific and Industrial Research (Ibnu Sina ISIR), Universiti Teknologi Malaysia, 81310 Skudai, Johor, Malaysia

M. A. A. Zaini et al. (eds.), *Proceedings of the 3rd International Conference on Separation Technology*, Lecture Notes in Mechanical Engineering,
https://doi.org/10.1007/978-981-16-0742-4_29

1 Introduction

The nature and behaviour of chromium, Cr released into the wastewater usually different from those present in natural waters due to the altered physicochemical conditions of the effluents which coming from numerous industrial sources. The concentration of Cr in the discharged effluents highly depends on the Cr compound used during the technology process. Hexavalent (VI) and trivalent (III) are the two most common oxidation states of Cr. Nevertheless, the hexavalent Cr(VI) is much more toxic compared to Cr(III) because of its strong oxidizing nature [1]. Normally, Cr(VI) is produced from the metallurgical, metal finishing, refractory, and electroplating industries. The removal of this hazardous Cr(VI) has become a necessary task for environmental safety.

Prior studies have introduced several methods for Cr(VI) removal including precipitation [2], adsorption [3], electrodialysis [4], electrodeposition [5], liquid–liquid extraction [5] and so forth. Mostly, these techniques provide several limitations such as sludge production as well as high costs and energy consumption. Liquid–liquid extraction (LLE) appears as one of the simplest methods for Cr removal with high separation factor. Through this technique, non-uniform distribution of substances between two immiscible liquid phases which are organic and aqueous phases is utilized. An enrichment of the targeted solute in one of the phases highly depends on several factors including pH, metal ion concentration and ionic strength in aqueous phase, carrier concentration in organic phase, and so forth. Under suitable conditions, a desired substance can be extracted to one of the phases, while undesired substances are remained in the other phases. The transfer of the solute from one liquid phase to another involves extraction reactions, which permit the establishment of liquid–liquid equilibrium [5]. The use of single carrier sometimes ineffective with low capacity loading and slow kinetic extraction [6].

Synergism in LLE process involves a cooperation of two carriers forming a binary mixtures system and improve the extraction efficiency of solute ion [7] (Chauhan and Patel, 2014). Through synergistic relationship, the solubility of the extracted solute in the organic phase is increased as well as inhibiting the emulsification and third phase formation. Binary mixed carrier system including acidic-basic, basic-basic, acidic-acidic, acidic-neutral and basic-neutral have been widely studied [8] (Tong et al. 2013). For instance, Zhang et al. (2016) [6] have stated the combination of D2EHPA and isooctanol for Cr(III) extraction. Isooctanol helps destroy the dimer of D2EHPA molecules, hence increasing the extraction site for Cr. Rajewski and Religa (2016) [9] reported the use of D2EHPA/Cyanex 272 system for Cr(III) extraction. Both Cyanex 272 and D2EHPA played a role as a carrier and synergist carrier, respectively. As for Cr(VI), most of carriers used for extraction are basic or neutral since it appears as anion complexes. In this study, no research has been claimed yet on the extraction of Cr(VI) using the mixtures of organophosphorus acidic carriers. Several variables have been investigated such as effect of different type of acidic carriers, type of mixed carrier system, different composition of mixed carrier system as well as the stripping of chromium from the loaded organic phase.

2 Experimental Procedures

2.1 Materials and Methods

Electroplating wastewater was taken from electroplating company in Pasir Gudang, Johor, Malaysia. The wastewater was in pH 2 with viscosity and density of 0.83 cP and 1.008 g/ml, respectively. Ions detected including cadmium, Cd (0.04 ppm), arsenic, As (0.04 ppm), chromium, Cr(VI) (41 ppm), iron, Fe (0.027) and sulfate, SO_4^{2-} (445 ppm) [10] (Noah et al. 2018). Organophosphorus acidic extractants including, Di-2,4,4-trimethylpentyl mono-thio-phosphinic acid (Cyanex 302) (99%) and Di-2-ethylhexyl phosphoric acid (D2EHPA) (≥95% assay) were purchased from Sigma Aldrich. Meanwhile, 5,8-diethyl-7-hydroxyldodecane-6-oxime (LIX63) was obtained from Cognis, Australia. Kerosene as diluent was purchased from Sigma Aldrich. Diphenylcarbazide (DPZ) supplied by Sigma Aldrich was used for the analysis of chromium concentration. Stripping agents such as sulfuric acid (H_2SO_4) (95% assay), nitric acid (HNO_3) (65% assay) and hydrochloric acid (HCI) (37% assay) were purchased from Merck. All chemicals and raw materials supplied were directly consumed without further purification.

2.2 Liquid-liquid Extraction

LLE process was done by mixing equal volumes of organic (carrier in diluent) and aqueous feed phases (Cr(VI) solution) which subsequently stirred by means of a mechanical shaker at 320 rpm for 18 h to attain the equilibrium [1]. After settling process for about 30 min, the aqueous phase containing Cr was diluted and mixed with Diphenylcarbazide and sulfuric acid solution, producing red-violet colour solution. This solution was then measured using UV–visible spectrophotometer (Jenway) at a wavelength of 540 nm. Meanwhile, the concentration of Cr in the organic phase was calculated via mass balance principle. As for the stripping process, the Cr loaded organic phase was shaken with equal volume of stripping solution using similar abovementioned conditions to attain the equilibrium. Similarly, after being separated, the Cr concentration in the aqueous phase was determined using UV–visible spectrophotometer. All experiments were performed at ambient temperature (26 ± 1°C) and performed in triplicates with standard deviations less than 5%.

2.3 Determinations and Calculations

In LLE, the extraction and stripping efficiencies were calculated using Eqs. (1) and (2). Then, the distribution ratio, D and synergistic factor, R were determined using Eqs. (3) and (4), respectively. D represents the ratio of the Cr concentration present in the organic phase to the aqueous phase at equilibrium.

$$Extraction = \frac{C_i - C_{aq}}{C_i} \times 100\% \quad (1)$$

$$Stripping = \frac{C_s}{C_{org}} \times 100\% \quad (2)$$

$$Distribution\ ratio,\ D = \frac{C_{org}}{C_{aq}} \quad (3)$$

$$Synergistic\ factor,\ R = \frac{D_{mix}}{(D_{LIX63} + D_{C302})} \quad (4)$$

where, C_i is the initial Cr (VI) concentration in aqueous feed phase (ppm), C_{aq} is the Cr(VI) concentration in aqueous feed phase after extraction (ppm), C_s is the Cr(VI) concentration in aqueous stripping phase after extraction(ppm), C_{org} is the Cr(VI) concentration in organic phase. Meanwhile, D_{mix}, D_{LIX63} and D_{C302} denote the distribution ratios of a chromium ion for the mixture of Cyanex 302/LIX63 system as well as single LIX63 and Cyanex 302, respectively.

3 Results and Discussion

3.1 Influence of Different Acidic Carriers Type Towards Extraction of Cr (VI)

Figure 1 presents the extraction percentages of Cr(VI) with respect to different types of acidic carriers. The carriers investigated including Cyanex 302, LIX63 and D2EHPA which are phosphinic, chelating and phosphoric acids, respectively. As observed, the extraction efficiency of Cr(VI) increased in the order: Cyanex 302 (68%) > LIX63 (43%) > D2EHPA (17%). The extraction involving acidic carriers is highly pH dependent [11]. Cyanex 302 contains sulfur which can reduce Cr(VI) to Cr (III) in the organic phase. As a result, the organic phase has changed from yellow to green colour. The sulfur increases the acidity of Cyanex 302, making it more suitable for the metal extraction from acidic media. This is supported by Sole and Hiskey [12] who indicated that the sulfur containing compound is much stronger acid and capable of extracting the metal at lower pH. Meanwhile, both LIX63 and D2EHPA contain an oxygen donor atom (hard base) which exhibits a strong tendency to form a complex with Cr(VI) (hard acid). According to Chauhan and Patel [7], LIX63 is a hydroxime carrier and capable of extracting metal at lower pH values efficiently but offers a slow kinetic of metal extraction. Then, D2EHPA is more preferrable for the Cr(III) extraction through cation exchange mechanism [6].

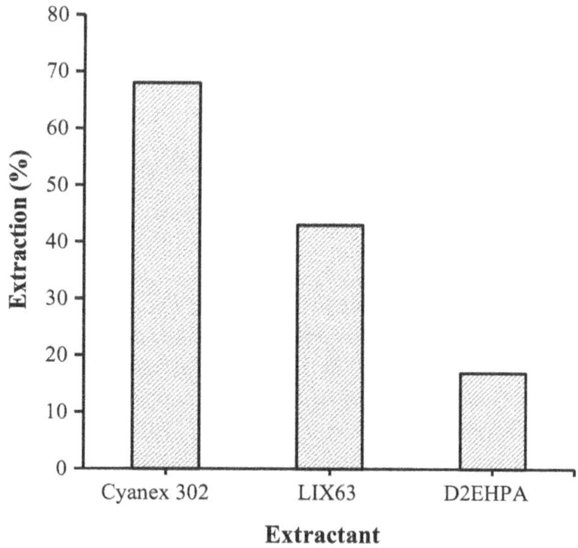

Fig. 1 Effect of acidic carriers on Cr (VI) extraction (Experimental conditions: [Cr]: 41 ppm; [Carrier]: 0.1 M; diluent: kerosene; agitation speed: 320 rpm; agitation time: 18 h; temperature = 26 ± 1 °C)

3.2 *Influence of Different Mixed Carrier System Towards Extraction of Cr (VI)*

Figure 2 shows an influence of several mixed acidic carrier systems towards the extraction performance of Cr (VI). Meanwhile, Table 1 illustrates the results of distribution ratio, D_{mix} and synergistic enhancement factor, R of mixed-extractant system of Cr (VI) when using the mixed carrier systems studied. It is noted that Cyanex 302/LIX63 (83%) system was found to be advantageous towards the extraction of Cr(VI) followed by LIX63/D2EHPA (51%) system. Additionally, although

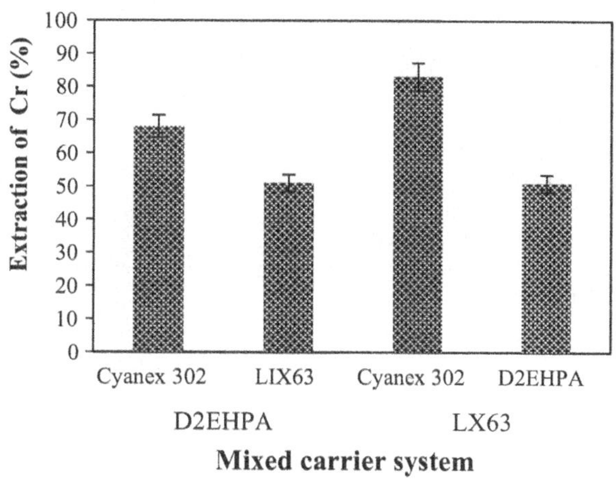

Fig. 2 Effect of mixed carrier system on Cr(VI) extraction. (Experimental condition: [Cr]: 41 ppm; [Based carrier]: 0.05 M; [synergist carrier]: 0.05 M; diluent: kerosene; agitation speed: 320 rpm; agitation time: 18 h; temperature = 26 ± 1 °C)

Table 1 Distribution ratio, D_{mix} and synergistic enhancement factor, R of mixed extractant system on Cr (VI) extraction

Mixed carrier system	Distribution ratio of mixture, D_{mix}	Synergistic enhancement factor, R
Cyanex 302/LIX63	4.85	2.10
LIX63/D2EHPA	1.05	2.14
Cyanex 302/D2EHPA	1.41	0.58

showing a comparable R values, the D_{mix} of Cyanex 302/LIX63 system was much higher than LIX63/D2EHPA system which were 4.85 and 1.05, respectively. Sarkar et al. [13] indicated that the mixed carrier system with D > 1 is pronounced to have a synergistic effect as well as high capacity loading of the targeted solute into the organic phase. Such result implies that LIX63/Cyanex 302 system has synergistically improved the Cr extraction, wherein high concentration of Cr (VI) was observed in the organic phase. Besides, LIX63/D2EHPA system only showed a slight increase in the D_{mix} value, hence insignificantly affect the carrier synergism for Cr (VI) extraction. According to Ayanda et al. [11], phosphinic acid (Cyanex 302) is a stronger acidic carrier compared to phosphoric acid (D2EHPA), thus capable of reducing the equilibrium pH, which is suitable for Cr extraction from acidic media. On the other hand, Fig. 3 displays the comparison among the single and mixed carrier system containing Cyanex 302 and LIX63 on Cr(VI) extraction. Considering a low extraction efficiency of Cr (VI) using LIX63 (43%), it turns out both LIX63 and Cyanex 302 play a role as carrier and synergist carrier, respectively. This is in accordance with Eyupoglu and Kumbasar [14] who claimed that the main carrier should have a slow kinetic extraction while the synergist carrier should accelerate the extraction process. Liu et al. [15] found that the synergistic extraction system predominantly affected by the nature of the metal ions and the phase medium. As previously mentioned, LIX63 contains oxygen donor atom which can form a strong bond with Cr(VI). Then, in term of the phase medium, Cyanex 302 can control the pH equilibrium so that it

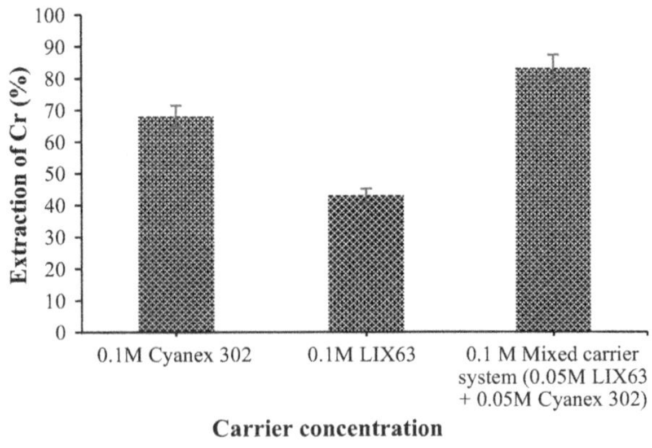

Fig. 3 Effect of single and mixed carrier system containing Cyanex 302 and LIX63 on Cr(VI) extraction. (Experimental conditions: [Cr]: 41 ppm; Diluent: kerosene; agitation speed: 320 rpm; agitation time: 18 h; temperature = 26 ± 1 °C)

Table 2 Effect of type of acid towards stripping performance of chromium from loaded organic phase containing LIX63/Cyanex 302 system. (Experimental conditions: $[Cr]_{org}$: 32 ppm; [acid]: 7.0 M; diluent: kerosene; agitation speed: 320 rpm; agitation time: 18 h; temperature = 26 ± 1 °C)

Type of acid	Stripping (%)
HCl	91
H_2SO_4	100
HNO_3	94

become acidic and able to extract Cr from acidic waste solution. Conversely, in the case of Cyanex 302/D2EHPA system, there is no change in the extraction efficiency of Cr (VI) (68%) when comparing with the single Cyanex 302 (68%). Such behaviour indicates that the addition of D2EHPA has provided insignificant effect in this mixed carrier system. Likewise, the low R value of 0.58 implies that an antagonism effect has occurred during the extraction. As a result, the mixed carrier system containing Cyanex 302 and LIX63 was used for further experiment.

3.3 Effect of Different Compositions of LIX63/Cyanex 302 System Towards Extraction of Cr (VI).

Figure 4 demonstrates an influence of several different compositions of LIX63/Cyanex 302 system towards extraction and synergistic enhancement factor of Cr (VI). Upon increasing the mole fraction of LIX63,X_{LIX63} from 0.2 to 0.5, both extraction efficiency and synergistic enhancement factor, R of Cr (VI) was increased from 73 to 83% and 0.70 to 3.32, respectively. A slight decrease in the extraction of Cr occured at X_{LIX63} of 0.4.

At low X_{LIX63}, it is believed that the LIX63 easily degraded at high concentration of Cyanex 302, hence reducing their complexation with the desired metal ion [16]. Previous work reported that the synergistic extraction take places when $R > 1$ whilst $R < 1$ represents antagonistic extraction [17]. As can be seen, the synergism has occured at X_{LIX63} of 0.5 with R value of 3.32, suggesting that the mixtures of equimolar concentration of Cyanex 302 and LIX63 is suitable for Cr(VI) extraction from acidic medium. However, further increase the mole fraction of LIX63 up to 0.6 started to decrease both the extraction and synergistic enhancement factor, R of Cr (VI) to 54% and 0.63, respectively. Such results indicate that beyond 0.05 M LIX63 insignificantly affected the extraction performance of Cr (VI). Likewise, below 0.05 M Cyanex 302 might reduce the acidity of the pH equilibrium between the organic and aqueous phases, hence lowering the extraction efficiency of Cr(VI). Therefore 0.05 M Cyanex 302 + 0.05 M LIX63 is employed for further experiments.

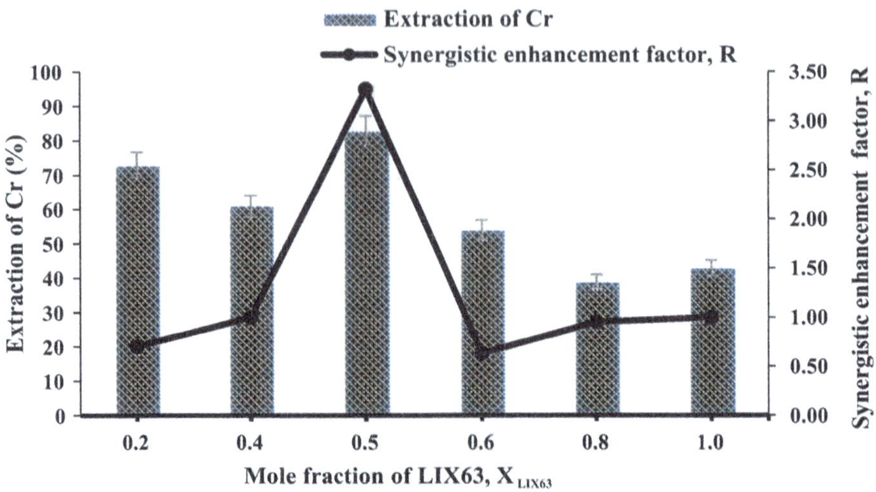

Fig. 4 Effect of composition of LIX63/Cyanex 302 system on Cr(VI) extraction. (Experimental conditions: [Cr]: 41 ppm; [LIX63 + Cyanex 302]: 0.10 M; diluent: kerosene; agitation speed: 320 rpm; agitation time: 18 h; temperature = 26 ± 1 °C)

3.4 Stoichiometry Study of LIX63 and Cyanex 302 Towards Synergistic Extraction of Cr (VI).

Figure 5 exhibits the dependency experiments which was performed to determine the effect of both LIX63 and Cyanex 302 concentrations on the synergistic extraction of Cr (VI). Using slope analysis method, the stoichiometry study of LIX63 was carried out by plotting the graph of log D_{mix} vs log [LIX63] at fixed concentration of Cyanex 302 as shown in Fig. 4a. Result indicated that the slope value obtained was about 5, implying that the mole ratio of LIX63: Cr(VI) in extracted complex was 5:1. Meanwhile, Fig. 4b shows that the mole ratio of Cyanex 302: Cr(VI) in extracted complex was about 0.3 (3:10). The reaction equation of Cr extraction using the synergistic mixtures of LIX63 and Cyanex 302 is given in Eq. (5) wherein both LIX 63 and Cyanex 302 represent (HL) and $(HR)_2$, respectively.

$$3HCrO_4^{-1}\,(aq) + 15(HL)\,(org) + 10(HR)_2\,(org) \rightarrow 3[(Cr(OH)_2.\,3L^{-1}.\,2(HR)_2)]\,(org) + 3(HL)_2.4(HR)_2\,(org) + 3O_2\,(g) + 6H^+ \quad (5)$$

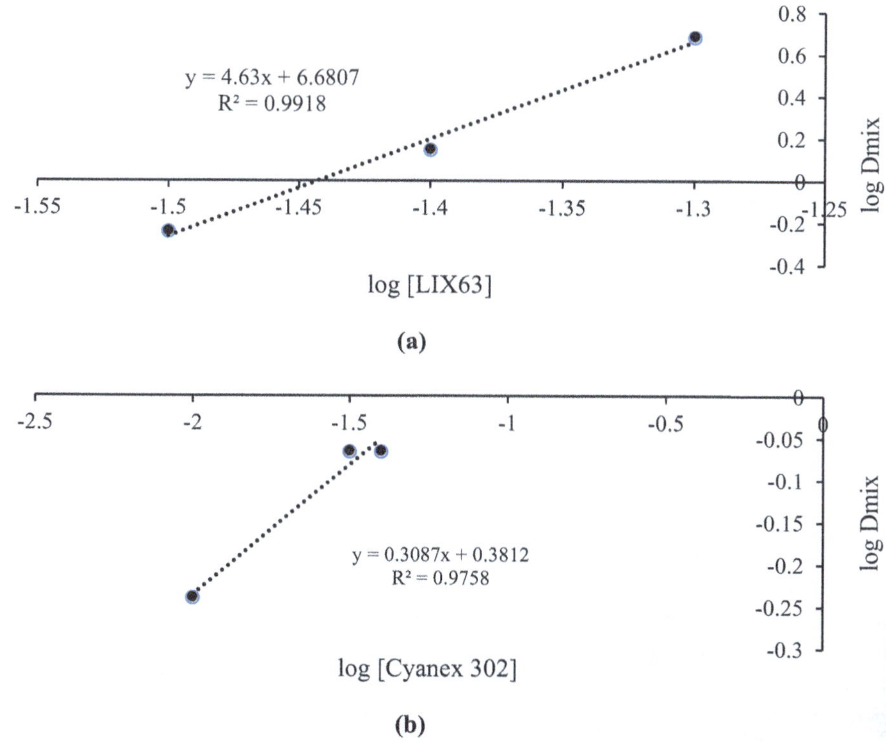

Fig. 5 Effect of LIX63 concentration (**a**) and Cyanex 302 concentration (**b**) on the distribution ratio. (Experimental conditions: [Cr]: 41 ppm; diluent: kerosene; agitation speed: 320 rpm; agitation time: 18 h; temperature = 26 ± 1 °C)

3.5 *Re-extraction of Chromium from Loaded Organic of LIX63/Cyanex 302 System*

Figure 6 presents the effect of different HCl concentrations towards stripping performance of chromium loaded with organic phase containing LIX63 and Cyanex 302. As observed, the stripping performance of chromium (III) increased gradually from 0 to 100% when increasing HCI concentrations from 1.0 to 8.0 M, respectively. Obviously, high strength of HCI was required to strip out the chromium from the highly stable complex of Cr-LIX63/Cyanex 302. Such behaviour is in agreement with Cerpa and Alguacil (2004) who observed that the extraction of metal ion with the mixture of chelating oxime (LIX63) and acidic carrier give rise to a very slow stripping of metal ion. Work by Eyupoglu and Kumbasar [14] claimed that the difficulty of stripping when using LIX63 is due to the polymeric species of LIX63 that hard to be destabilized. Above 7.0 M of HCI, the decomposition of organic phase has occurred due the noticeable brownish color in the stripping solution. Similarly, Eyupoglu et al. [18] also reported the decomposition of LIX63 when using up to 8 M

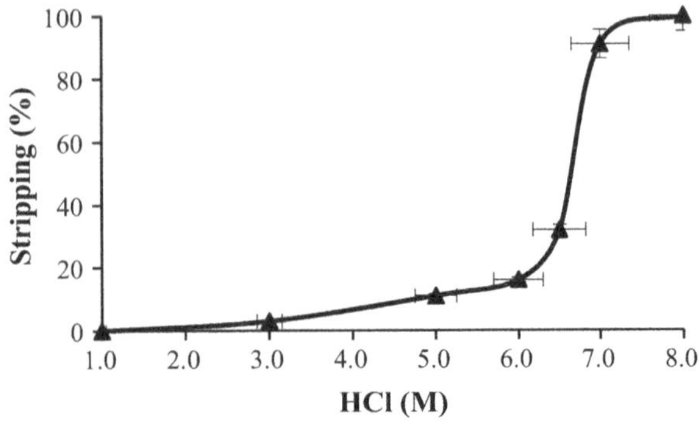

Fig. 6 Effect of different HCI concentrations on the stripping of chromium from. (Experimental conditions: $[Cr]_{org}$: 32 ppm; [HCI]: 1.0–8.0 M; diluent: kerosene; agitation speed: 320 rpm; agitation time: 18 h; temperature = 26 ± 1 °C)

HCl for metal ion extraction. Hence, 7.0 M HCI is enough to be used for the stripping of chromium to avoid the decomposition of carrier into the stripping phase. Then, further study was carried out to investigate the effect type of stripping agent towards stripping performance of chromium involving other acids of H_2SO_4 and HNO_3 at fixed concentration of 7.0 M as shown in Table 2. As can be seen, aside of HCI, both H_2SO_4 and HNO_3 also provide high potential as stripping agent for chromium (III) since they are categorized as strong acid. Nevertheless, the decomposition of carrier occured when applying 7.0 M HNO_3. Thus, 7.0 M H_2SO_4 seems to be more convenient as a stripping agent for chromium as it offers almost 100% of stripping efficiency as well as appearing as less strong acid compared with HCI.

4 Conclusion

In conclusion, about 98% of Cr (VI) was successfully extracted using synergetic binary mixed carrier system containing Cyanex 302 and LIX63. The maximum synergistic enhancement factor, R_{max} of 27.6 was found at 0.5 mol fraction of LIX63 (0.05 M Cyanex 302 + 0.05 M LIX63). Both LIX63 and Cyanex 302 played a role as a carrier and synergist carrier, respectively. To the best of our knowledge, Cr (VI) was extracted in the form of less toxic Cr (III) into the organic phase since Cyanex 302 is a sulfur reducing agent for chromium. In terms of the stripping performance of chromium from loaded organic phase containing LIX63/Cyanex 302 system, H_2SO_4 seems to be more preferable as a stripping agent. This new LLE formulation can be further used in liquid membrane technology application for chromium removal from wastewater.

Acknowledgements The grants from Universiti Teknologi Malaysia for making this research possible including (Professional Development Research University (PDRU) (Q. J130000.21A2.04E68), (Research University Grant (RUG) (Q. J130000.2451.08G02), and (Research University Grant (RUG) (Q. J130000.2451.08G03) are highly acknowledged.

References

1. Noah NFM, Sulaiman RNR, Othman N, Jusoh N, Rosly MB (2020) Extractive continuous extractor for chromium recovery: Chromium (VI) reduction to chromium (III) in sustainable emulsion liquid membrane process. J Cleaner Prod 247:119167. https://doi.org/10.1016/j.jclepro.2019.119167
2. Sun JM, Chang SY, Li R, Huang JC (2007) Factors affecting co-removal of chromium through copper precipitation. Sep Purif Technol 56:57–62. https://doi.org/10.1016/j.seppur.2007.01.013
3. Khalifa EB, Rzig B, Chakroun R, Nouagui H, Hamrouni B (2019) Application of response surface methodology for chromium removal by adsorption on low-cost biosorbent. Chemom Intell Lab Syst 189:18–26. https://doi.org/10.1016/j.chemolab.2019.03.014
4. Santos CSLD, Reis MHM, Cardoso VL, Resendel MMD (2019) Electrodialysis for removal of chromium (VI) from effluent: analysis of concentrated solution saturation. J Environ Chem Eng 7:103380. https://doi.org/10.1016/j.jece.2019.103380
5. Luo JH, Li J, Yang ZP, Liu XF (2013) Removal of chromium(III) from aqueous waste solution by predispersed solvent extraction. T Nonferr Met Soc China 23:524–529. https://doi.org/10.1016/S1003-6326(13)62494-2
6. Zhang G, Chen D, Zhao W, Zhao H, Wang L, Wang W, Qi T (2016) A novel D2EHPA-based synergistic extraction system for the recovery of chromium (III). Chem Eng J 302:233–238. https://doi.org/10.1016/j.cej.2016.05.063
7. Chauhan S, Patel T (2014) A review on solvent extraction of nickel. Int J Eng Res Technol 3(9):1315–1322. https://www.ijert.org/research/a-review-on-solvent-extraction-of-nickel-IJERTV3IS091089.pdf
8. Tong H, Wang Y, Liao W, Li D (2013) Synergistic extraction of Ce(IV) and Th(IV) with mixtures of Cyanex 923 and organophosphorus acids in sulfuric acid media. Sep Purif Technol 118:487–489. https://doi.org/10.1016/j.seppur.2013.07.039
9. Rajewski J, Paweł R (2016) Synergistic extraction and separation of chromium(III) from acidic solution with a double-carrier supported liquid membrane. J Mol Liq 218:309–315. https://doi.org/10.1016/j.molliq.2016.02.079
10. Noah NFM, Jusoh N, Othman N, Sulaiman RNR, Parker NAMK (2018) Development of stable green emulsion liquid membrane process via liquid-liquid extraction to treat real chromium from rinse electroplating wastewater. J Ind Eng Chem 66:231–241. https://doi.org/10.1016/j.jiec.2018.05.034
11. Ayanda OS, Adekola FA, Baba AA, Ximba BJ, Fatoki OS (2013) Application of Cyanex® extractant in Cobalt/Nickel separation process by solvent extraction. Int J Phy Sci 8:89–97. https://doi.org/10.5897/IJPS12.135
12. Sole KC, Hiskey JB (1995) Solvent extraction of copper by Cyanex 272, Cyanex 302 and Cyanex 301. Hydrometallurgy 37:129–147. https://doi.org/10.1016/0304-386X(94)00023-V
13. Sarkar R, Ray S, Basu S (2014) Synergism in solvent extraction and solvent extraction kinetics. J Chem Bio Phy Sci. 4: 3156–318
14. Eyupoglu V, Kumbasar RA (2015) Extraction of Ni(II) from spent Cr-Ni electroplating bath solutions using LIX 63 and 2BDA as carriers by emulsion liquid membrane technique. J Ind Eng Chem 21:303–310. https://doi.org/10.1016/J.JIEC.2014.02.037
15. Liu Y, Lee M, Senanayake G (2018) Potential connections between the interaction and extraction performance of mixed extractant systems: a short review. J Mol Liq 268:667–676. https://doi.org/10.1016/j.molliq.2018.07.097
16. Truong HT, Lee MS, Son SH (2017) Extraction of palladium(II) from hydrochloric acid solutions by solvent extraction with mixtures containing either Cyanex 301 or LIX 63. Metals 541:1–12. https://doi.org/10.20944/preprints201711.0059.v1
17. Zaheri P, Abolghasemi H, Maraghe MG, Mohammadi T (2015) Intensification of Europium extraction through a supported liquid membrane using mixture of D2EHPA and Cyanex272 as carrier. Chem Eng Process 92:18–24. https://doi.org/10.1016/j.cep.2015.03.004

18. Eyupoglu V, Kumbasar RA, Isik M (2015) Performance evaluation of nickel separation and extraction process from simulated spent Cr/Ni electroplating bath solutions by ELM process using LIX63 and PC88A. J Dispersion Sci Technol 1309–1319. https://doi.org/10.1080/01932691.2014.974813

The manufacturer's authorised representative in the EU is Springer Nature Customer Service Centre GmbH, Europaplatz 3, 69115 Heidelberg, Germany. If you have any concerns regarding our products, please contact ProductSafety@springernature.com

Printed and bound by CPI Group (UK) Ltd, Croydon, CR0 4YY
30/09/2026
02225251-0014